Mycobacteria and Human Disease

Mycobacteria and Human Disease

John M Grange, MSc, MD

Reader in Clinical Microbiology,
Cardiothoracic Institute,
Brompton Hospital,
London

Edward Arnold

First published in Great Britain 1988 by
Edward Arnold (Publishers) Ltd, 41 Bedford Square, London WC1B 3DQ

Edward Arnold (Publishers) Pty Ltd, 80 Waverley Road, Caulfield East,
Victoria 3145, Australia

Edward Arnold, 3 East Read Street, Baltimore, Maryland 21202, U.S.A.

British Library Cataloguing in Publication Data

Grange, John M.
    Mycobacteria and human disease.
    1. Mycobacterial diseases
    I. Title
    616.9'2   RC116.M8

ISBN 0-7131-4566-8

Text set in 10/11pt Times
by 𝓕\ Tek Art Ltd, Croydon
Printed and bound in Great Britain
by Butler & Tanner, Frome and London

Preface

The genus *Mycobacterium*, despite great advances in medical science, continues to be a major cause of misery and suffering throughout the world. Leprosy and tuberculosis attack the human race with undiminished vigour while several other species of mycobacteria are emerging as important causes of life-threatening disease.

It was originally envisaged that this book should be a revised version of my previous work *Mycobacterial Diseases*, published by Edward Arnold in 1980. In practice, there have been so many major developments in the subject over the last seven years that the text has been almost entirely rewritten. In particular, the drug treatment of leprosy, tuberculosis and other mycobacterial disease has become much more rational and considerable advances have been made in immunology, facilitated by the introduction of monoclonal antibodies, cell cloning techniques and modern 'genetic engineering' procedures. In addition, there have been advances in the ecology, biochemistry, epidemiology and classification of the mycobacteria. Sadly, though. there is an increasing gap between the 'high-tech' researchers and those responsible for the basic care of the victims of mycobacterial disease. I hope this book will help to bridge that gap.

This monograph provides a review of the mycobacteria themselves, their place in the environment, the way in which they interact with the living host, the nature of the diseases they cause and the available means of diagnosing, preventing and curing such disease. It is intended for both undergraduate and postgraduate students seeking a general account of the mycobacteria and the diseases they cause, for the clinician wishing to understand the underlying mechanisms of the pathogenesis of the diseases, for the epidemiologist and health care administrator wishing to appreciate the nature and magnitude of the public health problems posed by the diseases, and for the microbiologist providing a clinical service. The potential researcher will find an account of the exciting developments in the science of mycobacteriology and, more importantly, will become aware of the many gaps in our present-day knowledge!

London, 1987   JMG

Acknowledgements

During my time in the field of 'mycobacteriology' I have met so many delightful, erudite and friendly scientists who have contributed immensely to this book by imparting their wisdom. I shall mention none by name lest I upset any whom I might inadvertently omit, but I express my deep gratitude to them all.

Illustrations have been kindly provided by the Robert Koch Institute (Figs. 1.1 and 1.2), and Miss C.A. Dewar (Fig. 2.2), Mr B.W. Allen (Fig. 2.6), Dr J.L. Stanford (Fig. 3.1), Prof. J. Swanson Beck (Figs. 5.6 and 5.7), the Leprosy Unit of the World Health Organization (Fig. 6.2), Dr N.M. Samuel (Fig. 7.3), the late S.G. Browne (Figs. 7.5, 7.7 and 7.9), Dr K. Schopfer (Figs. 8.7, 8.8 and 9.7), Prof. W.C. Noble (Fig. 9.5) and Mr A.J. Prosser (Fig. 9.6).

Finally I owe a deep gratitude to my wife Helga for her unremitting support and encouragement.

Contents

1 Introduction

From time immemorial tuberculosis and leprosy have ranked amongst the most feared and dreaded of the numerous diseases that afflict mankind. The evangelist John Bunyan dubbed tuberculosis 'the Captain of all of these men of Death', and in India it was known as the King of Diseases. Leprosy may be termed the Disease of Kings, as Robert the Bruce, King of Scotland, and, according to legend, the emperor Constantine are numbered amongst its victims.

In the present century leprosy has become virtually extinct in the industrially developed nations (although a few new cases still occur in southern Europe and the USA) and the incidence of tuberculosis has dropped dramatically. Unfortunately only a quarter of the world's population lives in these regions: the other three-quarters live in the developing countries where both diseases, despite being preventable and curable, remain major public health problems. The fact that around 5000 individuals die of tuberculosis every day and millions of people are severely physically and socially handicapped by leprosy is 'a blot on the conscience of the world community'. Of all the chronic infectious diseases, none is associated with as much physical and mental suffering as leprosy. Thus, although leprosy is not nearly as prevalent as malaria and filarial infections, its very nature makes it a special case for compassionate attention. In the West, tuberculosis and, more especially, leprosy are often thought of as 'tropical' diseases, but it is important to remember that they were, and could again become, infections of worldwide distribution. Few people are aware that leprosy was prevalent in Scandinavia until the early part of this century.

Although the classical mycobacterial infections are relatively uncommon in the developed nations, disease due to other mycobacteria (the so-called atypical mycobacteria) is becoming an increasingly serious problem. The 'little red rods' seem determined to continue to inflict misery on the human race: they are most tenaceous and can only be dislodged from individual patients and from communities with great difficulty. Regrettably, their greatest ally is man's indifference to the sufferings of others.

The history of mycobacterial disease is divisible into three eras: those of ignorance, hope and enlightenment. The era of ignorance lasted from the dawn of recorded history until the discovery of the causative organisms in the latter part of the nineteenth century. The era of hope lasted from then until the introduction of effective therapy around 1950. In the present era, that of enlightenment, we are acquiring detailed knowledge of the mycobacteria and the diseases they cause. Can we hope that this will

Fig. 1.1 Robert Koch (1843–1910): discoverer of the tubercle bacillus in 1882

enable us to enter the fourth and final era – that of conquest?

The turning point in the history of tuberculosis occurred at a meeting of the Berlin Physiological Society on the evening of 24 March 1882, when Robert Koch (Fig. 1.1) described the isolation of the causative organism of that disease. This, in fact, was eight years after Armauer Hansen published his observation of the leprosy bacillus in lepromatous nodules. Hansen's work did not receive the adulation accorded to that of Koch as he was unable to isolate the organism in pure culture. Hansen attempted, though fortunately without success, to demonstrate the transmissibility of leprosy by incising the conjunctivae of uninfected patients with a contaminated cateract knife. Hansen was prosecuted for these highly unethical experiments and received a severe reprimand.

The work of Hansen and Koch did not occur in scientific isolation. The stage had been set by the clear establishment of the germ theory of communicable disease by Louis Pasteur and, in particular, by the experimental demonstration of the transmissibility of tuberculosis in rabbits by Jean-Antoine Villemin (Fig. 1.2), a French military surgeon, in 1868. It was therefore considered very likely that tuberculosis and leprosy

Fig. 1.2 Jean-Antoine Villemin (1827–92): pioneer of experimental studies on the transmissibility of tuberculosis

were caused by 'germs' and many workers attempted to isolate them. Koch's critics have remarked that he was only able to discover the tubercle bacillus because he used methods developed by other workers, namely Weigert's stains and Tyndall's inspissated serum medium. On the other hand, Koch's acknowledged industry, patience, tenacity and technical skill must have contributed greatly to his success. Indeed, Koch's detailed descriptions of his techniques enabled other workers to reproduce his findings and the few antagonists were rapidly silenced.

Koch's discovery heralded the era of hope. In 1908 Leonard Williams wrote: 'The riddle of the white plague, which had so long defied solution, had been read at last; the dreary watches of the night were over; and the dawn, with its promise of victory, peace, and purity, were really at hand'. Nevertheless, much work was required before practical benefits were to be reaped from the discoveries of Hansen and Koch. Such work started rapidly in three main directions: the isolation and culture of the bacillus for diagnostic purposes, the search for an effective cure, and the development of a vaccine. Diagnosis required specific stains and methods for the *in vitro* cultivation of mycobacteria. Koch stained his preparations with an alcoholic solution of methylene blue and used vesuvin as a counterstain. Very shortly afterwards, Paul Ehrlich discovered the 'acid-fastness' of the tubercle bacillus and introduced a staining technique which, with minor modifications by Ziehl and Neelsen whose names the method now bears, is still widely used today. Originally, tubercle bacilli were grown on heat-coagulated serum, then in glycerol–beef broth. Egg-based media were introduced by Dorset in 1902 and were modified by Lowenstein in 1930. Methods for 'decontaminating' clinical specimens were introduced by Petroff and others around 1915. No further significant developments were made so that, in 1954, Dubos remarked that tuberculosis bacteriology was based on 'primitive bacteriological techniques worked out decades ago'. Apart from the introduction of rapid, but very expensive, radiometric techniques, little has changed since.

Koch's discovery coincided with the birth of the discipline of immunology. Koch certainly did not regard himself as an 'immunologist'. Indeed, when Metchnikoff demonstrated the phenomenon of phagocytosis, Koch remarked 'I am a hygienist and it is of no interest to me where the microbes are, whether inside or outside the cells'. Nevertheless he attempted to develop an agglutination test for tuberculosis using the whole bacillus as antigen and also tried to attenuate a human strain for use as a vaccine for tuberculosis in cattle. But his main studies centered on the development of a cure for tuberculosis, and this led to the extraordinary saga of Old Tuberculin. This, as outlined in Chapter 5, followed a meticulous series of experiments on guinea pigs that led to the description of the necrotic hypersensitivity reaction termed the Koch Phenomenon. Unfortunately Koch was under considerable pressure from his political overlords to develop a cure for tuberculosis and his use of Old Tuberculin in patients proved disastrous and almost ruined his reputation. Nevertheless, this was the first attempt at 'immunotherapy' – a topic of great interest at the present time. Furthermore, Old Tuberculin and the reaction it elicited were used by Clemens von Pirquet to develop the tuberculin test – one of the most widely used and misunderstood of all diagnostic tests.

Another of Koch's errors that was to have far-reaching consequences was his announcement, at the British Congress on Tuberculosis in 1901, that bovine tuberculosis was of no danger to man. The veterinary surgeons present were so shocked by this pontification that they persuaded the Minister of Agriculture to convene a Royal Commission to investigate the issue. In a period of ten years the commissioners accrued an enormous amount of information on the epidemiology, bacteriology and pathology of bovine tuberculosis and this work paved the way for the virtual eradication of that disease in Great Britain. The Commission amply demonstrated the benefits of state-sponsored medical research and was the forerunner of the British Medical Research Council.

The principles of vaccination were well established by Pasteur, and many workers attempted to attenuate the tubercle bacillus for use as a vaccine. One of the first successful attempts was made by Edward Trudeau who attenuated a human strain by repeated passage on coagulated sheep serum for two years. Although this, the R1 strain, was attenuated for the guinea-pig, no further development was undertaken. Trudeau founded a tuberculosis research institute, which bears his name, at Saranac Lake, New York State. Years later, while working at that institute, George Mackaness laid the foundations to the study of the mechanisms of cell mediated immunity and the role of the macrophage in tuberculosis.

A vaccine was eventually produced by Calmette and Guerin after passaging a supposedly bovine tubercle bacillus 230 times on potato slices soaked in bile and glycerol over a period of 13 years. (It is sad to reflect that the present political and financial pressures on academic scientists to produce results and 'papers' would render such a procedure unthinkable nowadays.) This vaccine, Bacille Calmette–Guerin (BCG), was first used in 1921 as an oral vaccine for infants. Its early use was delayed by considerable controversy concerning its safety and by the 'Lubeck disaster' in 1930 when many children were accidentally vaccinated with a virulent strain of *M. tuberculosis* and 73 died. After a further delay caused by the Second World War, a freeze-dried vaccine was introduced and has been widely used since.

The next milestone in the history of mycobacterial disease occurred in the mid twentieth century with the discovery of the first effective drugs. As in the case of the discovery of the causative organism, leprosy preceded tuberculosis though with less fanfare and acclaim. Faget and his colleagues found that promin was effective against leprosy in 1943 and streptomycin was discovered, after an extensive search, by Waksman and his team in 1948. This, and the subsequent discoveries of isoniazid and other drugs, at last removed the fear of tuberculosis. The general public and some physicians were convinced that the disease was conquered and would soon be extinct. Others, including Waksman himself, were not so optimistic and doubted whether anti-tuberculosis drugs alone would solve the problem. Sadly they have been proved correct. Indeed the promises held out by drugs and BCG have probably done more to eradicate interest in the mycobacterial diseases than to eradicate the diseases themselves.

The 1950s were a time of great excitement owing not only to the introduction of effective chemotherapy and the early BCG trials, but also to the serious interest being taken in disease due to other mycobacterial species. The first to be described in detail were due to *M. ulcerans* (Buruli ulcer) and *M. marinum* (swimming pool granuloma). In 1954 Ernest Runyon published the first of his studies on the classification of 'anonymous' mycobacteria causing lung disease in man. This, together with the pioneering studies of Ruth Gordon, led to a renewed interest in the taxonomy of the mycobacteria, culminating in the extensive studies undertaken by the International Working Group on Mycobacterial Taxonomy (IWGMT) established by Dr Larry Wayne. At present there are 41 'approved' mycobacterial species (see Chapter 3), and a dozen or so others. About half the species are known to cause disease in animals or man.

In addition to interest in mycobacteria as pathogens, attention has been given to their place in the inanimate environment. It is now clear, contrary to earlier views, that the genus *Mycobacterium* is essentially one of environmental saprophytes and that pathogenicity is not their usual behaviour. Thus, the major pathogens *M. tuberculosis* and *M. leprae* are

atypical mycobacteria although, paradoxically, this term is usually applied to the typical saprophytic species! Ecological studies have proved to be of great relevance to disease as there is little doubt that immunologically effective contact with environmental mycobacteria has a profound influence on the way an individual responds to BCG vaccination or to infection by a pathogenic species.

The period from 1970 to the present time (the time spanned by the author's involvement in the subject) has been one of great fascination and excitement. Rifampicin was introduced and made it possible at last to contemplate short-course curative chemotherapy for both tuberculosis and leprosy. After a decade or so of extensive clinical trials organized by Professors Mitchison and Fox of the British Medical Research Council and their collaborators abroad, it has been clearly shown that it is possible to cure almost all patients with tuberculosis by an orally administered regimen in 64 doses over a 5-month period. This contrasts sharply with the 3000 doses of drugs over a 2-year period used in the early days of chemotherapy. Indeed it is salutory to note that safe, simple and effective anti-tuberculosis therapy has only been available for a very few years.

During the same period of time enormous strides have been made in mycobacterial ecology, taxonomy, structural and biochemical studies (especially on the lipid-rich cell wall), genetics and immunology. Technological advances in the latter two disciplines have been particularly exciting. It is now possible, by 'genetic engineering' to clone DNA, to insert genes into alternative hosts and to obtain gene products from the new hosts. By such means pure mycobacterial antigens, even those from the non-cultivable leprosy bacillus, are obtainable in sufficient quantities for diagnostic tests, immunological studies and possible incorporation into new vaccines. Very sensitive 'probes' are being produced for the detection of mycobacterial DNA in sputum and tissues and may prove to have great diagnostic applications in the future. Recombinant gene technology has also enabled some immunological mediators of possible therapeutic value to be prepared in quantity. 'Fine structure' analysis of the immune response in mycobacterial disease is possible by using monoclonal antibodies to identify and separate the various cells and mediators involved, while the availability of cell growth factors enables lymphocytes to be cultured, cloned and characterized *in vitro*.

Although these scientific advances are very exciting, they have contributed very little so far to the practical problems of the control of mycobacterial disease. Sadly, it seems unlikely that we will achieve, in the few remaining years of this millenium, the World Health Organization's Alma Ata Declaration: Health for all by the year 2000. There is no doubt that we have the technical ability to detect individuals with infectious forms of tuberculosis and leprosy, and we have chemotherapeutic agents that are capable of curing virtually every case. What we do not have is the motivation, finance and infrastructure necessary to detect and treat all patients. Short-course drug regimens for both diseases should, in principle, render supervision of therapy much easier. In practice, enormous difficulties are still experienced, as discussed in Chapter 6, and it is abundantly clear that the currently available regimens are still too long, too cumbersome to manage and too expensive. It is for this reason that interest is once again turning to the possibility of enhancing the patient's immune defences so that the very few bacilli remaining after a much shorter (and much cheaper) course of therapy would be effectively eliminated. There is also no doubt that BCG vaccination, under some circumstances, protects against both tuberculosis and leprosy. Some workers are attempting to use 'genetic engineering' to develop more effective vaccines, while others are seeking better ways using the currently available BCG vaccine. It remains to be seen which approach proves most beneficial, but certainly the whole issue of vaccination against mycobacterial disease needs, and is receiving, a thorough reappraisal.

In fine, a full scientific, medical and financial cooperation between the developed and developing nations will be required for the eventual conquest of mycobacterial disease. There is no doubt that these are diseases of the socio-economically underprivileged and that the relative freedom of the West from such ills is a direct result of its prosperity. Until the barriers of race, creed and nationality are broken down, and until mistrust and strife are replaced by brotherly love, compassion and cooperation, the tyrannical reign of the King of Diseases and the Disease of Kings will continue.

Publications of general and historical interest

Bishop, P.J. and Neumann, G. (1970) The history of the Ziehl–Neelsen stain. *Tubercle* **51**: 196–206.

Brothwell, D. and Sandison, A.T. (Eds.) (1967) *Diseases of Antiquity*. Springfield: Charles C. Thomas. (Contains chapters on leprosy, tuberculosis and diseases in the Bible and the Talmud.)

Browne, S.G. (1977) *Leprosy: New Hope and Continuing Challenge*. London: The Leprosy Mission.

Crawfurd, P. (1911) *The King's Evil*. Oxford: Clarendon Press.

Dubos, R. and Dubos, J. (1952) *The White Plague: Tuberculosis, Man and Society*. Boston: Little, Brown.

Francis, J. (1959) The work of the British Royal Commission on Tuberculosis, 1901–1911. *Tubercle* **40**: 124–32.

Grange, J.M. and Bishop, P. (1982) Über Tuberkulose – A tribute to Robert Koch's discovery of the tubercle bacillus, 1882. *Tubercle* **63**: 3–17.

Heifets, L. (1982) Metchnikoff's recollections of Robert Koch. *Tubercle* **63**: 139–41.

Keers, R.Y. (1978) Pulmonary tuberculosis: a journey down the centuries. London: Baillière Tindall.

Koch, R. (1882) Die Aetiologie der Tuberculose. *Berliner Klinische Wochenschrift* **19**: 221–38. Translated by Pinner, B. and Pinner, M. (1932) *American Review of Tuberculosis* **25**: 285–323.

Pallamary, P. (1955) Translation of Gerhard Armauer Hansen: *Spedalskheden Aarsager* (Causes of Leprosy). *International Journal of Leprosy* **23**: 307–9.

Rosenthal, S.R. (1957) *BCG Vaccination Against Tuberculosis*. London: Churchill. (Includes a historical chapter by Guerin.)

Villemin, J.A. (1868) Etudes experimentales et cliniqies sur tuberculose. Paris: Baillière et Fils.

Vogelsang, T.M. (1978) Gerhard Henrik Armauer Hansen (1841–1912), the discoverer of the leprosy bacillus. His life and work. *International Journal of Leprosy* **46**: 257–332.

Waksman, S.A. (1964) *The Conquest of Tuberculosis*. London: Cambridge University Press.

Williams, L. (1908) The worship of Moloch. *British Journal of Tuberculosis* **2**: 56–62.

2 The genus *Mycobacterium*

The generic name *Mycobacterium* was introduced by Lehmann and Neumann in the first edition of their 'Atlas of Bacteriology' published in German in 1896. At that time the genus contained only two species, *Mycobacterium tuberculosis* and *M. leprae*. The name *Mycobacterium*, meaning fungus-bacterium, was derived from the mould-like pellicular growth of the tubercle bacillus on liquid media. The name did not, and should not, imply that the mycobacteria are related to the fungi. The non-culturable leprosy bacillus was included in the genus because it shares a staining property with the tubercle bacillus; namely, resistance to decolouration by weak mineral acids after staining with one of the arylmethane dyes. This property, *acid-fastness*, is the basis of the widely used Ziehl–Neelsen stain, the history of which was reviewed by Bishop and Neumann (1970). Acid-fastness, although a useful distinguishing property, is not unique to the mycobacteria: bacterial spores, for example, are often strongly acid-fast and members of the related genus *Nocardia* are weakly acid-fast.

Shortly after the introduction of the generic name, acid-fast bacilli were cultured from birds and cold-blooded animals such as frogs, turtles and fish. Also, at that time, small but constant differences between tubercle bacilli isolated from man and cattle were described. Thus, four 'tubercle bacilli' were recognized; namely, human, bovine, avian and 'cold-blooded'. In addition, acid-fast bacilli were isolated from inanimate sources such as hay, compost and butter. As tuberculosis in man and cattle was such a serious problem, these other mycobacteria received scant attention, although there were a few reports of their involvement in human disease. Despite the lack of clinical interest, numerous supposedly new species were described and the 1966 edition of Index Bergeyana listed 128 validly published species. Paradoxically, this plethora of names made identification of individual isolates so difficult that mycobacteria other than *M. tuberculosis* were often termed 'anonymous mycobacteria'.

Interest in the classification of the genus was awakened in the 1950s by the descriptions of two new mycobacterial diseases of man – swimming pool granuloma and Buruli ulcer (see Chapter 9) – and by the pioneering studies of Ruth Gordon and Ernest Runyon.

Runyon (1959) drew attention to the role of 'anonymous' mycobacteria in human lung disease and placed the responsible strains into four groups

according to their speed of growth and pigmentation. These groups are:

I   photochromogens (yellow pigment formed in the light)
II   scotochromogens (yellow pigment formed in the dark)
III non-chromogens
IV   rapid growers

Though now virtually obsolete, this grouping was of great value in that era of taxonomic chaos. Since then, much effort has been devoted to the classification of the mycobacteria and, as a result, many species names have been reduced to synonymity. Indeed, only 16 of the 128 names in the 1966 edition of Index Bergeyana are now in use (Ratledge and Stanford, 1982). Before 1980, the correct name for a species was, by international agreement, the first one to be validly published after 1 May 1753, the publication date of Linne's *Species Planetarum.* It is now only necessary to refer back to the 'Approved lists of bacterial names' (Skerman *et al.*, 1980) published in the *International Journal of Systematic Bacteriology* on 1 January 1980. This list (see Table 3.1) contains 41 species of mycobacteria, but it omits a number of apparently distinct species, and several others have been described subsequently.

The variation of properties within the genus *Mycobacterium* is enormous and is reflected in the range of virulence, habitat, rate of growth, nutritional requirements and antigenicity. There are, in fact, relatively few properties that are common to all mycobacteria and yet clearly delinate this genus from related ones. Many of the unique characteristics of the mycobacteria are to be found in their very complex lipid-rich cell walls.

The mycobacteria appear to have evolved from the group of Gram-positive aerobic rods which includes the genera *Corynebacterium* and *Nocardia*. Indeed mycobacteria are Gram-positive although they are not easily stainable by this method. Mycobacteria are aerobic (although some such as bovine tubercle bacilli prefer low oxygen tensions), non-sporing and non-motile. They do not form capsules in the strict sense although some strains are very smooth, even slimy, owing to a thick coat composed of lipids termed mycosides (see page 16).

Mycobacteria appear to divide by simple binary fission, although some authors have postulated more complex life cycles, possibly including cell wall-free, or spheroplast, forms. Although such forms may be produced as laboratory artefacts, claims that they occur naturally or indeed that they are the causative agents of certain granulomatous diseases such as sarcoid or Crohn's disease require careful substantiation.

Antigenic structure

Mycobacteria, being complex unicellular organisms, contain many antigenic proteins, lipids and polysaccharides. The exact number of antigenic determinants is unknown. The sensitive technique of crossed immunoelectrophoresis reveals up to 90 antigens (Closs *et al.*, 1980) but this high number is probably still far short of the total.

The antigens are conveniently divided into cytoplasmic (soluble) and

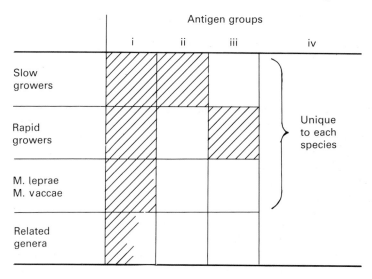

Fig. 2.1 The distribution of soluble antigens in the genus *Mycobacterium*

cell-wall lipid-bound (insoluble) antigens. Both have proved of value for classifying species and typing strains. Up to 15 precipitin lines are demonstrable when ultrasonicates of mycobacteria are tested against homologous antisera by double diffusion in agar gel. This technique has been extensively studied for taxonomic purposes by Stanford and his colleagues (see Stanford and Grange, 1974) who described four groups of soluble (diffusible) antigens (Fig. 2.1): those common to all mycobacteria (group i); those restricted to slowly growing species (group ii); those occurring in rapidly growing species (group iii); and those unique to each individual species (group iv). This antigenic distribution indicates a fundamental difference between the slowly growing and rapidly growing species, and suggests that these groups separated early in the evolution of the genus. Furthermore, some of the group iii antigens are also found in the genus *Nocardia*, suggesting a close relationship between this genus and the rapidly growing mycobacteria. Many of the common (group i) antigens are also found in the norcardiae and some are detectable in related genera such as *Corynebacterium* and *Listeria*. This intergeneric sharing of antigens is probably responsible for the notorious lack of specificity of serological tests for tuberculosis.

Electrophoretic analysis of culture filtrates of *M. tuberculosis* (Daniel and Janicki, 1978) revealed eleven precipitin arcs which, for reference purposes, have been numbered and some have been characterized. Antigens 1, 2 and 3 are, respectively, the polysaccharides arabinomannan, arabinogalactan and glucan and these are common to all mycobacteria. Antigens 6, 7 and 8 are widely distributed proteins, while antigen 5 is a glycoprotein with an antigenic determinant apparently unique to *M. tuberculosis*.

Many workers have attempted to isolate the species-specific (group iv) antigens for use in diagnostic tests, but this task has proved very difficult for two reasons. First, specific antigenic determinants often occur on the same protein molecule as shared antigens. Even purification methods based on binding to highly specific antibodies (affinity chromatography) cannot separate two determinants if they are on the same molecule. Secondly, a given determinant may be present on a range of molecules of differing physicochemical properties. Thus, preparative techniques based on such differences (gel filtration and ion-exchange chromatography) have not proved very useful. Recently, the development of 'genome libraries' and DNA cloning techniques have provided powerful tools for obtaining pure antigens (see page 25). Also, monoclonal antibodies have been used to detect antibodies to specific epitopes in competition immunoassays (see Chapter 5, page 83).

Insoluble cell-wall bound antigens are usually demonstrated by direct agglutination of whole bacilli by appropriate antisera. This technique is applicable to those species of mycobacteria that form stable, smooth suspensions and was used extensively by Schaefer and his colleagues (see Wolinsky and Schaefer, 1973) for typing *M. avium*, *M. intracellulare* and *M. scrofulaceum*. These serotypes are discussed in detail on page 37.

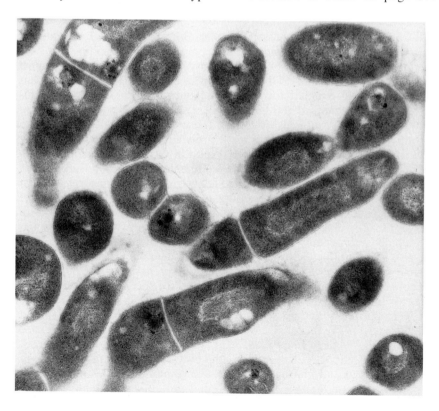

Fig. 2.2 Electron micrograph of a thin section of mycobacteria (BCG vaccine) showing nuclear bodies, cell walls, septa and lipid inclusion bodies (×42,000)

Serotypes are identifiable in several other species but not, unfortunately, in *M. tuberculosis* which is rough and readily auto-agglutinates. The responsible antigens have been identified as the sugar moieties on a group of peptidoglycolipids and phenolic glycolipids collectively termed mycosides (see page 16).

The structure of the cell

The mycobacteria consist of cytoplasm bounded by a plasma membrane and enclosed by a complex lipid-rich cell wall. The single chromosome is tightly wrapped into a nuclear body (Fig. 2.2) but is not bounded by a nuclear membrane. Thus, like other bacteria, the mycobacteria are prokaryots (higher unicellular and multicellular forms of life have nuclear membranes and are termed eukaryots). In common with many other bacteria, some mycobacteria contain additional small circles of DNA termed plasmids or episomes. The cell membrane consists of a bilayer of polar phospholipids with their hydrophobic ends facing inwards and their hydrophilic ends facing outwards. The membrane is closely associated with the enzymes and cofactors involved in energy production.

The bacterial cells vary in shape from species to species, and even within an individual strain according to the growth conditions. The cells of *M. avium* may be almost coccoid with those of *M. xenopi* may be filamentous with occasional branching. Cells of *M. kansasii and M. marinum* are often elongated and with a distinctive beaded or banded appearance (see page 35).

The cell wall

The mycobacterial cell wall is the most complex in all of nature. Its major characteristic is a very high lipid content. Indeed lipids account for about 60 per cent of the cell wall weight and they consist of a wide range of compounds, some being similar to those found in other organisms and others being unique to the mycobacteria. Freeze-fracturing techniques reveal that the cell wall has several distinct layers (Barksdale and Kim, 1977). These are shown diagrammatically in Fig. 2.3. The inner layer, overlying the cell membrane, is composed of peptidoglycan (murein). This, as in other bacteria, consists of long polysaccharide chains cross-linked by short peptide chains, thereby forming a net-like macromolecule that gives the cell its shape and rigidity. The polysacharide chains contain N-glycolyl muramic acid and N-acetyl glucosamine in alternating positions and the cross-linking peptide chains consist of the four amino acids L-alanine, D-isoglutamine, *meso*-diamino-pimelic acid and D-alanine. An exception is *M. leprae* which has glycine instead of L-alanine (Draper, 1976). The mycobacterial murein is very similar to that in other genera except that it contains N-glycolyl muramic acid instead of the more usual N-acetyl analogue.

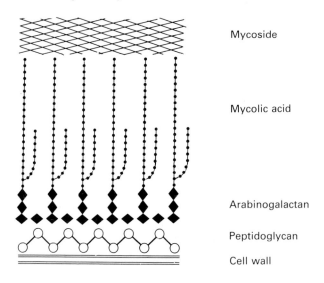

Mycoside

Mycolic acid

Arabinogalactan

Peptidoglycan

Cell wall

Fig. 2.3 Diagrammatic section of the mycobacterial cell wall

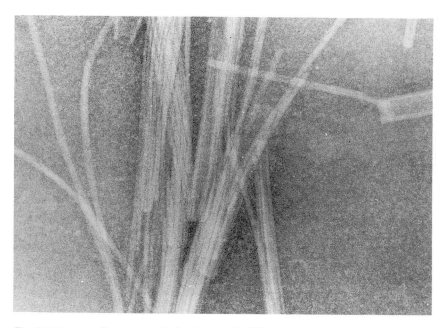

Fig. 2.4 Mycoside filaments of *M. fortuitum* (×90,000)

Mycobacteria are powerful adjuvants: Freund's complete adjuvant consists of killed mycobacteria in oil. This activity resides largely in the murein and also in small water-soluble fragments released from murein by digestion with lysosome. One such water-soluble adjuvant, N-acetyl-muramyl-L-alanyl-D-isoglutamine (muramyl dipeptide, MDP) has been synthesized (Lefrancier *et al.*, 1977) and is commercially available. External to the murein is a layer of arabinogalactan, a branded polysaccharide composed of arabinose and galactose (Lederer, 1971). The terminal arabinose units on the side chains are covalently linked to a group of long-chain fatty acids termed mycolic acids (see below). These form a dense pallisade, arranged in rope-like structures, which gives the cell wall its thickness and is largely responsible for acid-fastness. The outer layer is composed of a heterogeneous group of peptidoglycolipids or phenolic glycolipids termed mycosides. These often appear as ribbon-like filaments (Fig. 2.4) and, in the case of intracellular pathogens such as *M. avium*, *M. lepraemurium* and *M. leprae*, they appear as a capsule-like electron transparent zone around the bacilli. The cell wall also contains a range of other lipids as well as various proteins and polysaccharides.

Lipids

In recent years, mycobacterial lipids have been extensively investigated and several excellent and detailed reviews have been published (Goren and Brennan,1979; Minnikin, 1982).

Many of the lipids are based on long-chain fatty acids, i.e. compounds having the general formula $CH_3\text{-}(CH_2)_n\text{-}COOH$. These are often modified by the presence of unsaturated bonds, cyclopropane rings or methyl side chains. These lipids include tuberculostearic, mycoserosic, phthienoic, and mycolic acids.

Tuberculostearic acid, although found in many mycobacteria, has not been found in other genera. Thus its detection by mass spectroscopy has been proposed as a sensitive way to determine whether mycobacteria are present in a clinical specimen. Mycoserosic acids are found in some members of a class of surface lipids termed mycosides and also in the sulpholipids which may be associated with virulence.

The mycolic acids are a complex group of long-chain α-alkyl, β-hydroxy fatty acids. This term implies that there is an alkyl chain attached to the CH_2 group adjacent to the terminal carboxylic (–COOH) group. The general formula for the simplest type of mycolic acid is shown in Fig.

$$CH_3 - (CH_2)_x - CH.OH - CH - COOH$$
$$|$$
$$(CH_2)_y$$
$$|$$
$$CH_3$$

Fig. 2.5 The general formula of mycolic acid

Mycolic acids and related compounds occur in the corynebacteria and nocardiae as well as in the mycobacteria although the sizes of the molecules vary. The number of carbon atoms in the corynomycolic acids, nocardomycolic acids and mycobacterial mycolic acids are, respectively, 28 to 40, 40 to 60 and 60 to 90. This variation in size is of considerable value in defining the genera. It is thought that these long-chain molecules are formed by the fusion of a number of shorter-chain fatty acids, although the details are far from clear. The mechanism for the synthesis of mycolic acids is the target for the antituberculous drug isoniazid.

In addition to differences in chain length, the fatty acid component may be modified by the presence of unsaturated bonds, cyclopropane rings or methyl side chains. These are all termed α-mycolic acids. Others possess oxygen-containing groups and are named after the group, i.e. carboxy-, methoxy- or keto-mycolic acids.

Mycobacteria contain complex mixtures of these mycolic acids, but two-dimensional chromatography of their methyl esters shows that they fall into four main patterns represented by *M. tuberculosis*, *M. avium*, *M. fortuitum* and *M. chelonei* (Minnikin, 1982). An exception is *M. leprae* which possesses a simpler pattern of α-mycolic acids and keto-mycolic acids only.

Mycolic acids are often covalently linked to sugars. As mentioned above, those forming the main cell wall pallisade are linked to the arabinose residues of the structural polysaccharide arabinogalactan. A related class of lipids are the trehalose dimycolates which consists of two mycolic acids linked to the disaccharide trehalose. These lipids are also known as 'cord factors' and are discussed in relation to virulence on page 20.

A further biologically important class of mycobacterial lipids is represented by the mycosides. This is a collective name for a rather heterogeneous group of type-specific surface lipids which are responsible for agglutination serotype, colonial morphology and possibly virulence. They also serve as bacteriophage receptor sites. In view of these biological activities, Lederer remarked that the mycosides are 'in charge of the public relations of the mycobacteria'. In their structure and biological role, they are analogous to the *O* antigens of the Gram-negative enteric bacteria.

There are two major classes of mycosides, the peptidoglycolipids which contain lipids (mycoserosic acids), sugars and amino acids, and the phenolic glycolipids (phenol-phthiocerol dimycoserosates) which lack amino acids. These molecules often contain combinations of unusual sugars and therefore form unique antigenic determinants. A third and less frequent type is a sugar-free mycoside and includes the 'attenuation indicator lipid' described below and on page 21. Phenolic glycolipids have been isolated from *M. kansasii*, *M. bovis* and *M. marinum* and are termed mycosides A, B and G respectively. Large amounts of phenolic glycolipids similar to mycoside A have been isolated from *M. leprae* (Hunter and Brennan, 1981). One of these, phenolic glycolipid-I (PGL-I) is specific for this organism and a monoclonal antibody has been produced against it (Young *et al.*, 1984). This should prove of great value for diagnosis as well as for verifying claims for the *in vitro* cultivation of *M. leprae*.

The peptidoglycolipids, collectively termed mycoside C, are widely distributed throughout the genus and differ widely in their structure, particularly in their antigenic sugar groups. The C mycosides of *M. avium* and related organisms have been studied in great detail (Brennan *et al.*, 1979) and the nature of their sugar groups has been shown to determine the agglutination serotype of the strain.

Studies on a group of South Indian strains of *M. tuberculosis* that are attenuated in the guinea pig (see page 32) revealed the presence of a phenolic lipid similar to mycosides A and B except that the terminal sugar group was absent (Goren *et al.*, 1974). As this compound was not detected in strains virulent for the guinea pig, it was termed the 'attenuation indicator' (AI) lipid. The functional association with this lipid and attenuation is not known. Likewise the relationship of a companion substance, phthiocerol dimycoserosate (DIM), to virulence is uncertain, although a DIM-less mutant of *M. tuberculosis* strain H37Rv was found to be avirulent.

The sulpholipids are strongly acidic compounds consisting of mycoserosic acid covalently linked to trehalose sulphate. It was thought that this class of glycolipids was a major determinant of virulence of *M. tuberculosis*. Indeed, as this lipid strongly binds to the dye neutral red, such binding formed the basis of the 'cytochemical neutral red test for virulence' (Dubos and Middlebrook, 1948). This association now appears less likely as some fully virulent strains of *M. tuberculosis* contain only small amounts of these lipids (Goren *et al.*, 1982). They may, nevertheless, contribute to virulence by inhibiting phagosome–lysosome fusion (see page 66), neutralizing lysosomal enzymes and forming toxic complexes with trehalose dimycolate (cord factor).

Phospholipids occur mainly in the cell membrane rather than in the wall. The principal ones are cardiolipin, phosphatidyl ethanolamine and a group of phosphatidyl inositol mannosides. These lipids are antigenic and have been used in attempts to develop serodiagnostic tests for tuberculosis (Reggiardo *et al.*, 1980).

The ecology of the mycobacteria

Until recently, the major pathogens *M. tuberculosis* and *M. leprae* were the main foci of scientific attention within the genus while other species were regarded as being second-rate citizens. This attitude was reflected in the choice of *M. tuberculosis* as the type species of the genus and in the use of collective epithets such as 'atypical', 'tuberculoid', 'pseudotubercle' and 'MOTT' (mycobacteria other than typical tubercle) bacilli.

It is now realized that the genus *Mycobacterium* consists typically of free-living saprophytes and that only a small minority of species have adapted to a dependence on a living host.

The mycobacteria are very hydrophobic on account of their thick, waxy cell walls yet, paradoxically, their natural habitats are watery ones – mud, marshes, ponds, rivers and estuaries (Collins *et al.*, 1984). They are particularly abundant in sphagnum bogs (Kazda, 1979) and they prefer a slightly acidic environment (Stanford and Paul, 1973). In view of their water-repellent coats, they are often found at air–water interfaces: indeed they have been termed 'the ducks of the microbial world' (Grange, 1987). This situation enables them to derive oxygen from the atmosphere and nutrients from the water – such nutrients probably arising from decomposing vegetation.

In addition to their natural watery environments, some species have made their home in domestic, industrial and hospital water supplies from which they are easily transmissible to man. In view of this, great care must be taken to avoid contaminating clinical specimens with tap water. An apparent outbreak of infection due to the thermophilic species *M. xenopi* was traced to the practice of collecting sputum into metal pots that had been rinsed under a hot tap. Even distilled and deionized water may be contaminated; indeed, colonization of deionizer resins by mycobacteria has been reported (Azadian *et al.*, 1981). Consequently, if small numbers of acid-fast bacilli are seen in an inappropriate number of sputum smears, the staining reagents should be checked.

In recent years, considerable interest has developed in the effect of previous contact with environmental mycobacteria on challenge with – BCG vaccine or with virulent mycobacteria later in life. This topic is discussed in Chapter 5.

Water is probably the usual source and vector of environmental strains that cause immunological sensitization or overt disease in animals and man. Mycobacteria may thus enter the host through the alimentary tract, skin and lung by drinking, bathing and inhalation of aerosols respectively. In some cases, such as 'swimming pool granuloma', caused by *M. marinum*, there is direct evidence for infection of cuts and abrasions by water-borne strains. Likewise, cases of pulmonary *M. kansasii* infection amongst coalminers have occurred in those using showers supplied from tanks contaminated with this species. There is also evidence that mycobacteria are carried from rivers into the sea, aerosolized by waves, wafted inland by sea breezes and inhaled by dwellers in coastal regions (Gruft *et al.*, 1981).

Pigmentation

Many of the mycobacteria produce yellow, orange or, less frequently, salmon-pink pigments either in the dark (scotochromogens) or on exposure to light (photochromogens). The pigments are carotenoids, a class of polyterpene lipids that occur widely throughout nature. There are numerous different carotenoids and these are divisible into two groups, the oxygen-free carotenes and the oxygen-containing xanthophylls. The biosynthesis and structure of mycobacterial carotenoids has been reviewed by Minnikin (1982) and David (1984). Chromatographic analysis shows that the distribution of the pigments varies considerably from species to species (Tarnok and Tarnok, 1970) although those of the photochromogens *M. kansasii* and *M. marinum* appear identical. The most widely distributed of the pigments are α-carotene and leprotene, the latter so-named as it was first isolated from a strain thought to be a cultivable form of *M. leprae*. Pigmentation is of value in identification, although pigmentary variants within species are occasionally encountered.

Carotenoids are situated in the hydrophobic interior of the plasma membrane bilayer and it is probable that their natural function is to protect the cells against photodynamic injury. They are found in close association with the menaquinones, a photoreactive class of lipids involved in electron transfer and energy generation.

Acid-fastness

Acid-fastness is defined as the ability of the bacterial cell to resist decolouration by weak mineral acids after staining with one of the basic arylmethane dyes. The property is not confined to the mycobacteria: nocardiae, some corynebacteria and related organisms and bacterial spores are weakly acid-fast. Nevertheless, the property is widely used for the microscopic detection of mycobacteria in clinical or environmental specimens (see page 52).

Despite numerous investigations, the chemical basis of acid-fastness is poorly understood. Mycolic acids are certainly involved and the degree of acid-fastness is related to the size of the acids. Thus the corynomycolic acids, nocardomycolic acids and mycobacterial mycolic acids are progressively larger (see page 16) and are associated with progressively more intense staining. Chemical binding of the dye to the mycolic acid occurs but this is not the whole explanation as disruption of the cell wall by any means reduces its acid-fastness considerably. It has therefore been postulated that the mycolic acids are arranged in certain configurations that cause a trapping of the dye. This view is supported by the finding that acid-fastness is associated with the mycolic acid that is covalently bound to the layer of arabinogalactan rather than that lying free within the cell wall (Goren, 1972).

Pathogenicity and virulence

Pathogenicity is the ability of a micro-organism to cause disease. Clearly this property depends on the susceptibility of the host as well as the aggressiveness of the invading organism. Some micro-organisms are obligate pathogens, having developed a total dependence on a living host for their continued existence. In the case of the mycobacteria this includes the human pathogens *M. tuberculosis* and, possibly, *M. leprae*. Many other mycobacteria are opportunist pathogens, normally existing as harmless saprophytes but becoming pathogens under certain permissive conditions.

Virulence is a quantitative measure of pathogenicity and may vary considerably according to the host species. Thus, although virulence may be quantitated in a standard, preferably inbred, animal, care must be taken when interpreting such findings in relation to other animals or man. This is evident within *M. tuberculosis* and related strains from the following examples: (a) the vole tubercle bacillus (*M. microti*) is virulent for voles and some other small animals yet attenuated in man; (b) the bovine tubercle bacillus is much more virulent than the human type for the rabbit; (c) a South Indian variant of the human tubercle bacillus is attenuated in the guinea pig but is virulent in man; and (d) bovine tubercle bacilli are virulent for both cattle and man yet human tubercle bacilli rarely cause progressive disease in cattle.

It has long been recognized that mycobacteria owe their pathogenicity to their ability to invade and survive within the macrophages, as described in detail in Chapter 5. Although some strains certainly liberate toxic compounds, their virulence is not primarily associated with such substances.

There have been several claims that virulence of *M. tuberculosis* is related to the presence of certain toxic lipids in the cell wall. These include cord factor (dimycolyl trehalose; Bloch, 1950) and the sulpholipids (Middlebrook *et al.*, 1959). Cord factor derives its name from the erroneous assumption that it is reasonable for the characteristic 'serpentine cords' seen in microcolonies of tubercle bacilli (Fig. 2.6). Although fairly toxic for mice, this lipid occurs throughout the genus and it is unrelated to virulence.

The association of sulpholipid content with virulence appears to be fortuitous. Thus, although attenuated South Indian strains contain only small amounts of sulpholipids, such small amounts are also found in a minority of fully virulent tubercle bacilli (Grange *et al.*, 1978). In addition, tubercle bacilli of phage type B (see page 34) contain very little sulpholipid

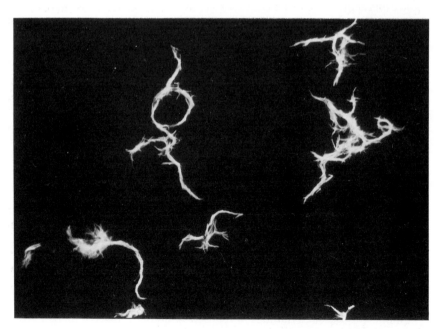

Fig. 2.6 Microcolonies of *M. tuberculosis* on a slide culture showing serpentine cords. Courtesy B.W. Allen

yet they are as virulent as strains of phage type A which contain large amounts (Goren *et al.*, 1982). Thus it is probable that sulpholipid content is not a prime cause of virulence. On the other hand, sulpholipid binds to cord factor and the complex is highly toxic for mitochondria (Kato and Goren, 1974).

A further lipid of possible relevance to virulence of *M. tuberculosis* is the attenuation indicator lipid, described on page 17. This is characteristic of the South Indian strains but is not found in isoniazid-resistant classical strains which, in common with the former, are often of diminished virulence for the guinea pig, or in laboratory attenuated strains. Thus its association with attenuation may also be a fortuitous one.

The nature of virulence of *M. leprae* is even less well understood. This species is remarkably non-toxic: lepromatous leprosy patients with huge numbers of bacilli throughout their bodies often appear remarkably well. The virulence of this organism may well be attributable to its ability to enter non-immunological cells and to escape into the cytoplasm of macrophages. A further property of the leprosy bacillus that may be associated with its virulence is its ability to suppress DNA synthesis within an infected cell (Samuel *et al.*, 1987).

Strains of *M. kansasii* isolated from cases of human disease tend to be more resistant to hydrogen peroxide than strains isolated from the environment, suggesting an association similar to that postulated for *M. tuberculosis*. *Mycobacterium ulcerans* may owe its virulence to a cytotoxic substance (see page 141), while virulence of *M. avium* appears to be related to colony morphology, suggesting a protective role for surface mycosides (Kuze and Uchihira, 1984). Apart from these examples, the mechanisms of virulence of the opportunist mycobacteria have received scant attention.

There is an evident need for more studies on mycobacterial virulence. In particular, an identification of the determinant(s) of virulence could be of relevance to the development of new vaccines. On the other hand, virulence may be multifactorial and may depend more on the presence of antigens that elicit inappropriate immune responses than on any aggressive properties of the bacilli. This aspect of pathogenicity is discussed in Chapter 5.

Nutrition and metabolism

Despite the relatively slow growth of mycobacteria and the complexity of their lipid-rich cell walls, most species have very simple nutritional requirements. They do, nevertheless, show an enormous diversity in the substrates that they are able to use as nitrogen and carbon sources – a diversity exploited by taxonomists. Nutritional requirements include oxygen, carbon, nitrogen, phosphorus, sodium, potassium, sulphur, iron and magnesium. A typical simple medium that supplies all these nutrients is Sauton's medium, composed of asparagine, glucose, glycerol, Na_2HPO_4, K_2HPO_4, $MgSO_4$ and ferric ammonium citrate.

Oxygen is derived from the atmosphere. All mycobacteria are aerophilic, although some strains, such as bovine tubercle bacilli, are micro-aerophilic, i.e. they grow preferentially at reduced oxygen tensions. Carbon dioxide is essential for growth and may be derived from the atmosphere or from carbonates or bicarbonates in the media. The vitamin biotin is a cofactor involved in CO_2 fixation and is included in some media.

Trace elements are probably also required and sufficient quantities are available as impurities in the constituents of the media (Ratledge, 1982).

Carbon is obtained from sugars or organic acids. Variation in saccharolytic activity is used as a taxonomic tool (Gordon and Smith, 1953). Glycerol is a particularly well utilized carbon source and is metabolized to pyruvate. Accordingly, glycerol is incorporated in most media except those for the culture of bovine tubercle bacilli which grow much better in the presence of pyruvate.

Nitrogen is derived from ammonia, organic amides, amino acids, nitrate and nitrite. Variation in the range of amides from which ammonia is liberated has been widely used for classification and identification (Bönicke, 1962). Likewise, variation in amino acid utilization may be used to identify and type certain strains (Grange, 1976; Barrow, 1986). Asparagine is a particularly well utilized nitrogen source and is included in Löwenstein–Jensen and many other media.

In most respects, mycobacterial metabolism is very similar to that of bacteria in other genera. The reason for a particular preference for glycerol and asparagine as carbon and nitrogen sources is unknown. Likewise, the biochemical basis for differences in growth rate between the rapid and slow growers remains a mystery. Possible explanations include differences in the rates of DNA replication, diffusion of nutrients, cell wall synthesis and assembly, and the presence or absence of key energy-generating pathways.

There are two aspects of mycobacterial metabolism that are of particular interest, namely the synthesis and assembly of the various lipids and the acquisition of iron. Lipid synthesis has been reviewed in detail by Ratledge (1982) and Minnikin (1982). This fascinating subject is, perhaps, rather esoteric for the clinician, although mycolic acid synthesis is of relevance as this is the target for the widely used antituberculous agent isoniazid.

Acquisition of iron: exochelins and mycobactins

Iron is an essential requirement for bacterial growth and metabolism. As there is very little free soluble iron in the inanimate environment, and even less in the living host, bacteria require mechanisms for the capture and solubilization of iron and for its transport to the interior of the cell. In the living host, these mechanisms must compete with the various iron-binding proteins in serum and tissue fluids. It is now well recognized that the ability of micro-organisms to acquire iron from their hosts is an important determinant to pathogenicity (Griffiths, 1983). Acquisition of iron by bacteria is mediated by chelating agents collectively termed siderophores. Mycobacteria are unique in synthesizing two distinct classes of siderophores: exochelins and mycobactins. The synthesis, structure and function of these have been reviewed in detail by Ratledge (1982, 1984).

Exochelins are secreted into the environment while mycobactins are found within the cell wall. Exochelins are low-molecular-weight peptides and two major types have been described: the MB-type, which occurs principally in slowly growing species, and the MS-type, isolated mainly from the rapidly growing species (Macham *et al.*, 1977). Both are water-soluble, but the MB-type is also chloroform-soluble when complexed to iron. Each mycobacterial strain produces several different exochelins within one of the two types.

Mycobactins are water-insoluble lipids and are synthesized by all mycobacteria except *M. paratuberculosis* and some strains of *M. avium* (see page 40). In order to cultivate these strains *in vitro* it is necessary to add heat-killed mycobacteria or pure mycobactin to the medium. Mycobactins show a wide variation in structure both between and within species (Snow, 1970), but exploitation of this variation for identification and typing is restricted by the complexity of the analytical methods.

The interaction between exochelins and mycobactins is shown in Fig. 2.7. Iron is solubilized and chelated from the external environment by exochelin and, in the case of the MB-type, the iron is then transferred to mycobactin for transport across the cell wall. The MS-type exochelin–iron complex, on the other hand, appears to be transported directly across the cell wall. In this case the mycobactin serves as an iron store rather than as an active transport mechanism (Stephenson and Ratledge, 1980).

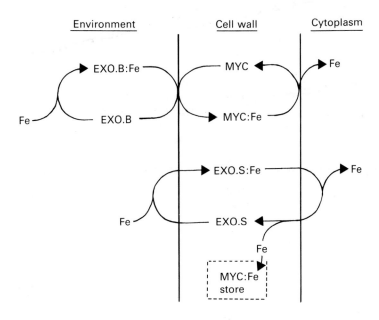

Fig. 2.7 The role of exochelins of the MB- and MS-type and mycobactins in the acquisition of iron by mycobacteria. Fe = iron; EXO.B = MB-type exochelin; EXO.S = MS-type exochelin; MYC = mycobactin

The mycobacterial genome

In common with all other bacteria, the mycobacteria contain a single circular chromosome (the genome) and some strains also contain additional small circular units of DNA termed plasmids (Crawford *et al.*, 1981). The molecular weights of mycobacterial genomes range from 3×10^9 to 5.5×10^9 (Baess and Mansa, 1978). For comparison the molecular weight of the genome of *Escherichia coli* is 2.5×10^9.

The ratio of the base pairs adenine (A) and thymine (T) to guanine (G) and cytosine (C) varies considerably between bacterial genera. The G+C content of the mycobacteria is high, from 66 to 71 per cent of the total base content (Baess and Mansa, 1978).

The degree of relation between mycobacterial (or any other) genomes may be determined by the technique of DNA hybridization. This technique is based on the ability of single strands of DNA to associate into double strands provided that the sequence of the base pairs of the two strands complement each other. Thus the extent of hybridization between fragments of DNA from two different strains or species gives an indication of the similarity of the sequence of the bases. Technical details of this powerful taxonomic tool are given by Baess and Bentzon (1978). Hybridization studies have confirmed the species boundaries as determined by other methods. They also reveal a low degree of homology between slowly growing and rapidly growing strains, thus supporting the serological evidence (see page 11) that these are evolutionary divergent subgenera.

The tendency for single-stranded DNA to bind to very similar or identical single strands has been exploited in the development of 'DNA probes'. Thus a radio-labelled single-stranded DNA fragment of a base sequence unique to a species can, in principle, be used to detect minute amounts of homologous DNA in a clinical specimen. At the time of writing several workers are attempting to develop such probes and it is hoped that these will help to confirm, or deny, that granulomatous diseases such as sarcoid or Crohn's disease have a mycobacterial aetiology.

A further method of comparing DNA from different mycobacteria is that of restriction endonuclease analysis. Many bacteria produce enzymes that split DNA chains but only at sites where there is a certain characteristic sequence of bases. Such enzymes are therefore termed restriction endonucleases. If a genome is digested by one of these enzymes, many fragments of DNA of different sizes will be produced; the number and sizes depending on the number and position of digestible sites. The fragments are separable according to their molecular weights by high-resolution electrophoresis and demonstrable as bands by fluorescent staining with ethidium bromide. This technique may reveal differences between closely related strains. Thus Collins *et al.* (1986) showed that bovine tubercle bacilli isolated in New Zealand could be separated into 33 restriction types by using three different endonucleases.

Restriction endonucleases have also proved of great value for DNA cloning and the production of 'genomic libraries' in *Escherichia coli*. The principles and techniques were reviewed in detail by Dahl *et al.* (1981). In brief, the free ends of DNA split by a restriction endonuclease have a 'stickiness' for each other and may be reconnected by the enzyme ligase. Likewise, a cleaved DNA fragment can be inserted into a chain of 'foreign' DNA by this process. In the most widely used technique, DNA is inserted into a non-essential site of a vector genome, usually the temperate bacteriophage lambda of *E. coli*. This hybrid vector is introduced into the host bacterium by transfection. When the phage enters the lytic cycle, hundreds of copies of the DNA of the phage together with the inserted fragment are produced. Genomic libraries therefore consist of nothing more than little bottles containing many phages, each bearing a small fragment of the cloned genome.

If the DNA inserted into a phage represents a whole gene, the gene product may be synthesized by the host bacterium. If these products are antigenic, they may be detected by staining phage plaques with a suitable labelled antibody. Lysogenic bacteria within such plaques are then isolated and used as a source for that antigen. Genome libraries have been produced from *M. leprae* (Clark-Curtiss *et al.*, 1985), *M. tuberculosis* (Young *et al.*, 1985; Eisenach *et al.*, 1986) and BCG (Thole *et al.*, 1985). Practical applications of these libraries are awaited, but in view of the great difficulties experienced in purifying mycobacterial antigens, they may prove of great value in obtaining pure antigens for diagnostic purposes and for immunological studies.

Bacteriophages of the mycobacteria

Since the first description of a phage lytic for a mycobacterium by Gardner and Weiser (1947), many 'mycobacteriophages' have been isolated. Most were isolated from environmental sources, but a few were found in naturally occurring lysogenic strains. Mycobacteriophages usually have hexagonal or oval heads and long, non-contractile tails (Fig. 2.8(a)), an exception being phage I3 which has a contractile tail (Fig. 2.8(b)). Little is known about the dynamics of the host–phage interactions, although in some cases it has been shown that the phage receptors on the bacteria are the mycosides (see page 16). Many of the phages have very wide host-ranges that appear to ignore the usual species boundaries. Accordingly, some phages isolated and propagated in strains of *M. smegmatis*, are lytic for *M. tuberculosis*. For details of the biology of mycobacteriophages, see Grange and Redmond (1978).

In contrast to the closely related genus *Corynebacterium*, there is no evidence that mycobacterial virulence is associated with lysogeny. Interest in the mycobacteriophages has therefore centred on their use as vectors for gene transfers and for typing of mycobacteria, particularly tubercle bacilli, for epidemiological purposes (see page 34).

Genetic transfers in mycobacteria

Numerous attempts have been made to transfer genes from one mycobacterium to another by direct cell contact (conjugation), by free DNA

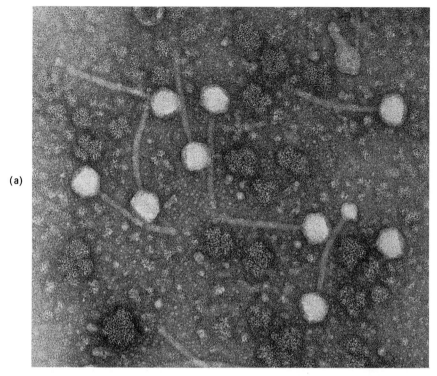

Fig. 2.8 Mycobacteriophages: (a) *M. kansasii* phage (×120,000); (b) *M. smegmatis* phage I3 (×200,000)

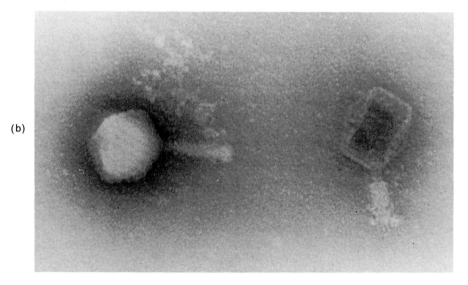

(transformation), and by use of phages as vectors (transduction). With few exceptions, these attempts, reviewed by Grange (1982), have met with little success. These exceptions include transduction of *M. smegmatis* with phage I3 (Sundar Raj and Ramakrishnan, 1970), conjugation of *M. smegmatis* (Mizuguchi *et al.*, 1976), and recombination of mutants of *M. aurum* by producing spheroplasts and fusing them by treatment with polyethylene glycol (Rastogi *et al.*, 1983). Genetic transfers have not yet been achieved in the pathogenic species.

The entry of free-phage DNA into mycobacterial cells (transfection) has been clearly demonstrated (Nakamura, 1970). Studies on transfection may therefore help to establish the optimal conditions for the transfer of bacterial DNA.

A technique for gene transfers in the major pathogenic species and BCG would be of value now that libraries of cloned mycobacterial DNA are available. This could lead to the development of new vaccines, including ones containing *M. leprae* antigens. Whether such vaccines would prove superior to BCG remains to be seen.

References

Azadian, B.S., Beck, A., Curtis, J.R., Cherrington, L.E., Gower, P.E., Phillips, M., Eastwood, J.B. and Nicholls, J. (1981) Disseminated infection with *Mycobacterium chelonei* in a haemodialysis patient. *Tubercle* **62**: 281–4.

Baess, I. and Bentzon, M.W. (1978) Deoxyribonucleic acid hybridization between different species of mycobacteria. *Acta Pathologica et Microbiologica Scandinavica* **86B**: 71–6.

Baess, I. and Mansa, B. (1978) Determination of genome size and base ratio of deoxyribonucleic acid from mycobacteria. *Acta Pathologica et Microbiologica Scandinavica* **86B**: 309–12.

Barksdale, L. and Kim, K.S. (1977) Mycobacterium. *Bacteriological Reviews* **41**: 217–372.

Barrow, P.A. (1986) Physiological characteristics of the *Mycobacterium tuberculosis–M. bovis* group of organisms with particular reference to heterogeneity within *M. bovis*. *Journal of General Microbiology* **132**: 427–30.

Bishop, P.J. and Neumann, G. (1970) The history of the Ziehl–Neelsen stain. *Tubercle* **51**: 196–206.

Bloch, H. (1950) Studies on the virulence of tubercle bacilli: isolation and biological properties of a constituent of virulent organisms. *Journal of Experimental Medicine* **91**: 197–217.

Bönicke, R. (1962) Report on identification of mycobacteria by biochemical methods. *Bulletin of the International Union Against Tuberculosis* **32**: 13–68.

Brennan, P.J. and Goren, M.B. (1979) Structural studies on the type specific antigens and lipids of *Mycobacterium avium, Mycobacterium intracellulare, Mycobacterium scrofulaceum*. *Journal of Biological Chemistry* **254**: 4205–11.

Clark-Curtiss, J.E., Jacobs, W.R., Docherty, M.A., Ritchie, L.R. and Curtiss, R. (1985) Molecular analysis of DNA and construction of genome libraries of *Mycobacterium leprae*. *Journal of Bacteriology* **161**: 1093–102.

Closs, O., Harboe, M., Axelson, N.H., Bunch-Christensen, K. and Magnusson, M. (1980) The antigens of *Mycobacterium bovis* strain BCG studied by crossed immunoelectrophoresis: a reference system. *Scandinavian Journal of Immunology* **12**: 249–63.

Collins, C.H., Grange, J.M. and Yates, M.D. (1984) Mycobacteria in water. *Journal of Applied Bacteriology* **57**: 193–211.

Collins, D.M., De Lisle, G.W. and Gabric, D.M. (1986) Geographic distribution of restriction types of *Mycobacterium bovis* isolates from brush-tailed possums (*Trichosuris vulpecula*) in New Zealand. *Journal of Hygiene* **96**: 431–8.

Crawford, J.T., Cave, M.D. and Bates, J.H. (1981) Characterization of plasmids from strains of *Mycobacterium avium-intracellulare*. *Reviews of Infectious Diseases* **3**: 949–52.

Dahl, H.H., Flavell, R.A. and Grosveld, F.G. (1981) The use of genomic libraries for the isolation and study of eukaryotic genes. In: *Genetic Engineering*, Vol. 2, pp. 49–127. Edited by R. Williamson. London: Academic Press.

Daniel, T.M. and Janicki, B.W. (1978) Mycobacterial antigens: a review of their isolation, chemistry and immunological properties. *Microbiological Reviews* **42**: 84–113.

David, H.L. (1984) Carotenoid pigments of the mycobacteria. In: *The Mycobacteria: A Sourcebook*, Part A, pp. 537–45. Edited by G.P. Kubica and L.G. Wayne. New York: Mercel Dekker.

Draper, P. (1976) Cell walls of *Mycobacterium leprae*. *International Journal of Leprosy* **44**: 95–8.

Dubos, R.J. and Middlebrook, G. (1948) Cytochemical reaction of virulent tubercle bacilli. *American Review of Tuberculosis* **58**: 698–9.

Eisenach, K.D., Crawford, J.T. and Bates, J.H. (1986) Genetic relatedness among strains of the *Mycobacterium tuberculosis* complex. *American Review of Respiratory Disease* **133**: 1065–8.

Gardner, G.M. and Weiser, R.S. (1947) A bacteriophage for *Mycobacterium smegmatis*. *Proceedings of the Society for Experimental Biology and Medicine* **66**: 205–6.

Gordon, R.E. and Smith, M.M. (1953) Rapidly growing acid-fast bacteria. *Journal of Bacteriology* **66**: 41–8.

Goren, M.B. (1972) Mycobacterial lipids: selected topics. *Bacteriological Reviews* **36**: 33–64.

Goren, M.B. and Brennan, P.J. (1979) Mycobacterial lipids: chemistry and biologic activities. In: *Tuberculosis*, pp. 63–193. Edited by G.P. Youmans. Philadelphia: W.B. Saunders.

Goren, M.B., Brokl, O. and Schaefer, W.B. (1974) Lipids of putative relevance to virulence in *Mycobacterium tuberculosis*: phthiocerol dimycoserosate and the attenuation indicator lipid. *Infection and Immunity* **9**: 150–8.

Goren, M.B., Grange, J.M., Aber, V.R., Allen, B.W. and Mitchison, D.A. (1982) Role of lipid content and hydrogen peroxide susceptibility in determining the guinea-pig virulence of *Mycobacterium tuberculosis*. *British Journal of Experimental Pathology* **63**: 693–700.

Grange, J.M. (1976) Enzymic breakdown of amino acids and related compounds by suspensions of washed mycobacteria. *Journal of Applied Bacteriology* **41**: 425–31.

Grange, J.M. (1982) The genetics of mycobacteria and mycobacteriophages. In: *The Biology of the Mycobacteria*, Vol. 1, pp. 309–51. Edited by C. Ratledge and J.L. Stanford. New York: Academic Press.

Grange, J.M. (1987) Infection and disease due to the environmental mycobacteria. *Transactions of the Royal Society of Tropical Medicine and Hygiene* **81**: 179–82.

Grange, J.M., Aber, V.R., Allen, B.W., Mitchison, D.A. and Goren, M.B. (1978) The correlation of bacteriophage types of *Mycobacterium tuberculosis* with guinea pig virulence and *in vitro* indicators of virulence. *Journal of General Microbiology* **108**: 1–7.

Grange, J.M. and Redmond, W.B. (1978) Host–phage relationships in the genus

Mycobacterium and their clinical significance. *Tubercle* **59**: 203–25.

Griffiths, E. (1983) Adaptation and multiplication of bacteria in host tissues. *Philosophical Transactions of the Royal Society* **303B**: 85–96.

Gruft, H., Falkinham, J.O. and Parker, B.C. (1981) Recent experience in the epidemiology of disease caused by atypical mycobacteria. *Review of Infectious Diseases* **3**: 990–6.

Hunter, S.W. and Brennan, P.J. (1981) A novel phenolic glycolipid from *Mycobacterium leprae* possibly involved in immunogenicity and pathogenicity. *Journal of Bacteriology* **147**: 728–35.

Kato, M. and Goren, M.B. (1974) Synergistic action of cord factor and mycobacterial sulfatides on mitochondria. *Infection and Immunity* **10**: 733–41.

Kazda, J., Muller, K. and Irgens, L.M. (1979) Cultivable mycobacteria in sphagnum vegetation of moors of South Sweden and coastal Norway. *Acta Pathologica et Microbiologica Scandinavica* **87B**: 97–101.

Kuze, F. and Uchihira, F. (1984) Various colony formers of *Mycobacterium avium–intracellulare*. *European Journal of Respiratory Diseases* **65**: 402–10.

Lederer, E. (1971) The mycobacterial cell wall. *Pure and Applied Chemistry* **25**: 135–65.

Lefrancier, P., Choay, J., Derien, M. and Lederman, I. (1977) Synthesis of N-acetylmuramyl-L-alanyl-D-isoglutamine, an adjuvant of the immune response, and of some N-acetyl muramyl peptide analogs. *International Journal of Peptide and Protein Research* **9**: 249–57.

Lehmann, K.B. and Neumann, R. (1896) *Atlas und Grundriss der Bakteriologie und Lehrbuch der speciellen bakteriologischen Diagnostik*, 1st edn. Munich: J.F. Lehmann.

Macham, L.P., Stephenson, M.C. and Ratledge, C. (1977) Iron transport in *Mycobacterium smegmatis*: the isolation, purification and function of exochelin MS. *Journal of General Microbiology* **101**, 41–9.

Middlebrook, G., Coleman, C.M. and Schaefer, W.B. (1959) Sulfolipid from virulent tubercle bacilli. *Proceedings of the National Academy of Science of the USA* **45**: 1801–4.

Minnikin, D.E. (1982) Complex lipids: their chemistry, biosynthesis and roles. In: *The Biology of the Mycobacteria*, Vol. 1, pp. 95–184. Edited by C. Ratledge and J.L. Stanford. New York: Academic Press.

Mizuguchi, Y., Suga, K. and Tokunaga, T. (1976) Multiple mating types of *Mycobacterium smegmatis*. *Japanese Journal of Microbiology* **20**: 435–43.

Nakamura, R.M. (1970) Transfection of *Mycobacterium smegmatis* in an acidic medium. In: *Host–Virus Relationships in Mycobacterium, Nocardia and Actinomyces*, pp. 166–78. Edited by S.E. Juhasz and G. Plummer. Illinois: Charles C. Thomas.

Rastogi, N., David, H.L. and Rafidnarivo, E. (1983) Spheroplast fusion as a mode of genetic recombination in mycobacteria. *Journal of General Microbiology* **129**: 1227–37.

Ratledge, C. (1982) Nutrition, growth and metabolism. In: *The Biology of the Mycobacteria*, Vol. 1, pp. 185–271. Edited by C. Ratledge and J.L. Stanford. New York: Academic Press.

Ratledge, C. (1984) Metabolism of iron and other metals by mycobacteria. In: *The Mycobacteria: A Sourcebook*, Part A, pp. 603–27. Edited by G.P. Kubica and L.G. Wayne. New York: Marcel Dekker.

Ratledge, C. and Stanford, J.L. (1982) Introduction. In: *The Biology of the Mycobacteria*, Vol. 1, pp. 1–6. Edited by C. Ratledge and J.L. Stanford. New York: Academic Press.

Reggiardo, Z., Vazquez, E. and Schnaper, L. (1980) ELISA tests for antibodies against mycobacterial glycolipids. *Journal of Immunological Methods* **34**: 55–60.

Runyon, E.H. (1959) Anonymous mycobacteria in pulmonary disease. *Medical Clinics of North America* **43**: 273–90.

Samuel, N.M., Jessen, K.R., Grange, J.M. and Mirsky, R. (1987) Gamma interferon, but not *Mycobacterium leprae*, induces major histocompatibility class II (Ia) antigens on cultured rat schwann cells. *Journal of Neurocytology* **16**: 281–7.

Skerman, V.D.B., McGowan, V. and Sneath, P.H.A. (1980) Approved lists of bacterial names. *International Journal of Systematic Bacteriology* **30**: 225–420.

Snow, G.A. (1970) Mycobactins: iron chelating growth factors from mycobacteria. *Bacteriological Reviews* **34**: 99–125.

Stanford, J.L. and Grange, J.M. (1974) The meaning and structure of species as applied to mycobacteria. *Tubercle* **55**: 143–52.

Stanford, J.L. and Paul, R.C. (1973) A preliminary report on some studies of environmental mycobacteria in Uganda. *Annales de la Societe Belge de Medicine Tropicale* **53**: 389–93.

Stephenson, M.C. and Ratledge, C. (1980) Specificity of exochelins for iron transport in three species of mycobacteria. *Journal of General Microbiology* **116**: 521–3.

Sundar Raj, C.V. and Ramakrishnan, T. (1970) Transduction in *Mycobacterium smegmatis*. *Nature* **228**: 280–1.

Tarnok, I. and Tarnok, Zs. (1970) Carotenes and xanthophylls in mycobacteria. I: technical procedures: thin layer chromatographic patterns of mycobacterial pigments. *Tubercle* **51**: 305–12.

Thole, J.E.R., Dauwerse, H.G., Das, P.K., Groothuis, D.G., Schouls, L.M. and van Embden, J.D.A. (1985) Cloning of *Mycobacterium bovis* BCG DNA and expression of antigens in *Escherichia coli*. *Infection and Immunity* **50**: 800–6.

Young, D.B., Khanolkar, S.R., Barg, L.L. and Buchanan, T.M. (1984) Generation and characterization of monoclonal antibodies to the phenolic glycolipid of *Mycobacterium leprae*. *Infection and Immunity* **43**: 183–8.

Young, R.A., Bloom, B.R., Grosskinsky, C.M., Ivanyi, J., Thomas, D. and Davis, R.W. (1985) Dissection of *Mycobacterium tuberculosis* antigens using recombinant DNA. *Proceedings of the National Academy of Science* **82**: 2583–7.

Wolinsky, E. and Schaefer, W.B. (1973) Proposed numbering scheme for mycobacterial serotypes by agglutination. *International Journal of Systematic Bacteriology* **23**: 182–3.

3 The species of mycobacteria

In contrast to many other groups of bacteria, the boundaries of the species within the genus *Mycobacterium* are usually very distinct. There are now few mycobacteria that cannot be assigned to one of the well-described species. Indeed, this may well now be the best classified of all bacterial genera.

The science of classification and speciation is known as taxonomy and it embraces several principles and schools of thought. One of the most widely used systems is Adansonian taxonomy, in which a large number of cultural and biochemical properties are used to group strains into clusters or 'taxa' according to their similarity. This technique has been applied successfully to the mycobacteria by the International Working Group on Mycobacterial Taxonumy (Meissner *et al.*, 1974; Saito *et al.*, 1977; Wayne *et al.*, 1971, 1978). Other techniques such as antigenic analysis (see page 11) and DNA hybridization (page 24) have also been used and give results that are in close accord with those of the numerical taxonomic studies.

From the taxonomic viewpoint, mycobacteria are divisible into the rapid growers, slow growers and those not yet cultivated *in vitro*. The clinical microbiologist finds it more useful to divide them into the major pathogens of man (the tubercle and leprosy bacilli), those species that frequently cause opportunist infections, and those that rarely or never do so. Whether the microbiologist identifies all isolates at the species level depends on local resources, workload and interest. Some consider it sufficient to allocate strains to clusters or complexes of similar clinical significance.

The 'approved lists of bacterial names' (Skerman *et al.*, 1980) contain 41 species of mycobacteria (Table 3.1). A few apparently distinct species, such as *M. diernhoferi*, were omitted and several have been described subsequently. In addition, some of the listed species are considered to be variants of other species by some authorities. Indeed most of the difficulties encountered in mycobacterial classification are associated with nomenclature rather than with the underlying scientific principles.

In this chapter the main features of the species are outlined, with particular emphasis on those that are pathogenic in man.

The *M. tuberculosis* group

For historical reasons the nomenclature of the causative organisms of tuberculosis is confusing. Koch's tubercle bacillus, later called *Mycobacterium tuberculosis* (Lehmann and Neumann, 1896) is divisible into human

Table 3.1 Approved names for mycobacteria

M. africanum	M. gordonae	M. phlei
M. asiaticum	M. haemophilum	M. scrofulaceum
M. aurum	M. intracellulare	M. senegalense
M. avium	M. kansasii	M. simiae
M. bovis	M. komossense	M. smegmatis
M. chelonei*	M. leprae	M. szulgai
M. chitae	M. lepraemurium	M. terrae
M. duvalii	M. malmoense	M. thermoresistibile
M. farcinogenes	M. marinum	M. triviale
M. flavescens	M. microti	M. tuberculosis
M. fortuitum	M. nonchromogenicum	M. ulcerans
M. gadium	M. neoarum	M. vaccae
M. gastri	M. parafortuitum	M. xenopi
M. gilvum	M. paratuberculosis	

*Also termed *M. chelonae*

and bovine types which, for many years, were called *M. tuberculosis* var *hominis* and *M. tuberculosis* var *bovis* respectively. In 1970, the separate scientific name *M. bovis* was proposed for the latter (Karlson and Lessel, 1970). Also, at about the same time, a heterogeneous group of tubercle bacilli with properties intermediate between the human and bovine strains were isolated from man in equatorial Africa (Castets *et al.*, 1969). Although the authors accepted that these were variants of the human tubercle bacillus, they used the separate species name *M. africanum*. To complicate the issue further, the rarely encountered vole tubercle bacillus of Wells (1946) was named *M. microti* by Reed (1957).

These four names all appear in the approved lists but there is ample evidence that if the same criteria of speciation is applied to these organisms as to other mycobacteria, they would clearly emerge as variants of a single species (Grange, 1982). Thus, despite the official nomenclature, *M. bovis*, *M. africanum* and *M. microti* may be regarded as variants of *M. tuberculosis*. A simple classification of those variants that cause disease in man on the basis of four tests has been proposed (Collins *et al.*, 1982) and is shown in Table 3.2. This classification is primarily for epidemiological use: the human, bovine and African types are of similar clinical significance and many bacteriologists, in order to avoid confusion, report them all as *M. tuberculosis*.

The human tubercle bacilli are divisible into two major types: the classical type originally isolated by Koch and the South Indian or Asian type. The latter was originally isolated in the Madras area of South India and differs from the classical type in being attenuated in the guinea pig and susceptible to killing by hydrogen peroxide *in vitro* (Mitchison, 1970). This variant was subsequently found to be susceptible to the isoniazid analogue thiophen-2-carboxylic acid hydrazide (TCH), to contain the characteristic attenuation indicator lipid (see page 17), and to be of phage type I (Grange *et al.*, 1978). About 60 per cent of tubercle bacilli isolated in the Madras region are of this type and it is also found in Asian communities who have settled in other parts of the world. Despite its low virulence for the guinea

Table 3.2 Subdivision of the *Mycobacterium tuberculosis* group for epidemiological purposes

Species and variant	TCH* susceptibility	Pyrazinamide susceptibility	Nitratase	Oxygen preference
M. tuberculosis				
Classical	Resistant	Sensitive	Positive	Aerobic
Asian	Sensitive	Sensitive	Positive	Aerobic
M. africanum				
African I	Sensitive	Sensitive	Negative	Micro-aerophilic
African II	Sensitive	Sensitive	Positive	Micro-aerophilic
M. bovis				
Bovine	Sensitive	Resistant	Negative	Micro-aerophilic

*TCH = thiophen-2-carboxylic acid hydrazide

pig, it appears fully virulent for man and causes similar disease to the classical strains.

Bovine strains differ from human strains in a few but important properties. Originally these strains were differentiated by their growth characteristics on Lowenstein–Jensen medium. Human strains produce luxuriant, heaped up, breadcrumb-like 'eugonic' colonies, while bovine strains grow as small, flat 'dysgonic' colonies (Royal Commission on Tuberculosis, 1907). Growth of bovine strains is stimulated by incorporating pyruvate in place of glycerol in the media. As shown in Table 3.2, bovine strains are distinguishable from human strains in being micro-aerophilic, nitratase negative and resistant to pyrazinamide. They resemble the Asian type in being sensitive to TCH.

The African type (*M. africanum*) was originally isolated from man in equatorial Africa. The strains resemble the bovine type in being micro-aerophilic and sensitive to TCH but resemble the human type in being sensitive to pyrazinamide. Those isolated from West Africa (Dakar, Mauretania and Yaounde) more closely resemble bovine strains and are nitratase negative, while those from East Africa (Rwanda and Burundi) are nitratase positive and may have other properties in common with human strains. African strains have also been isolated from patients, mostly of African origin, resident in Europe. For epidemiological purposes, African strains are divisible into two groups according to their nitratase activity (Table 3.2; Collins *et al.*, 1982).

Mycobacterium microti, a rarely encountered pathogen of voles and other small mammals, is not included in Table 3.2 as it is not naturally virulent for man. It was once used in man as a vaccine but was abandoned in favour of BCG as some vaccinees developed lupus-like lesions at the site of injection. This variant has not been well characterized but appears to have properties intermediate between the human and bovine types.

The vaccine strain, BCG, was reputedly derived from a bovine strain but differs from that type in several respects, particularly in being eugonic and aerobic. As BCG occasionally causes severe local or widespread lesions

Table 3.3 The major phage types of *M. tuberculosis*

Phage type	DS$_6$A	GS$_4$E	Susceptibility to: BG1	PH	BK1	33D
A	+	−	−	−	−	+
I	+	+	+	+	−	+
B	+	+	+	+	+	+
BCG	+	−	−	−	−	−

(Lotte *et al.*, 1984) there is a need to distinguish it from virulent tubercle bacilli. This may be achieved by the tests shown in Table 3.2 and by its characteristic resistance to cycloserine.

Members of the *Mycobacterium tuberculosis* group are divisible into three clear-cut phage types, A, B and C (Table 3.3). These may be further subdivided, although with less reliability (Grange and Redmond, 1978). Bovine, African and the majority of human strains are of type A. The Asian type, and a few other strains, are type I (intermediate), while strains of type B contain low levels of sulpholipids but resemble those of type A in other respects, including virulence (Goren *et al.*, 1982). Strains of BCG differ from other tubercle bacilli in being resistant to phage 33D (Yates *et al.*, 1978).

Slow growing photochromogens

This group contains four species – *M. kansasii*, *M. marinum*, *M. simiae* and *M. asiaticum*. All are opportunist human pathogens and *M. simiae*, as the name suggests, also causes disease in monkeys.

Mycobacterium kansasii

This species, named by Hauduroy (1955), is the most frequent cause of opportunist mycobacterial disease in Great Britain and is second only to the *M. avium* group in the USA. This species has been isolated on several occasions from piped water supplies (Collins *et al.*, 1984) but is rarely encountered in the natural environment.

M. kansasii is one of the faster growing of the slow growers – a good growth is obtained within 14 days of subculture on to Löwenstein–Jensen medium. The bacterial cells in clinical material and on subculture tend to be elongated and to show distinct beading (Fig. 3.1).

Mycobacterium marinum

This species is responsible for a granulomatous skin disease of man described by Linell and Norden (1954) and known as swimming-pool

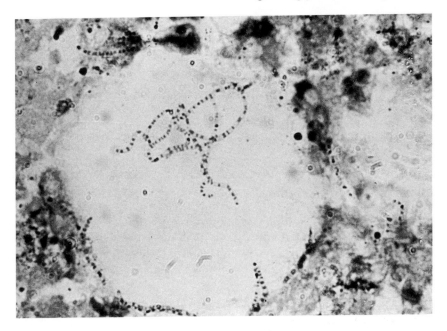

Fig. 3.1 *M. kansasii* in sputum, showing characteristic elongated cells with distinct banding. Courtesy J.L. Stanford

granuloma, fish-tank granuloma and fish-fancier's finger (see Chapter 9). It was originally isolated from diseased fish (Aronson, 1926). It is very similar in its colonial and cellular appearance to *M. kansasii* and the pigments of the two species appear to be identical (Tarnok and Tarnok, 1970). It has a lower temperature range of growth than *M. kansasii* and grows poorly, or not at all, at 37°C. It is also differentiated from *M. kansasii* by its antigenic structure and lack of nitratase activity. A characteristic peculiar to this species is a very strong α-L-fucosidase activity (Grange and McIntyre, 1979).

Mycobacterium simiae

This species, originally isolated from monkeys (Karasseva *et al.*, 1965) is identical to *M. habana* (Valdivia-Alvarez *et al.*, 1971). In common with *M. tuberculosis*, it synthesizes niacin and may thus be identified as the latter unless tested for pigment production in the light. Although nitratase-negative, it differs from *M. marinum* in growing well at 37°C and in failing to hydrolyse Tween 80.

Mycobacterium asiaticum

This was also isolated from monkeys by Karasseva and colleagues (1965) but was later identified as a separate species (Weiszfeiler *et al.*, 1971).

Although photochromogenic, it closely resembles *M. gordonae* biochemically and differs from *M. simiae* in its amidase activity and in its failure to produce niacin. It is a very rare cause of lung disease in man.

Slow growing scotochromogens

This group includes the species *M. gordonae*, *M. scrofulaceum* and *M. szulgai*, although occasional strains of *M. avium* are also scotochromogenic.

Mycobacterium gordonae

Named after Dr Ruth Gordon (Bojalil *et al.*, 1962), this species is found in water and has also been termed *M. aquae* and the 'tap-water scotochromogen'. Although a frequent contaminant of clinical specimens, it also appears to be a rare cause of disease in man.

Mycobacterium scrofulaceum

This species (Prissick and Masson, 1956) was originally termed *M. marianum* but this name, although the earlier of the two, was abandoned because of its frequent confusion with *M. marinum*. Although differing in antigenic structure, *M. scrofulaceum* superficially resembles pigmented strains of *M. avium* and some workers group these, and *M. intracellulare*, into the MAIS complex (see page 37). As suggested by its name, *M. scrofulaceum* is a cause of cervical lymphadenitis but it also causes pulmonary disease. It is differentiated from *M. avium* by its urease activity and by agglutination serology.

Mycobacterium szulgai

This species was first characterized by its lipid chromatography pattern (Marks *et al.*, 1972) and named after Dr T. Szulga, from Poland, who helped to develop the technique. Although a scotochromogen at 37°C, it is photochromogenic when incubated at 27°C. Although a relatively uncommon pathogen, cases of pulmonary disease, tenosynovitis, bursitis and lymphadenopathy have been reported (Collins *et al.*, 1986)

Slow growing non-chromogens

This group includes the '*M. avium* group', *M. farcinogenes*, *M. gastri*, *M. malmoense*, the '*M. terrae* group', *M. ulcerans* and *M. xenopi*, although the latter two species may produce a pale yellow pigment on prolonged incubation. The '*M. terrae* group' contains three phenetically similar species, *M. terrae*, *M. triviale* and *M. nonchromogenicum*, that superficially resemble *M. avium*. *Mycobacterium shimoidei* was omitted from the approved lists but it may nevertheless be a distinct species. Mention must also be made of *M. haemophilum* which, despite being an 'approved' species, has not yet been clearly shown to be a mycobacterium.

The *M. avium* group

This is a group of closely related strains whose classification is still somewhat confusing. The 'official' species within the group are *M. avium* (the avian tubercle bacillus), *M. intracellulare*, *M. lepraemurium* (the cause of rat leprosy) and *M. paratuberculosis* (the cause of Johne's disease, or chronic hypertrophic enteritis, of cattle). This is therefore an important group from the point of view of both human and animal disease.

In 1901 Chester gave the name *Mycobacterium avium* to the avian tubercle bacillus. In 1967 Runyon found that an organism called *Nocardia intracellularis* was a mycobacterium and renamed it *M. intracellulare*. He also found that the 'Battey bacillus', an acid-fast organism isolated at the Battey State Hospital, Georgia, belonged to this species. Since that time, there has been much debate as to whether *M. avium* and *M. intracellulare* are really distinct species. From the results of an international cooperative taxonomic study (Meissner *et al.*, 1974) it was concluded that they were variants of a single species to be termed simply *M. avium*. A minority of the participants favoured the expression *M. avium–intracellulare* (MAI) complex. Others couple this group with the phenetically similar but genetically distinct species *M. scrofulaceum* in the *M. avium–intracellulare–scrofulaceum* (MAIS) complex.

The *M. avium* group of strains is an heterogeneous one and may be subdivided in several ways. Schaefer and his colleagues described 21 agglutination serotypes which have been numbered (Wolinsky and Schaefer, 1973). Serotypes 1, 2 and 3 are classical bird-pathogenic *M. avium*, with the type species being serotype 2. The other serotypes were regarded as *M. intracellulare*, with the type species being serotype 16.

In 1967 Kazda described an apparently new species and termed it *M. brunense*. This was found to be identical to strains of serotype 8, previously known as the Davis serotype.

Immunodiffusion analysis (Stanford, 1983) revealed a close antigenic relationship between the types described above, as shown in Fig. 3.2. All types share three antigens which do not occur in any other mycobacterial species. There are also seven antigens which delinate variants within the group. Applying the principles of classification by immunodiffusion analysis (see page 11) this group of mycobacteria appear to be one species, *M. avium*, with four subspecies, namely, *avium*, *brunense*, *intracellulare* and *lepraemurium*. The first three subspecies have also been termed *M. avium* A, B and C (McIntyre and Stanford, 1986a). The *avium* subspecies contains the Schaefer agglutination serotypes 1, 2 and 3, the *brunense* subspecies contains serotypes 4–12, 20 and 21, and the *intracellulare* subspecies contains serotypes 13–19 (McIntyre and Stanford, 1986a). Strains of *M. paratuberculosis* and other mycobactin dependent strains,

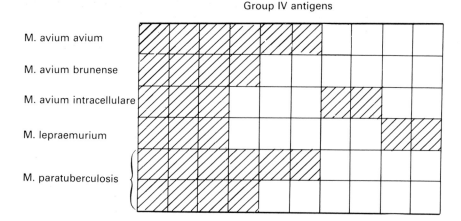

Fig. 3.2 Distribution of the group iv (species-specific) antigens in the *M. avium* group. Data from McIntyre and Stanford (1986a,b)

including those from wood pigeons, were found to be either immunodiffusion type A or B (McIntyre and Stanford, 1986b). The wood pigeon isolates were serotype A while the type strain of *M. paratuberculosis* and most cattle isolates were type B. Thus, according to this system, strains previously identified as *M. paratuberculosis* would be either *M. avium avium* (type A) or *M. avium brunense* (type B).

An almost identical subdivision of *M. avium* and *M. intracellulare* to that outlined above was obtained by Meissner and Anz (1977) on the basis of virulence for the chicken and sensitin typing: strains were divided into *M. avium*, 'intermediate' and *M. intracellulare*. There are a few differences between the two classifications described above: for example, serotype 7 strains were classified as *M. intracellulare* by Meissner and Anz. Some of the serotypes were examined by Baess (1983) for DNA homology by the hybridization technique and two groups were delineated. The first group contained the *M. avium* serotypes 1, 2 and 3 and the *brunense* types 4–6 and 8. The other group contained the *intracellulare* serotypes 12, 14, 16 and 18 and, in agreement with the findings of Meissner and Anz, serotype 7. Thus there are several ways of dividing this group of strains, but none of the ways gives completely concordant results (see Table 3.4).

The taxonomic investigations described above raise a number of nomenclatural problems. Most clinical laboratories do not subdivide members of this group by serological or other means and will therefore continue to refer simply to *M. avium*, *M. avium–intracellulare* or to the *M. avium–intracellulare* (MAI) complex. Furthermore, as the behaviour of *M. lepraemurium* and *M. paratuberculosis* are so different in respect to the diseases they cause and in their cultural properties, it is unlikely that their reduction to subspecific status will be readily accepted. We will therefore adhere to the 'official' nomenclature; namely, *M. avium*, *M. intracellulare*, *M. lepraemurium* and *M. paratuberculosis*.

Table 3.4 Subdivision of *M. avium–intracellulare* by agglutination, immunodiffusion, sensitin typing and conventional taxonomic tests. Data from Stanford (1983)

Agglutination serology	Immunodiffusion analysis	Sensitin testing	Conventional taxonomic
1 ⎫			
2 ⎬ *M. avium*	*M. avium*	*M. avium*	*M. avium*
3 ⎭			
4			'Intermediate'
5		*M. intracellulare*	
6	*M. avium*		*M. intracellulare*
7	*brunese*		
8		*M. avium*	'Intermediate'
9			
10			
11			
12			
13 *M. intracellulare*			
14		*M. intracellulare*	
15			
16	*M. avium*		*M. intracellulare*
17	*intracellulare*		
18			
19			
20	*M. avium*		
21	*brunese*		
22			
23	Not tested	Not tested	Not tested
24			

Mycobacterium avium and *M. intracellulare*

Worldwide, these species are the most prevalent causes of the 'opportunist' mycobacterial diseases in man. They are of increasing importance in view of their association with the acquired immune deficiency syndrome (see page 148). They usually grow as smooth non-pigmented colonies, although rough or pigmented variants are occasionally encountered. There is a relation between colony form and virulence, especially amongst the bird-pathogenic strains (Kuze and Uchihira, 1984). The two species are not readily subdivided by standard biochemical tests and are therefore, as mentioned above, often reported as *M. avium–intracellulare*. Strains of *M. avium* usually grow at 44°C while those of *M. intracellulare* do not, but this difference is not entirely reliable for identification purposes. Strains of both species are usually resistant to almost all the antituberculosis drugs *in vitro*, although infections often respond to therapy with these drugs (see page 165).

Mycobacterium paratuberculosis

This is the cause of chronic hypertrophic enteritis, or Johne's disease, of cattle and, less often, of sheep, goats and other ruminants. The disease was described by Johne and Frothingham (1895) who, interestingly, thought that it was caused by the 'avian tubercle bacillus'. Characteristically, strains fail to synthesize the lipid-soluble iron-binding compound mycobactin (see page 23). Thus, for its cultivation, media must be supplemented with killed mycobacteria or purified mycobactin. Even on such medium, growth is very slow and incubation for up to 14 weeks is indicated in some cases.

Some mycobacteria with the properties of *M. avium* were found to be mycobactin-dependent, particularly on primary isolation (McDiarmid, 1948). Such strains were isolated principally from wood pigeons but also from deer, pigs and hares. Experimentally, some of wood pigeon strains cause hypertrophic enteritis in calves. As discussed above, such strains are serologically identical on immunodiffusion serology to bird-pathogenic *M. avium*, while cattle isolates were mainly *avium* type B (*brunense* subspecies).

Mycobacterium lepraemurium

This is the causative organism of rat leprosy, described by Stefansky (1903), who found that about 5 per cent of rats caught in the Black Sea port of Odessa were infected. The disease, which involves the skin and lymph nodes, has subsequently been described in many countries and a similar disease, caused by bacilli with the same properties as *M. lepraemurium*, occurs in cats (Mori and Kohsaka, 1987). The organism has proved very difficult to culture *in vitro*, although limited growth is obtainable in Ogawa egg medium (Mori, 1975). *Mycobacterium lepraemurium* is quite unrelated to *M. leprae* and the corresponding diseases are quite distinct.

Mycobacterium farcinogenes (and M. senegalense)

The name *Mycobacterium farcinogenes* was given by Chamoiseau (1973) to strains (previously thought to be nocardiae) responsible for farcy – a chronic disease of the lymphatics – in zebu cattle in Chad and Senegal. Two subtypes were recognized, *tchadense* and *senegalense*, but the latter was found to be a separate, and rapidly growing, species and was termed *M. senegalense* (Chamoiseau, 1979). The exact taxonomic status of these species and their relationship to other mycobacteria requires further elucidation. Neither species is known to cause disease in man.

Mycobacterium gastri

Although originally isolated from gastric washings, this species, described by Wayne (1966), has never been directly associated with human disease. It has much in common with non-pigmented variants of *M. kansasii*: indeed its separate species status remains in doubt. Although *M. gastri* is separable from *M. kansasii* by numerical taxonomy, the soluble antigens of the two species are identical (Stanford and Grange, 1974).

Mycobacterium malmoense

This very slow growing species was first isolated on Malmo in Sweden by Schröder and Juhlin (1977). On primary isolation, growth may not be visible until ten weeks incubation, so that it may be missed if cultures are not kept for this length of time. Since 1980, the incidence of human disease due to this species has shown a significant increase in Great Britain and some other countries.

The *Mycobacterium terrae–triviale–nonchromogenicum* group

These three species are superficially similar, rarely cause disease and may be mistaken for *M. avium–intracellulare*. Their nomenclatural history is confusing and doubts still exist as to whether these, and *M. novum*, are really separate species. *Mycobacterium nonchromogenicum* was described by Tsukamura (1965), but in the following year he became dissatisfied with this name and changed it to *M. terrae*. By coincidence, that name had, just a few months previously, been given by Wayne (1966) to a new species isolated from soil and, originally, from radishes. Accordingly Tsukamura's new species reverted to its original name, *M. nonchromogenicum*. The third member of this group, *M. triviale*, was described by Kubica *et al.* (1970). Although isolated on one occasion from a case of septic arthritis in a baby, its name reflects its unimportance in clinical microbiology.

Mycobacterium shimoidei

This species is a very rare cause of pulmonary disease in Japan (Tsukamura, 1982) and a case has also been reported in Germany. It is biochemically rather unreactive and grows at 45°C but not at 25°C.

Mycobacterium ulcerans

This is the cause of Buruli ulcer (Chapter 9) and was first described in Australia (MacCallum *et al.*, 1948). It is a very slow growing organism, has a very restricted temperature range of growth – 31 to 34°C – and is biochemically very inactive. Epidemiological evidence (Barker, 1973) suggests that it is an environmental saprophyte although attempts to cultivate it from inanimate sources have so far failed.

Mycobacterium xenopi

Originally isolated from a toad (*Xenopus laevis*) by Schwabacher (1959), this species is the most frequent cause of opportunist mycobacterial disease in South London. The growth rate is slower than that of *M. tuberculosis* and, on primary isolation, colonies may not be visible until after six to eight weeks incubation. Colonies are off-white or pale-lemon coloured and may show small ariel hyphae. Bacterial cells are often filamentous.

Mycobacterium xenopi is thermophilic and grows well at 45°C. It has, on occasions, been isolated from hot water systems. It may therefore contaminate specimen containers that are washed but not sterilized (see page 18).

Mycobacterium haemophilum

The name of this species is derived from its requirement for a high concentration of iron which is conveniently supplied as blood, haem or ferric ammonium citrate. It was originally described by Sompolinsky *et al.* (1978) in Israel and almost all isolations have been from skin lesions of immunosuppressed individuals. In common with *M. marinum*, it grows well at 30°C but not at 37°C: this may explain its association with skin lesions.

The rapidly growing mycobacteria

There are 17 species of rapidly growing mycobacteria in the approved lists and a few additional ones that were excluded from those lists or were described subsequently. With the exception of *M. chelonei* and *M. fortuitum*, rapidly growing mycobacteria cause human disease with extreme rarity.

Mycobacterium chelonei and *M. fortuitum*

These are both members of the 'cold-blooded tubercle bacilli', having been isolated from the turtle in 1903 and the frog in 1905 respectively. *Mycobacterium chelonei* was used as a vaccine and therapeutic agent for many years until it became apparent that it was not only ineffective but that it frequently caused injection abscesses. The taxonomic history of this species is somewhat confusing: synonyms include *M. abscessus*, *M. friedmannii*, *M. runyonii* and *M. borstelense*. This species is divisible into two major variants on the basis of antigenic differences, salt tolerance and citrate utilization. These are *M. chelonei chelonei* and *M. chelonei abscessus*. The former is the prevalent type in Europe while the latter is the usual type in Africa and the USA. Both types appear to be equally virulent and cause injection abscesses, wound infections, pulmonary lesions and disseminated disease (for further details see Grange, 1981).

The taxonomic history of *M. fortuitum* is no less confusing. This, being the original frog tubercle bacillus, was termed *M. ranae* but, as there has been some confusion due to the mislabelling of strains in culture collections, the name was changed to *M. fortuitum* in 1974. Synonyms include *M. giae*, *M. peregrinum* and *M. minetti*. There are major variants

Table 3.5 Properties of the pathogenic rapidly growing species *M. fortuitum* and *M. chelonei*

	Acid produced from mannitol	isonitol	Growth at 42°C	Nitratase	Citrate utilization	Growth on 5% NaCl
M. fortuitum biotype A	–	–	+	+	+	–
M. fortuitum biotype B	+	–	–	+	+	–
M. fortuitum biotype C	+	+	–	+	+	–
M. chelonei abscessus	–	–	–	–	–	+
M. chelonei chelonei	–	–	–	–	+	–

detectable by biochemical properties, agglutination or immunodiffusion serology and by lipid chromatography, namely the *fortuitum* and *peregrinum* (or *giae*) types. Although the latter has been isolated from the sputum of individuals with dust-associated disease, only the former type is regularly encountered as a pathogen. (For further bacteriological details see Grange and Stanford, 1974). *Mycobacterium fortuitum* causes injection abscesses and wound infections: pulmonary and disseminated disease also occurs but is rare.

These two species were often grouped in the '*fortuitum* complex', thus obscuring the fact that *M. chelonei* is the more virulent. Although superficially similar in being non-pigmented and strongly arylsulphatase positive, they are quite unrelated species and are easily differentiated from each other (Table 3.5).

Thermophilic rapid growers

This group contains three species: *M. phlei*, *M. smegmatis* and the aptly-named *M. thermoresistibile*. The importance of the first two of these species have been elevated out of all proportion by their frequent use in biochemical and genetic studies and by the groundless belief that *M. smegmatis* is a frequent contaminant of urine specimens. (Mycobacterial contaminants of urine do occur but are usually rapidly growing chromogens.) In fact, all three species are very rarely encountered. *Mycobacterium phlei*, also known as the Timothy grass bacillus, and *M. thermoresistibile*, both survive heating at 60°C for four hours. These species vary in their pigmentation. Strains of *M. smegmatis* may be chromogenic or colourless, *M. phlei* is salmon-pink coloured on Löwenstein–Jensen medium, while *M. thermoresistibile* is of a rusty-orange colour.

Other rapidly growing mycobacteria

The remainder of the rapid growers are mostly bright-yellow or orange in colour and, with very rare exceptions, only occur in clinical specimens as contaminants. The 'approved' species include the chromogenic *M. aurum*, *M. duvalii*, *M. flavescens*, *M. gadium*, *M. gilvum*, *M. komossense*,

M. neoaurum,   *M. parafortuitum*,   *M. rhodesiae*,   *M. senegalense*, *M. sphagni*, *M. vaccae* and the non-chromogenic *M. chitae*. Only *M. flavescens* has been implicated as a cause of human disease, but very rarely (Collins *et al.*, 1986).

In addition Tsukamura has reintroduced six rapidly growing species that were omitted from the approved lists, namely *M. agri* (Tsukamura, 1981), *M. aichense*, *M. chubuense*, *M. obuense*, *M. rhodesiae* and *M. tokaiense* (Tsukamura *et al.*, 1981). All, with the exception of *M. agri*, are chromogenic.

Mycobacterium diernhoferi, a rapidly growing non-chromogen described by Bönicke and Juhasz (1965), is a further species that was omitted from the list but probably merits reintroduction.

Mycobacterium leprae

This species, the causative organism of leprosy, was termed 'the bacteriologists's enigma' in an excellent comprehensive review by Stewart-Tull (1982). The leprosy bacillus has, despite many claims to the contrary, stubbornly eluded all attempts to cultivate it *in vitro*. This does not mean that it is unable to replicate under certain conditions in the environment. Among the intriguing aspects of *M. leprae* is that it appears to be an obligate pathogen with a host range virtually limited to man, in whom it is the cause of a relatively recent affliction. Although there have been a few claims of a leprosy-like disease in animals (see page 102), there is no extensive animal reservoir of *M. leprae*, and taxonomic studies have failed to relate it to any known culturable mycobacterium. This raises the fascinating question of what or where was *M. leprae* before it infected man.

Intriguing experiments performed by Kazda *et al.* (1980) have shown that sphagnum vegetation contains mycobacteria that could be propagated by serial passage in footpads of mice, but bacilli from the footpads could not be cultured *in vitro*. Although it cannot be assumed these organisms are *M. leprae*, it indicates that 'non-cultivable' mycobacteria may exist in the environment, possibly in symbiosis with plants or other living organisms such as bacteria or protozoa.

Mycobacterium leprae possesses the common (group i) mycobacterial antigens and some species-specific ones but not those that distinguish the slow growers (group ii) or the rapid growers (group iii). The growth rate-associated antigens are also absent in the culturable species *M. vaccae* and *M. nonchromogenicum*, but there is no evidence that *M. leprae* is more closely related to these than to any other mycobacterial species. The growth rate of *M. leprae* is, in fact, unknown. Although it appears to divide very slowly – about once every two weeks – it is also possible that it divides much more rapidly but that many of the progeny die. In one respect, *M. leprae* resembles the rapidly growing mycobacteria. It has been shown that the iron-binding exochelins from rapid growers are usually water-soluble while those from the slow growers are mostly chloroform-soluble. Although *M. leprae* does not produce detectable amounts of exochelin, it accepts iron from the water-soluble, but not from the chloroform-soluble, type (Hall and Ratledge, 1987).

Although non-cultivable, the metabolism of *M. leprae* has, to some extent, been studied in incubating tissue-derived bacilli with radio-labelled substrates and detecting enzyme reaction products, such as ^{14}C carbon dioxide (Wheeler, 1984). Such studies have, so far, failed to detect a metabolic defect that would explain the failure of this species to grow *in vitro*.

One of the unique characteristics of *M. leprae* is the antigenic component of a surface lipid termed phenolic glycolipid I (see page 16) The availability of this antigen, and its synthetic analogue, should assist in the diagnosis of leprosy and in the confirmation of any claimed success at culturing the bacillus *in vitro*.

References

Aronson, J.D. (1926) Spontaneous tuberculosis in salt water fish. *Journal of Infectious Diseases* **39**: 315–20

Baess, I. (1983) Deoxyribonucleic acid relationships between different serovars of *Mycobacterium avium*, *Mycobacterium intracellulare* and *Mycobacterium scrofulaceum*. *Acta Pathologica et Microbiologica Scandinavica* **91B**: 201–3.

Barker, D.J.P. (1973) Epidemiology of *Mycobacterium ulcerans* infection. *Transactions of the Royal Society of Hygiene and Tropical Medicine* **67**: 43–50.

Bojalil, L.F., Cerbon, J. and Trujillo, J. (1962) Adansonian classification of mycobacteria. *Journal of General Microbiology* **28**: 333–46.

Bönicke, R. and Juhasz, S.E. (1965) Beschreibung der neuen Species *Mycobacterium diernhoferi* n. sp. eine in der Umgebung des Rindes haufig vorkommende neue *Mycobacterium* Species. *Zentralblatt fur Bakteriologie, Parasitenkunde, Infektionskrankheiten und Hygiene, Abt 1.* **197**: 292–4.

Castets, M., Rist, N. and Boisvert, H. (1969) La variete africaine du bacille tuberculeux humain. *Medicine d'Afrique Noir* **16**: 321–2.

Chamoiseau, G. (1973) *Mycobacterium farcinogenes*: agent causal du farcin du boeuf en Afrique. *Annals De Microbiologie* (Institut pasteur) **124A**: 215–22.

Chamoiseau, G. (1979) Etiology of farcy in African bovines: nomenclature of the causal organisms *Mycobacterium farcinogenes* Chamoiseau and *Mycobacterium senegalense* (Chamoiseau) comb. nov. *International Journal of Systematic Bacteriology* **29**: 407–10.

Chester, F.D. (1901) *A Manual of Determinative Bacteriology*, p. 356. New York: Macmilan.

Collins, C.H., Grange, J.M. and Yates, M.D. (1984) Mycobacteria in water. *Journal of Applied Microbiology* **57**: 193–211.

Collins, C.H., Grange, J.M. and Yates, M.D. (1986) Unusual opportunist mycobacteria. *Medical Laboratory Sciences* **43**: 262–8.

Collins, C.H., Yates, M.D. and Grange, J.M. (1982) Subdivision of *Mycobacterium tuberculosis* into five variants for epidemiological purposes: methods and nomenclature. *Journal of Hygiene* **89**: 235–42.

Goren, M.B., Grange, J.M., Aber, V.R., Allen, B.W. and Mitchison, D.A. (1982) Role of lipid content and hydrogen peroxide susceptibility in determining the guinea-pig virulence of *Mycobacterium tuberculosis*. *British Journal of Experimental Pathology* **63**: 693–700.

Grange, J.M. (1981) *Mycobacterium chelonei*. *Tubercle* **62**: 273-6.

Grange, J.M. (1982) Koch's tubercle bacillus: a centenary reappraisal. *Zentralblatt fur Bakteriologie, Parasitenkunde, Infektionskrankheiten und Hygiene, Abt 1.* **251**: 297–307.

Grange, J.M., Aber, V.R., Allen, B.W., Mitchison, D.A. and Goren, M.B. (1978) The correlation of bacteriophage types of *Mycobacterium tuberculosis* with guinea pig virulence and *in vitro* indicators of virulence. *Journal of General Microbiology* **108**: 1–7.

Grange, J.M. and McIntyre, G. (1979) Fluorogenic glycosidase substrates: their use in the identification of some slow growing mycobacteria. *Journal of Applied Bacteriology* **47**: 285–8.

Grange, J.M. and Redmond, W.B. (1978) Host–phage relationships in the genus *Mycobacterum* and their clinical significance. *Tubercle* **59**: 203–25.

Grange, J.M. and Stanford, J.L. (1974) A re-evaluation of *Mycobacterium fortuitum* (synonym *Mycobacterium ranae*). *International Journal of Systematic Bacteriology* **24**: 320–9.

Hall, R.M. and Ratledge, C. (1987) Exochelin-mediated iron acquisition by the leprosy bacillus, *Mycobacterium leprae*. *Journal of General Microbiology* **133**: 193–9.

Hauduroy, P. (1955) *Derniers Aspects du Monde des Mycobacteries.* Paris: Masson.

Johne, H.H. and Frothingham, L. (1895) Ein eigentumlicher Fall von Tuberculose beim Rind. *Deutsche Zeitschrift fur Tiermedizin und Vergleichende Pathologie* **21**: 438–54.

Karasseva, V., Weiszfeiler, J. and Krasznay, E. (1965) Occurrence of atypical mycobacteria in *Macacus rhesus. Acta Microbiologica Academicae Scientiarum Hungarica* **12**: 275–82.

Karlson, A.G. and Lessel, E.F. (1970) *Mycobacterium bovis* nom. nov. *International Journal of Systematic Bacteriology* **20**: 273–82.

Kazda, J. (1967) Mycobakterien in Trinkwasser als Ursache der parallergie gegenuber Tuberculinen bei Tieren. III Mitteilung: Taxonomische Studie einiger rasch wachsender Mykobacterien und Beschreibung einer neuen Art: *Mycobacterium brunense. Zentralblatt fur Bakteriologie, Parasitenkunde, Infektionskrankheiten und Hygiene, Abt 1.* **203**: 199–201.

Kazda, J., Irgens, L.M. and Muller, K. (1980) Isolation of non-cultivable acid-fast bacilli in sphagnum moss vegetation by footpad technique in mice. *International Journal of Leprosy* **48**: 1–6.

Kubica, G.P., Silcox, V.A., Kilburn, J.O., Smithwick, R., Beam, R.E., Jones, W.D. and Stottmeier, K.D. (1970) Differential identification of mycobacteria. VI: *Mycobacterium triviale* Kubica sp. nov. *International Journal of Systematic Bacteriology* **20**: 164–74.

Kuze, F. and Uchihira, F. (1984) Various colony formers of Mycobacterium *avium-intracellulare. European Journal of Respiratory Diseases* **65**: 402–10.

Lehmann, K.B. and Neumann, R. (1896) *Atlas und Grundriss der Bakteriologie und Lehrbuch der speciellen bakteriologischen Diagnostik*, 1st edn. Munich: J.F. Lehmann.

Linell, F. and Norden, A. (1954) *Mycobacterium balnei*: a new acid-fast bacillus occurring in swimming pools and capable of producing skin lesions in humans. *Acta Tuberculosea Scandinavica Supplement*. **33**: 1–84.

Lotte, A., Wasz-Hockert, O., Poisson, N., Dumitrescu, N., Verron, M. and Couvet, E. (1984) BCG complications. *Advances in Tuberculosis Research* **21**: 107–93.

MacCallum, P., Tolhurst, J.C., Buckle, G. and Sissons, H.A. (1948) A new mycobacterial infection of man. *Journal of Pathology and Bacteriology* **60**: 93–122.

McDiarmid, A. (1948) The occurrence of tuberculosis in the wild wood-pigeon.

Journal of Comparative Pathology **58**: 128–33.

Marks, J., Jenkins, P.A. Tsukamura, M. (1972) *Mycobacterium szulgai* – a new pathogen. *Tubercle* **53**: 210–14.

McIntyre, G. and Stanford, J.L. (1986a) The relationship between immunodiffusion and agglutination serotypes of *Mycobacterium avium* and *Mycobacterium intracellulare*. *European Journal of Respiratory Diseases* **69**: 135–41.

McIntyre, G. and Stanford, J.L. (1986b) Immunodiffusion analysis shows that *Mycobacterium paratuberculosis* and other mycobactin-dependent mycobacteria are variants of *Mycobacterium avium*. *Journal of Applied Bacteriology* **61**: 295–8.

Meissner, G. Anz, W. (1977) Sources of *Mycobacterium avium* complex infection resulting in human disease. *American Review of Respiratory Disease* **116**: 1057–64.

Meissner, G. and 21 others (1974) A comparative numerical analysis of nonscoto- and nonphotochromogenic slowly growing mycobacteria. *Journal of General Microbiology* **83**: 207–35.

Mitchison, D.A. (1970) Regional variation in the guinea pig virulence and other characteristics of tubercle bacilli. *Pneumonology* **142**: 131–7.

Mori, T. (1975) Biochemical properties of cultivated *Mycobacterium lepraemurium*. *International Journal of Leprosy* **43**: 210–17.

Mori, T. and Kohsaka, K. (1987) Identification of cat leprosy bacillus grown in mice. *International Journal of Leprosy* **54**: 584–95.

Prissick, F.H. and Masson, A.M. (1956) Cervical lymphadenitis in children caused by chromogenic mycobacteria. *Canadian Medical Association Journal* **3**: 91–100.

Reed, G.B. (1957) Family 1 Mycobacteriaceae Chester 1897. In: *Bergey's Manual of Determinative Bacteriology*, 7th edn. Edited by R.S. Breed, E.G.D. Murray and N.R. Smith. Baltimore, Williams and Wilkins.

Royal Commission on Tuberculosis (1907) *Second Interim Report*. London: HMSO.

Runyon, E.H. (1967) *Mycobacterium intracellulare*. *American Review of Respiratory Disease* **95**: 861–7.

Saito, H. and 12 others (1977) Cooperative numerical analysis of rapidly growing mycobacteria. *International Journal of Systematic Bacteriology* **27**: 75–85.

Schroder, K.H. and Juhlin, I. (1977) *Mycobacterium malmoense* sp. nov. *International Journal of Systematic Bacteriology* **27**: 241–6.

Schwabacher, H. (1959) A strain of mycobacterium isolated from skin lesions of a cold-blooded animal *Xenopus laevis*, and its relation to atypical acid-fast bacilli occurring in man. *Journal of Hygiene* **57**: 57–67.

Skerman, V.D.B., McGowan, V. and Sneath, P.H.A. (1980) Approved lists of bacterial names. *International Journal of Systematic Bacteriology* **30**: 225–420.

Sompolinsky, D., Lagziel, A., Naveh, D. and Yankilevitz, T. (1978) *Mycobacterium haemophilum* sp. nov.: a new pathogen of humans. *International Journal of Systematic Bacteriology* **28**: 67–75.

Stanford, J.L. (1983) Immunologically important constituents of mycobacteria: antigens. In: *The Biology of the Mycobacteria*, Vol. 2, pp. 85–127. Edited by C. Ratledge and J.L. Stanford. New York: Academic Press.

Stanford, J.L. and Grange, J.M. (1974) The meaning and structure of species as applied to mycobacteria. *Tubercle* **55**: 143–52.

Stefansky, W.K. (1903) Ein lepraahnliche Erkrankung der Haut und der Lymphdrusen bei Wanderratten. *Zentralblatt fur Bakteriologie, Parasitenkunde, Infektionskrankheiten und Hygiene, Abt 1*. **33**: 481–7.

Stewart-Tull, D.E.S. (1982) *Mycobacterium leprae* – the bacteriologist's enigma. In: *The Biology of the Mycobacteria*, Vol. 1, pp. 274–307. Edited by C. Ratledge and J.L. Stanford. New York: Academic Press.

Tarnok, I. and Tarnok, Zs. (1970) Carotenes and xanthophylls in mycobacteria. I:

technical procedures: thin layer chromatographic patterns of mycobacterial pigments. *Tubercle* **51**: 305–12.

Tsukamura, M. (1965) A group of mycobacteria from soil sources resembling nonphotochromogens (Group III): a description of *Mycobacterium nonchromogenicum*. *Medicine and Biology* **71**: 110–13.

Tsukamura, M. (1981) Numerical analysis of rapidly growing nonphotochromogenic mycobacteria, including *Mycobacterium agri* sp. nov., nom. rev. *International Journal of Systematic Bacteriology* **31**: 247–58.

Tsukamura, M. (1982) *Mycobacterium shimoidei* sp. nov., nom. rev.: a lung pathogen. *International Journal of Systematic Bacteriology* **32**: 67–9.

Tsukamura, M., Mizuna, S. and Tsukamura, S. (1981) Numerical analysis of rapidly growing scotochromogenic mycobacteria, including *Mycobacterium obuense* sp. nov., nom. rev., *M. rhodesiae* sp. nov., nom. rev., *M. aichiense* sp. nov., nom. rev., *M. chubuense* sp. nov., nom. rev. and *M. tokaiense* sp. nov., nom. rev. *International Journal of Systematic Bacterology* **31**: 263–75.

Valdivia-Alvarez, J., Suarez-Mendez, R. and Echemendia-Font, M. (1971) *Mycobacterium habana*: probable neuva especie dentro de la micobatterias. *Boletin de Higiene Epidemiologia* **8**: 65–73.

Wayne, L.G. (1966) Classification and identification of mycobacteria III. Species within Group III. *American Review of Respiratory Disease* **93**: 919–28.

Wayne, L.G. and 14 others (1971) A co-operative numerical analysis of scotochromogenic slowly growing mycobacteria. *Journal of General Microbiology* **66**: 255–71.

Wayne, L.G., Andrade, L.B., Froman, S., Kappler, W., Kubala, E., Meissner, G. and Tsukamura, M. (1978) A co-operative numerical analysis of *Mycobacterium gastri*, *Mycobacterium kansasii* and *Mycobacterium marinum*. *Journal of General Microbiology* **109**: 319–27.

Weiszfeiler, J., Karasseva, V. and Karczag, E. (1971) A new mycobacterium species: *Mycobacterium asiaticum* n. sp. *Acta Microbiologica Academicae Scientiarum Hungarica* **18**, 247–52.

Wells, A.Q. (1946) The murine type of tubercle bacillus. *Medical Research Council Special Report 259*. London: HMSO.

Wheeler, P.R. (1984) Metabolism in *Mycobacterium leprae*: its relation to other mycobacteria and intracellular parasites. *International Journal of Leprosy* **52**: 208–30.

Wolinsky, E. and Schaefer, W.B. (1973) Proposed numbering scheme for mycobacterial serotypes by agglutination. *International Journal of Systematic Bacteriology* **23**: 182–3.

Yates, M.D., Collins, C.H. and Grange, J.M. (1978) Differentiation of BCG from other variants of *Mycobacterium tuberculosis* isolated from clinical material. *Tubercle* **59**: 143–6.

4 Diagnostic mycobacteriology

Diagnostic mycobacteriology is a complex and technically demanding branch of medical microbiology. The organization of diagnostic services and the technical procedures have been reviewed in detail elsewhere (Collins *et al.*, 1985). The aim of this chapter is to enable the clinician to make the most of the available services.

The examination of clinical specimens suspected of containing mycobacteria consists of four steps:

1. microscopical examination;
2. isolation of mycobacteria in culture;
3. identification of the organism;
4. sensitivity testing (where relevant).

In addition, reference laboratories play an important role in the collection of epidemiological information, staff training, quality control and research as well as giving clinical and technical advice.

In many countries, laboratories are arranged in a hierarchical fashion. Microscopy and culture are performed in peripheral laboratories, larger centres identify *Mycobacterium tuberculosis* while the identification of other species as well as sensitivity testing are usually performed in designated reference centres. In poorer countries, emphasis is placed on microscopy services so that open or infectious cases of tuberculosis are detected and treated.

Laboratory safety

The handling of specimens and cultures exposes laboratory staff to a serious risk of infection. Workers in post-mortem rooms are at particular risk. Safety measures are essential and, in some countries, mandatory. These include containment laboratories, approved safety cabinets and centrifuges, protective clothing, hand washbasins and facilities for the safe disposal of contaminated waste. All staff should be vaccinated with BCG, unless they are tuberculin positive or have a BCG scar, and must have regular clinical and radiological examinations as determined by local policies.

Health and safety measures are no substitute for adequate training of staff in the basic technical procedures, their constant awareness of potential hazards and their familiarization with local safety rules. For a full account of laboratory safety see Collins (1983) and Collins *et al.* (1985).

The clinician has a duty to consider the welfare of the laboratory staff by submitting specimens in a safe condition and by ensuring that specimen containers are not contaminated on the outside. Plastic bags containing tracheal catheters, Lukan traps and other objects contaminated with sputum are particularly hazardous and should be disposed of immediately without being opened.

The collection of specimens

The first, and most important, step in any microbiological investigation is the collection of suitable specimens. Much valuable information is lost, and time is wasted, by the submission of inadequate or mislabelled specimens or by delays in their delivery to the laboratory.

It is particularly important to collect all specimens into sterile containers. It is a common misassumption that, as specimens for mycobacterial investigations are 'decontaminated' before culture, cleanliness of the container is unimportant. Unfortunately, unsterilized containers may be contaminated with environmental mycobacteria. An apparent hospital outbreak of pulmonary infection due to the thermophilic species *M. xenopi* was traced to the practice of collecting sputum in metal pots that were washed under the hot tap.

Specimens

Sputum.   The most suitable containers for sputum are wide-mouthed, screw-capped disposable plastic pots, into which sputum may be directly expectorated. Such pots for use in developing countries are available from Unicef.

Most sputum specimens submitted for examination are 'spot' samples taken at clinics, but it is preferable to obtain at least three early-morning specimens. Sputum specimens are inevitably contaminated with non-acid-fast bacteria which multiply rapidly. Specimens should therefore be conveyed to the laboratory as quickly as possible. If delays in transport are unavoidable, sputum may be stored for up to one week at 0°C. If specimens are to be transported for long distances in hot climates and (as is often the case) refrigerated transport is not available, sputum may be preserved by the addition of an equal volume of 1% cetyl pyridinium chloride in 2% saline.

Other respiratory tract specimens.   Sputum is by far the best material for the diagnosis of pulmonary mycobacterial disease and all efforts should be made to obtain a specimen. Laryngeal swabbing and gastric aspiration are relatively ineffectual means of obtaining material suitable for culture, as well as being unpleasant for the patient. A more satisfactory alternative, if the equipment is available, is the use of the fibreoptic bronchoscope.

Laryngeal swabs are specially prepared for the purpose: they consist of stiff wires bent at one end at an angle of 35° and tipped with cotton wool. Some laboratories issue swabs tipped with alginate wool which can subsequently be dissolved so as to liberate the mycobacteria. Swabbing should only be done by staff trained in the technique: blind swabbing of the pharynx is useless. The operator should wear a visor as the patients inevitably cough violently during the procedure.

Gastric aspiration is performed in the early morning before food or drink is taken. The patient is requested to cough and swallow several times before the stomach contents are aspirated through a nasogastric tube. The aspirate should be transported without delay to the laboratory or, if this is not possible, it should be neutralized with sodium hydroxide.

The fibreoptic bronchoscope may be used to take biopsies of radiologi-cally visible lesions, to sample material on the bronchial walls by means of a small extendable brush, or to rinse out sections of the bronchial tree with saline (broncho-alveolar lavage). After use, the bronchoscope is sterilized with a 2% solution of glutaraldehyde.

Urine.   Three early-morning midstream specimens should be obtained in 28 ml glass or plastic ('Universal') containers. Some laboratories isolate mycobacteria by membrane filtration of larger quantities of urine, such as total early-morning specimens.

Other fluids.   Cerebrospinal fluid and pus should be placed directly in sterile containers and sent to the laboratory without delay. Examination of cerebrospinal fluid is an emergency investigation and the bacteriologist should be contacted by the clinician. Specimens of pleural and peritoneal fluids may contain enough fibrinogen to cause them to coagulate. This is preventable by adding sterile sodium citrate to the specimen.

Tissues.   Tissue is much more suitable for culture than necrotic material or pus, probably as the latter contain free fatty acids that are toxic to mycobacteria. For this reason the largest of a group of excised lymph nodes may not be the best for culture as it may be the most necrotic. Biopsies are usually examined histologically as well as bacteriologically.

Microscopy

As the isolation of *M. tuberculosis* and most other pathogenic mycobacter-ia by cultural methods takes several weeks, the use of microscopy to reach a preliminary diagnosis is of great importance. In addition, microscopy of sputum is of great value in the detection of open or infectious cases. It is well established that if no acid-fast bacilli are seen in sputum on a standard, yet competently performed, examination, the patient is unlikely to be infectious. Thus the establishment of good sputum microscopy services is of prime importance in developing countries where the first priority in tuberculosis control is the detection and treatment of the open cases.

Sputum is examined either directly or after liquefaction and centrifuga-tion (see page 53). Direct examination is performed by selecting a purulent-looking portion of sputum and spreading it thinly on a glass slide with a bacteriological loop or a wooden throat swab stick. If centrifuged deposits are examined, great care must be taken to make sure that the reagents and diluents are not contaminated with environmental mycobacteria in order to avoid false-positive smears.

Smears are stained by the Ziehl–Neelsen (ZN) method, or one of its various modifications. In the traditional ZN method heat-fixed slides are flooded with carbol fuchsin (a phenol/water solution of basic fuchsin) and heated until steam rises. Boiling must be avoided as this dislodges the smear from the slide. After five minutes the slide is washed under the tap and then flooded with a dilute mineral acid such as 3% sulphuric or hydrochloric acid. Some workers use acidified alcohol rather than acid alone as this tends to give a cleaner film. It does not, contrary to a popular belief, distinguish tubercle bacilli from other mycobacteria.

After five minutes the slide is washed again and a counterstain is applied. The counterstain may be green or blue, depending on the microscopist's preference: red–blue colour-blind workers may use a yellow stain such as picric acid. Various 'cold' ZN staining techniques have been introduced but are not widely used.

An alternative to the ZN method is staining with auramine or rhodamine (or both) for fluorescence microscopy (FM). Although the equipment is more expensive, FM is much less tiring to the technical staff. Smears are examined at a lower magnification, thereby increasing the chance of detecting low numbers of acid-fast bacilli. Some authorities consider that fluorescent staining is less specific than ZN staining and advise that all smears found to be positive by FM should be confirmed by the ZN method.

Biopsies may be homogenized by grinding with sand in a Griffith tube or examined histologically by use of a modified ZN staining procedure such as the Wade–Fite or TRIFF methods. For details see Ridley (1977). Urines, cerebrospinal fluids and other fluids are centrifuged and the deposits are stained.

Culture media

The tubercle bacillus was originally isolated on heat-coagulated serum. Nowadays the most popular media contain egg and are solidified by heating at 80°C for one hour, a process termed inspissation. The widely-used Löwenstein–Jensen medium contains eggs, asparagine, glycerol and some mineral salts. Stonebrink's medium is very similar to the former but contains sodium pyruvate instead of glycerol and is therefore suitable for the isolation of bovine tubercle bacilli. Egg-based media usually contain a dye such as malachite green which, in addition to being inhibitory to certain bacteria, gives a better background colour against which to see colonies of mycobacteria. The nutritional role of egg is uncertain. The yolk may provide lipid precursors but it is doubtful if the albumin protein is utilized. It is more likely that the role of the protein is to absorb free fatty acids which are toxic for mycobacteria, although they stimulate growth if present in small quantities.

There are a number of clear broths or agar-based solid media but these are used more for sensitivity testing and research work than for primary isolation of strains from clinical specimens. These include Middlebrook–Dubos 7H9 broth and 7H10 agar, Sauton's medium and Kirchner's broth. For details see Collins et al. (1985).

The above media support the growth of almost all mycobacteria (except, of course, *M. leprae*) but a few species require special media. The use of pyruvate instead of glycerol for the isolation of bovine tubercle bacilli has already been mentioned. *Mycobacterium paratuberculosis* and some strains of *M. avium* require media supplemented with mycobactin (see page 40) for growth. Such media are also preferable to standard media for the isolation of *M. avium* from specimens likely to contain only very few bacilli, such as porcine heart valves used in cardiac surgery. Media containing iron supplements (haem or ferric ammonium citrate) are required for the isolation of *M. haemophilum*.

Decontamination of specimens

Most specimens submitted for culture of mycobacteria contain other bacteria and possibly fungi which, if not destroyed, will rapidly overgrow the medium. It is therefore necessary to treat the specimens with reagents that selectively kill non-acid-fast organisms. Fortunately mycobacteria are relatively more resistant to acids, alkalis and certain disinfectants than other micro-organisms. The difference in resistance is a relative one and no agent can be relied upon to kill all other organisms without also killing a subtantial number of mycobacteria. Thus a decontamination procedure that permits a good isolation rate of mycobacteria will also result in a minority of cultures becoming contaminated – a fact that must be accepted by clinicians.

The choice of decontaminating agent depends on the specimen. 'Soft' methods are suitable for fresh specimens and for those likely to contain few contaminants, while 'hard' methods are used for heavily contaminated material. Soft reagents include trisodium phosphate and benzalkonium chloride, while hard reagents include sodium hydroxide and oxalic acid. The latter is particularly useful for the decontamination of urine specimens that contain *Pseudomonas aeruginosa* which is relatively resistant to alkali. In the case of sputum a liquefying agent, such as N-acetyl L-cysteine (NALC), may be added to the decontaminating reagent to aid concentration of acid-fast bacilli by centrifugation.

A widely used decontamination procedure is that of Petroff. Sputum is shaken with an equal volume of 4% NaOH (or 2% NaOH with 1% NALC) for 30 minutes then centrifuged. The deposit is neutralized with dilute hydrochloric acid or with 14% monopotassium phosphate, using neutral red as an indicator. The neutralization step may be avoided by adding an exact amount of deposit (0.2 ml) to an acid-buffered egg medium.

The 'sputum swab' method, although inferior to techniques in which sputum is concentrated, is suitable for laboratories without centrifuges. A throat swab dipped into sputum is placed first into a test tube containing 5% oxalic acid and is then transferred to another tube containing phosphate buffer. The swab is then used to inoculate a slope of Löwenstein–Jensen medium.

An alternative to decontamination is the use of a medium containing a 'cocktail' of antimicrobial agents that kill organisms other than mycobac-

teria. One of the most widely used media is Middlebrook–Dubos 7H10 agar containing polymyxin, trimethoprim, carbenicillin and amphotericin B (Mitchison *et al.*, 1983). Fewer than 1 per cent of plates seeded with untreated sputum became contaminated and the colony count of tubercle bacilli was 2.4 times higher than that on drug-free plates seeded with sputum decontaminated by treatment with NaOH.

Incubation and reading the cultures

There is a considerable variation in the temperature range of growth of the mycobacterial species – a feature utilized for identification purposes. Most mycobacteria grow at 35–37°C, but three species associated with skin disease, namely *M. marinum*, *M. haemophilum* and *M. ulcerans*, grow at a lower temperature. Thus all material from skin lesions should be incubated at 33°C as well as at a higher temperature. Mycobacteria are aerobic but their growth is enhanced by an atmosphere of 5–10 per cent CO_2 in air. As CO_2 incubators are costly to obtain and maintain, they are rarely used in clinical practice.

The slopes should be examined weekly for at least 8 weeks and preferably for up to 12 weeks as some species, such as *M. xenopi* and *M. malmoense*, may take this time to appear on primary isolation. An even longer incubation, up to 14 weeks, is required in veterinary practice for the isolation of *M. paratuberculosis*.

Contamination is usually indicated by a softening or discolouration of the media. Mycobacterial colonies can usually, with experience, be distinguished from those of contaminating bacteria but confirmation should always be made by Ziehl–Neelsen staining.

Animal inoculation for the isolation of mycobacteria

Guinea-pig inoculation was once a popular way of diagnosing tuberculosis but should now be regarded as obsolete. It has been clearly demonstrated that the use of this animal offers no advantages over *in vitro* culture (Marks, 1972; Pallen, 1987). In addition to humane considerations, animal inoculation is costly and generates many biohazards. Likewise, the use of animals to identify BCG and bovine strains of tubercle bacilli is obsolete as there are now very adequate *in vitro* tests for this purpose (Yates and Collins, 1979).

Radiometric methods

In view of the slow rate of growth of tubercle bacilli and most other pathogenic mycobacteria, attempts have been made to develop rapid methods. The only rapid method in routine clinical use is radiometry. This technique, originally introduced for the detection of bacterial growth in blood cultures, is based on the release of radioactive carbon dioxide from a labelled precursor by bacterial metabolism. The released gas is detected by periodic sampling of the air space over the medium: this is done automatically in a commercially available instrument (BACTEC, Beckton Dickenson). The specimen is added to a bottle of broth containing ^{14}C–palmitic acid and a cocktail of antimicrobial agents to inhibit growth of any organism other than mycobacteria (see page 54). Growth of mycobacteria may be detectable within two or three days, although the isolation rate is no higher than by standard culture. This technique is also used for rapid sensitivity testing by incorporating antituberculous drugs in the media. Although radiometry is undoubtedly more rapid than conventional techniques, it is much more costly and this will severely restrict its use in routine laboratories. For further details of radiometric methods, see Laszlo *et al.* (1983) and Roberts *et al.* (1983).

Identification of mycobacteria

By far the most frequently isolated mycobacterium in clinical practice is *M. tuberculosis*. The first step in identification is therefore to determine whether or not an isolate is of this species or the closely related *M. bovis* and *M. africanum*. This can be done in several ways; for example, the following properties together clearly distinguish these strains from all other mycobacteria: slow growth rate, no pigment produced in the light or dark, no growth at 25°C and no growth on egg media containing 500 mg/l of p-nitrobenzoic acid or p-nitro-α-acetylamino-β-hydroxypropriophenone (NAP).

Some workers identify the human tubercle bacillus by its production of large quantities of niacin, detectable by a simple chemical test (Konno, 1956) or by the use of commercially available test strips (Difco). Reliance should not be placed on this test alone as, in addition to bovine strains, a few human strains are negative. Conversely, positive reactions are given by *M. simiae* and by a few strains of *M. avium* and of the rapidly growing pathogen *M. chelonei*.

Typing tubercle bacilli

The tubercle bacilli that cause human disease are *M. tuberculosis*, *M. bovis* and *M. africanum* (see page 32). The former is divisible into the classical and South Indian or Asian types (Collins *et al.*, 1982). In addition BCG occasionally causes localized or even widespread disease and may need to be distinguished from the virulent types. The subdivision of the virulent tubercle bacilli is for epidemiological purposes only. The clinical management of all these types is identical except that bovine strains are naturally resistant to the antituberculosis drug pyrazinamide.

Table 4.1 Properties of the slowly growing mycobacteria

	Pigmentation	Growth at (°C)					Nitratase	Arylsulphatase	Tween 80 hydrolysis	Tellurite reduction	α-L-fucosidase	Urease	Acid phosphatase
		20	25	33	42	44							
M. tuberculosis	N	–	–	+	–	–	+	–	–	–	–	–	–
M. bovis	N	–	–	+	–	–	–	–	–	–	–	–	–
M. kansasii	P	–	+	+	v	–	+	+	+	–	–	+	+
M. marinum	P	+	+	+	–	–	+	+	+	–	+	+	+
M. asiaticum	P	–	v	+	–	–	–	(+)	+	–	–	–	+
M. simiae	P/N	–	+	+	v	–	–	(+)	(+)	–	–	+	–
M. scrofulaceum	S	v	+	+	+	–	–	–	–	–	–	+	–
M. szulgai	S	–	+	+	–	–	+	+	+	–	+	+	–
M. gordonae	S	+	+	+	–	–	–	+	+	–	–	+	+
M. avium–intracellulare	N	v	+	+	+	v	–	–	–	+	0	–	–
M. malmoense	N	–	+	+	–	–	–	–	(+)	–	0	–	–
M. ulcerans	Nx	–	–	+	–	–	–	–	–	–	–	–	–
M. xenopi	Nx	–	–	+	+	+	–	+	–	–	–	–	–
M. haemophilum	N	0	+	+	–	–	–	0	–	–	0	–	+
M. terrae	N	v	+	+	–	–	+	+	+	–	0	–	+
M. triviale	N	v	+	+	–	–	+	+	+	–	0	–	+
M. nonchromogenicum	N	v	+	+	–	–	–	+	+	–	0	–	+

N = nonchromogen; P = photochromogen; S = scotochromogen; Nx = light lemon-yellow colour; + = positive reaction in >85 per cent of strains;
– = negative reaction in >85 per cent of strains; (+) = weak or late reaction; v = variable growth; 0 = limited or no data

Identification of other mycobacterial species

There is no universally accepted protocol for the identification of mycobacteria other than the tubercle bacilli. Ideally each isolate should be identified at the species level and, in some cases, at the subspecies level so as to increase our knowledge of the types of mycobacteria that cause disease, and to determine the most appropriate therapy. Often, though, facilities and finance prevent such a thorough investigation.

Some laboratories are in the fortunate position of being able to perform a wide range of tests on each isolate and possibly to use highly discriminatory test systems. Others must be content with using a few simple tests which will identify most species that are encountered in clinical practice. Marks (1976) divided clinical isolates into 15 groups according to growth at 25, 37, 42 and 45°C, pigment production, oxygen preference and hydrolysis of Tween 80. Additional simple and useful tests include detection of nitratase and arylsulphatase activity and reduction of tellurite. Glycosidase activity is also of value, especially α-L-fucosidase activity which is particularly strong in *M. marinum* and also distinguishes *M. szulgai* from other slowly growing scotochromogens. The use of these and other cultural and biochemical tests for the identification of most of the slowly growing mycobacteria is shown in Table 4.1.

Rapidly growing species – those that give a good growth on subculture from a small inoculum on Löwenstein–Jensen medium within 7 days – may be identified by their liberation of ammonia from amides (Bönicke, 1962), by their production of acid from sugars (Gordon and Smith, 1953), and by their utilization of organic acids for growth. In this respect, ammonia production from allantoin, acid production from mannitol, inositol and xylose and utilization of citrate are of particular value, as shown in Table 4.2. From the clinical point of view the important rapid growers are *M. chelonei* and *M. fortuitum* as, with very rare exceptions, these are the only human pathogens in this group. These are characterized by their lack of pigment, strong arylsulphatase activity and limited saccharolytic activity and are distinguished from each other by the nitratase test and other properties shown in Table 3.5.

More sophisticated identification tests include characterization of mycosides (page 16) by lipid chromatography, immunodiffusion analysis of cytoplasmic antigens, protein electrophoresis, gas–liquid chromatography and pyrolysis mass spectroscopy.

Sensitivity testing

As outlined in Chapter 10, drug-resistant mutants continuously arise at a low rate in any mycobacterial population. Any culture will therefore inevitably contain a few such mutants. The purpose of sensitivity testing is to determine whether the great majority of bacilli in the culture are sensitive to the antituberculous drugs currently in use. In other words, sensitivity tests are designed to inform the clinician whether or not an isolate is as susceptible to a given drug as other known sensitive strains.

Table 4.2 Properties of some rapidly growing mycobacteria

	Pigment	Growth at 45°C	Growth after heating to 60°C for 4 hours	Arylsulphatase (3 days)	Nitrate reductase	Citrate utilization	Allantoinase	Acid from mannitol	Acid from inositol	Acid from xylose
M. fortuitum type A	–	+	–	+	+	+	+	–	–	–
M. fortuitum type B	–	–	–	+	+	+	+	+	–	–
M. fortuitum type C	–	–	–	+	+	+	+	+	+	–
M. chelonei abscessus	–	–	–	+	–	–	–	–	+	–
M. chelonei chelonei	–	–	–	+	–	+	–	–	–	–
M. smegmatis	–	+	–	–	+	+	–	+	+	+
M. phlei	+	+	+	–	+	+	–	+	–	+
M. diernhoferi	–	–	–	–	+	+	–	+	+	+
M. gilvum	+	–	–	+	+	+	–	+	+	–
M. duvalli	+	–	–	–	+	–	–	+	–	–
M. flavescens	+	–	–	+	+	–	–	V	–	–
M. vaccae	+	–	–	–	V	V	+	+	+	+

+ = positive; – = negative; V = variable result

Table 4.3 Examples of the resistance ratio method

	Increasing drug concentration (tube no.)						*Resistance ratio*
	1	2	3	4	5	6	
Modal resistance	C	D	O	O	O	O	–
Test strain no.							
1	C	C	O	O	O	O	1
2	C	D	O	O	O	O	1
3	C	C	D	O	O	O	2
4	C	C	C	D	O	O	4
5	C	C	C	D	+	O	4
6	C	C	C	C	C	O	8
7	C	C	C	D	+	O	8
8	C	C	C	C	D	+	8+

C = confluent growth; D = numerous discrete colonies; + = 20–100 colonies;
O = <20 colonies

There are three major techniques for sensitivity testing: the resistance ratio method, the absolute concentration method and the proportion method. Furthermore, testing may be direct or indirect (i.e. performed on the original specimen or on a subculture respectively).

The resistance ratio is determined by inoculating standardized suspensions of the test strain and a number of known sensitive strains on to media containing doubling dilutions of the drug. After incubation, the endpoint for each strain (i.e. the slope with 20 or less colonies) is determined. Test strains are then compared with the average or 'modal' resistance of the set of known sensitive strains. If the endpoint of test and controls is equal, the strain has a resistance ratio of 1. As doubling dilutions of drugs are used, 1, 2 or 3 tube differences in the endpoints of test and control strains give resistance ratios of 2, 4 and 8 respectively. Strains with resistance ratios of 4 or more are reported as resistant. Examples are shown in Table 4.3.

The absolute concentration method is very similar to the resistance ratio method, being based on a titration of the test strain, along with adequate control strains, on slopes of media containing known quantities of the drug in doubling dilutions. The difference is that the results are expressed in the actual endpoint concentration of drug. In practice, the activity of the drug may be less than its concentration in the medium owing to its denaturation during medium preparation. Accordingly, this method has no real advantage over the resistance ratio method.

In the proportion method the number of colonies growing from a standard inoculum on a drug-containing medium is compared with the colony count from the same sized inoculum on a drug-free medium. A strain is considered resistant to a given concentration of drug if the number of colonies growing on the drug-containing medium is 1 per cent or more of the number growing on the drug-free medium.

The three methods give essentially similar results. The resistance ratio method is used in Great Britain while the proportion method is widely used in the USA. More recently, radiometry (see page 54) has proved to be a rapid, reliable but expensive method for performing sensitivity testing.

Pyrazinamide sensitivity tests pose a particular problem as this drug is only active in acid media and there is a narrow pH range, around 5.2, at which the drug is active and the bacilli are able to grow. A reliable technique is described by Yates (see Collins *et al.*, 1985, p. 97).

Laboratory reports

A report of the microscopical examination may be issued soon after receipt of the specimen. It is difficult and arguably unwise to distinguish tubercle bacilli from 'atypical' mycobacteria on the basis of microscopy. Isolation of a mycobacterium usually takes from two to five weeks and at that stage it should be possible to say with a high degree of certainty whether or not the isolate is a tubercle bacillus. A firm identification of the strain together with results of sensitivity testing is available after a further three or four weeks. Thus sensitivity results are usually available about two months after submission of the specimen, unless direct testing or radiometric methods are used.

Laboratory methods in leprosy

As *Mycobacterium leprae* cannot be cultured *in vitro*, routine investigations are limited to the microscopical demonstration of the bacilli in clinical specimens. Leprosy bacilli may be propagated in the footpads of mice and, by incorporating drugs in the animals' food, sensitivity testing may be performed. This is a laborious procedure that is only carried out in a limited number of reference and research centres.

The usual specimens for the microscopical diagnosis of leprosy are slit-skin smears and nasal scrapings which are examined for the presence of acid-fast bacilli. These diagnostic procedures are described in Chapter 7.

References

Bönicke, R. (1962) Report on identification of mycobacteria by biochemical methods. *Bulletin of the International Union Against Tuberculosis* **32**: 13–68.

Collins, C.H. (1983) *Laboratory Acquired Infections.* London: Butterworths.

Collins, C.H., Grange, J.M. and Yates, M.D. (1985) *Organization and Practice in Tuberculosis Bacteriology.* London: Butterworths.

Collins, C.H., Yates, M.D. and Grange, J.M. (1982) Subdivision of *Mycobacterium tuberculosis* into five variants for epidemiological purposes: methods and nomenclature. *Journal of Hygiene* **89**: 235–42.

Gordon, R.E. and Smith, M.M. (1953) Rapidly growing acid-fast bacteria. *Journal of Bacteriology* **66**: 41–8.

Konno, K. (1956) New chemical method to differentiate human type tubercle bacilli from other mycobacteria. *Science* **124**: 985.

Laszlo, A., Gill, P., Handzel, V., Hodgkin, M.M. and Helbeque, D.M. (1983) Conventional and radiometric drug susceptibility testing of *Mycobacterium tuberculosis* complex. *Journal of Clinical Microbiology* **18**: 1335–9.

Marks, J. (1972) Ending the routine guinea-pig test. *Tubercle* **53**: 31–4.

Marks, J. (1976) A system for the examination of tubercle bacilli and other mycobacteria. *Tubercle* **57**: 207–25.

Mitchison, D.A., Allen, B.W. and Manickavasagar, D. (1983) Selective Kirschner medium in the culture of specimens other than sputum for mycobacteria. *Journal of Clinical Pathology* **36**: 1357–61.

Pallen, M. (1987) The inoculation of tissue specimens into guinea-pigs in suspected cases of mycobacterial infection – does it aid diagnosis and treatment? *Tubercle* **68**: 51–8.

Ridley, D.S. (1977) *Skin Biopsy in Leprosy.* Basel: Ciba-Geigy.

Roberts, G.D., Goodman, N.L., Heifets, L., Larsh, H.W., Lindner, T.H., McLatchy, K., McGinnis, M.R., Siddiqi, S.H. and Wright, P. (1983) Evaluation of the BACTEC radiometric method for the recovery of *Mycobacterium tuberculosis* from acid-fast smear-positive specimens. *Journal of Clinical Microbiology* **18**: 689–96.

Yates, M.D. and Collins, C.H. (1979) Identification of tubercle bacilli. *Annales de Microbiologie* **130B**: 13–19.

5 Immunology of mycobacterial disease

The clinical and histological features of mycobacterial disease are more the result of the immune reactions of the host than of the invasive powers of the pathogens. In the case of leprosy in particular these features enable a very accurate assessment of the nature of the patient's immune response to be made. The subject of mycobacterial immunology has become a very extensive one in recent years. For detailed reviews of the fundamentals of the subject the reader is referred to Rook (1983), Lagrange (1984) and Ottenhoff and de Vries (1987). The purpose of this chapter is to describe those areas of immunology that are relevant to an understanding of the pathogenesis of mycobacterial disease, the nature and significance of tuberculin reactivity and the mode of action of BCG vaccine.

Mycobacteria, in common with other intracellular parasites, owe their virulence to their ability to survive with macrophages. Protective immune reactions in mycobacterial disease are of the so-called cell-mediated type and serve to enhance the ability of the macrophage to inhibit or destroy the invaders. Humoral immune responses, i.e. antibody production, certainly occur and many attempts have been made to utilize these for diagnostic purposes; but there is no evidence that they play any major role in host defence.

The immunological phenomena seen in mycobacterial disease consist, as in other infections, of recognition, response and reaction. In the first step, the invading mycobacteria are recognized as being 'foreign'. In the second step the necessary defence mechanisms are alerted and recruited, while in the third the actual intracellular struggle between mycobacterium and macrophage takes place.

Antigen recognition

Antigen recognition is a property of the lymphocytes which are divided into two main sets: the B cells (bursa or bursa-equivalent dependent) and the T cells (thymus dependent) The former mature into antibody-secreting plasma cells, while the latter are involved in the initial recognition of antigen, the regulation of the immune response, cell-mediated cytotoxicity and the secretion of hormone-like molecules that affect other cells of the immune system. The T cells are divisible into subsets according to their functions and by the presence of certain markers (differentiation antigens) detectable by monoclonal antibodies. The main T cell subsets are those with the T4 or CD4 markers, which include the 'helper' cells, and those with the T8 or CD8 markers, which include the 'suppressor' cells. The association of differentiation antigens and functional activity is, in fact, not so clear-cut as originally thought and it is better to refer to CD4 and CD8 cells as 'putative' helper and suppressor cells respectively.

A lymphocyte bears receptors that bind to one of the thousands of possible antigenic determinants or epitopes. Accordingly, there are many thousands of subpopulations within each subset, each specific for just one epitope. An important stage of the immune response is that of clonal expansion, in which a small number of antigen-specific lymphocytes proliferate to form a clone of cells of sufficient numbers to mediate an effective immune response.

Before an antigen is able to induce such clonal proliferation it must be presented to the lymphocytes in a special way. This is mediated by *antigen presenting cells* (APCs) which are of the monocyte/macrophage series and include the dendritic cells of lymph modes and scattered lymphoid tissues and the Langerhans cells of the dermis. The process is shown in Fig. 5.1. The APCs engulf particulate and soluble antigens which usually contain a multitude of epitopes. The antigens are partially digested by enzymes within lysosomes and within the cell membrane, and the remaining epitopes are presented on the cell's surface in close relation to the immune-associated (Ia) antigens. The Ia antigens are class II (HLA-D) products of the major histocompatibility complex and lymphocytes will only proliferate if they 'see' the foreign epitope/Ia complex. The linkage between the lymphocyte and the APC induces the latter to secrete a mediator – interleukin I (IL1) – which in turn causes the lymphocyte to secrete

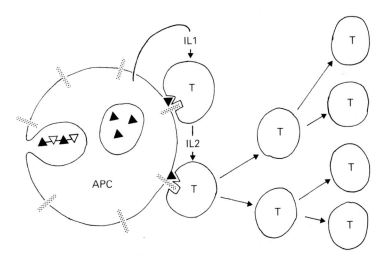

Fig. 5.1 Antigen presentation. A complex antigen is phagocytosed and degraded by the antigen processing cell (APC). Individual epitopes are presented in the cell membrane together with the cell's Ia antigens. Antigen-specific T-lymphocytes (T) cause the APC to secrete interleukin 1 (IL1). This induces the T-cell to secrete interleukin 2 (IL2) which causes the T-cells to proliferate into a clone

interleukin 2 (IL2) which mediates lymphocyte division and clonal expansion.

The immune response: macrophage activation

The next cells to enter the scenario are the macrophages which belong to the same cell lineage as the blood-borne monocytes. Macrophages are phagocytic cells but, unlike the other major class of phagocytes, the polymorphonuclear leucocytes, they are long-lived cells which settle in a given tissue and organ and adapt to the local environment. Thus, alveolar macrophages, osteoclasts, Kupffer cells of the liver and Schwann cells of the nerves are all specialized macrophages. Further, macrophages do not express their full microbial potential unless they are 'activated' (Fig. 5.2). Such activation is a major function of the clonally expanded lymphocytes which secrete a class of hormone molecules termed lymphokines. One group of lymphokines is collectively termed macrophage activating factor (MAF) and one of the principal members of this group is gamma interferon (INF-γ).

The activated macrophage differs from its resting counterpart in several respects. The cell membrane is much more motile – a phenomenon termed membrane ruffling. Random migration, glass adherence and the ability to phagocytose and kill micro-organisms are all increased. Activation is not a single-step process. Although INF-γ is an important macrophage activator, it is not the only one. It has been shown that macrophages activated by

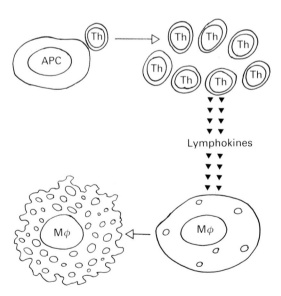

Fig. 5.2 Macrophage activation. In response to antigen presented by the APC and subsequent clonal expansion of the T-helper cells (Th), lymphokines are released by the Th cells and these activate the resting macrophage (MØ)

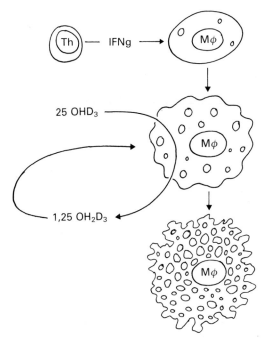

Fig. 5.3 Role of vitamin D in macrophage activation and maturation: After activation by gamma interferon (IFNg) the activated macrophage converts vitamin D_3 (25 OHD_3) into the more active 1,25 OH_2D_3 metabolite which causes further macrophage activation. Data from Rook (1986)

INF-γ have a limited anti-mycobacterial effect but that, by means of an autocrine positive feedback mechanism (Fig. 5.3), vitamin D induces further enhancement of anti-mycobacterial effect (Rook, 1986). This may well explain the success of vitamin D therapy in the treatment of lupus vulgaris.

In addition to its microbicidal activity, the macrophage synthesizes and secretes many important compounds that affect the pathogenesis of mycobacterial disease. These include some of the acute-phase reactant proteins, vasoactive peptides, and proteases that liquify necrotic tissue and contribute towards the formation of the tuberculosis cavity. In addition, macrophages secrete a substance known as cachectin which is thought to be responsible for the extreme wasting associated with advanced tuberculosis. This substance is identical, or extremely similar to, tumour necrosis factor (Beutler *et al.*, 1985) which may contribute to the necrotic reactions seen in delayed hypersensitivity.

The immune reaction: events within the macrophage

In order to survive within a macrophage, a bacterium must be able to resist destruction by the wide range of oxygen-dependent and non-oxygen-

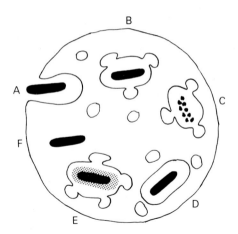

Fig. 5.4 Phagocytosis. A: Bacilli are engulfed by the cell membrane. B: The bacilli lie in a membrane vesicle – the phagosome – which fuses with lysosomes containing bactericidal substances. C: These substances destroy the bacilli. D: *M. tuberculosis* inhibits phagosome/lysosome fusion. E: Members of the *M. avium* group are protected by a thick capsule-like outer layer. F: *M. leprae* escapes from the phagosome and lies freely within the cytoplasm

dependent killing mechanisms of the infected cell. The former mechanisms include the generation of superoxide radicals, hydrogen peroxide and hypochlorite ions; the latter include enzymes such as lysozyme, lipases and phosphatases. The process of phagocytosis is shown in Fig. 5.4. An engulfed bacterium lies within a vesicle formed by invagination of the surface membrane. This vesicle, the phagosome, then fuses with the lysosomes which contain the bacteroicidal agents referred to above. There are three main strategies by which bacteria or other pathogens survive within the phagosome. First, phagosome/lysosome fusion may be inhibited. Secondly, the pathogen may cover itself with a protective layer that absorbs or neutralizes the bacteriocidal agents or, thirdly, it may escape from the vesicle and lie freely in the cytoplasm of the cell.

Mycobacterium tuberculosis, in common with the protozoal parasite *Toxoplasma gondii*, inhibits phagosome–lysosome fusion but neither the mechanism nor the significance of this activity are clearly understood. The inhibition of membrane fusion is presumably due to one or more mediators secreted by the bacilli: sulpholipids, polyglutamic acid, ammonium ions and cyclic adenosine monophosphate being possible candidates (Draper, 1981). It is not known to what extent this activity is responsible for intracellular survival: antibody-coated tubercle bacilli do not inhibit membrane fusion yet antibody appears to have no protective role in tuberculosis. It is possible that fusion-inhibition isolates the bacillus from nutrients as well as from the lysosomal contents, leading to bacterial stasis or dormancy, a condition in which the pathogen is relatively resistant to the effect of antibacterial agents.

Mycobacterium leprae is able to escape from the phagosome and lie freely within the cytoplasm of the macrophage, where it is not only free from attack by the lysosomal contents but is also bathed in nutrients.

Some other pathogens, notably members of the *M. avium* group, appear to survive the effects of phagosome–lysosome fusion, probably on account of the thick outer layer of mycosides which, on electron microscopy, appears as an electron transparent zone surrounding the bacilli.

The mechanism by which macrophages actually kill mycobacteria, as opposed to merely inhibiting their growth, remains shrouded in mystery. The major barrier to our understanding is our inability, so far, to achieve a convincing and repeatable demonstration of mycobacterial killing in macrophages *in vitro*. The available evidence, although largely derived from work with mouse or guinea pig cells *in vitro*, strongly suggests that hydrogen peroxide is the major mycobacteriocidal agent produced by the macrophage (Lowrie, 1983). An association of virulence for the guinea pig and resistance to hydrogen peroxide is suggested by the finding that both the South Indian variant and isoniazid-resistant classical tubercle bacilli are susceptible to hydrogen peroxide and are attenuated in this animal. It is uncertain whether this association applies to man, in whom these peroxide-susceptible strains are virulent.

The early immunological events

The events summarized above underlie the pathogenesis of primary pulmonary tuberculosis. The inhaled bacilli lodge in the periphery of the lung and are soon engulfed by alveolar macrophages. Some bacilli are transported within macrophages to the local lymph nodes, where antigen is processed and presented to lymphocytes which undergo clonal expansion. In the meantime the bacilli at the initial site of infection and in the lymph nodes proliferate and some are borne further afield in the lymphatics and blood stream. When the immune response has developed, activated macrophages and lymphocytes cluster around the invading mycobacteria. In the presence of lymphokines the activated macrophages mature into the so-called epithelioid cells and some fuse to form giant cells. This forms the characteristic lesion of chronic infection known as the granuloma. In tuberculosis the granuloma consists of a central area of cheese-like necrosis, or caseation, surrounded by epithelioid and giant cells which in turn are surrounded by lymphocytes. In many cases this protective immune response is sufficient to arrest the disease and destroy most of the mycobacteria. Collagen is then laid down by fibroblasts and the foci heal by scarring.

Similar resolution of early disease also occurs in many cases of leprosy and other mycobacterial infections, although it has not been possible to examine the sequence of events in the same detail as in pulmonary tuberculosis.

Protective cell-mediated reactions as described above are not the only immune phenomena associated with mycobacterial disease. Others, such as immunosuppression and delayed hypersensitivity, also occur and make the subject much more complex. Indeed these other reactions are of great relevance to the pathogenesis of the diseases.

Delayed hypersensitivity

A hypersensitivity reaction may be defined as one which causes tissue damage. Four main types of hypersensitivity are recognized; namely, anaphylactic (Type I), antibody-dependent cytotoxic (Type II); immune complex-mediated (Type III) and delayed (Type IV). The first three involve antibody but delayed hypersensitivity is 'cell-mediated'. The tuberculin reaction is often cited as the classical example of the delayed hypersensitivity (DH) or Type IV reaction but, in fact, the mechanism of this reaction and its relation to protective cell-mediated immune reactions is the subject of a long and unresolved controversy.

In order to understand DH it is necessary to look back to the original studies of Robert Koch which were carried out, under considerable pressure from his political masters, in an attempt to discover a cure for tuberculosis. Koch (1891) inoculated tubercle bacilli into the flanks of guinea-pigs. After a week or two a small, firm nodule developed and subsequently ulcerated. Viable tubercle bacilli were isolated from the ulcer which remained open until the animal died. About a month later local lymph nodes were enlarged and disseminated disease developed, leading to death three or four months later. When Koch gave a similar inoculation of bacilli into the opposite flank a month after the original infecting dose, a quite different lesion developed. After a day or two the skin at the inoculation site became black and necrotic and then sloughed off leaving a shallow ulcer from which no bacilli could be isolated and which soon healed. Koch then found that he could elicit a similar reaction, the Koch Phenomenon, by injecting either killed tubercle bacilli or a heat-concentrated filtrate of the medium in which the bacilli had been grown. The important point to notice is that, although this reaction clearly led to the elimination of bacilli inoculated into the skin, the animals were nevertheless dying of systemic tuberculosis. This implies that either the reaction does not occur in the presence of tubercle bacilli in the deep tissues or, if it does, it is either non-protective or positively harmful. In fact, Koch himself put the matter to the test by administering Old Tuberculin systemically to patients with tuberculosis. This proved disasterous although there were a few remarkable cures in individuals with disease of the skin or larynx. Thus it appears that if the necrotic Koch phenomenon occurs on the surface, the bacilli-laden tissue easily sloughs off. If, on the other hand, it occurs in the lung or other internal organ, the bacilli and necrotic tissue remain *in situ*.

Conversion to tuberculin positivity in man occurs about six to eight weeks after the initial infection. By then the primary lesion is, in most cases, resolving. Post-primary disease occurs months, years or even decades later and is the result of either reactivation of old foci of disease or of exogenous reinfection (see page 93). Post-primary lesions, which are frequently in the apex of the lung, are characterized by large amounts of caseous necrotic tissue. This material is liquified by proteases liberated by macrophages and, if the lesion erodes into a bronchus, the softened contents are coughed out leaving a cavity. The cavity wall is a perfect breeding ground for tubercle bacilli and large numbers enter the sputum, rendering the patient open or infectious. Thus the difference in immunological reactivity before and after tuberculin conversion has a profound effect on the pathogenesis of the disease and is an important factor in determining infectivity.

Tubercle bacilli escaping from cavities may cause secondary lesions in the lower lobes of the lung, in the upper respiratory tract and, if swallowed, in the alimentary tract (Chapter 8, page 124). On the other hand, lymphatic and haematogenous dissemination of disease is, in contrast to primary tuberculosis, most unusual. This is probably due to necrosis of the draining lymphatics and capillaries by the DH reaction. An important component of the protection afforded by BCG vaccine is a prevention of serious forms of primary tuberculosis that result from haematogenous dissemination, and this may be due to the induction of DH (Ladefoged *et al.*, 1976).

For many years the relationship between protective cell-mediated immunity (CMI) and delayed hypersensitivity (DH) has been the subject of much controversy. Some workers claim that these are essentially similar phenomena but differ in degree while others argue that they are separate reactions.

During the two decades following the introduction of the tuberculin test into routine clinical use by Clemens von Pirquet in 1907, it was widely asserted that tuberculin reactivity was a sign and measure of immunity. This view was seriously challenged by Rich and McCordock (1929) who observed that infected individuals with large tuberculin reactions were more likely to develop reactivation tuberculosis than those with less reactivity. Similar conclusions were reached in animal experiments by Wilson *et al.* (1940) and in clinical studies by Turner (1953). Thus it appeared that a moderate amount of tuberculin reactivity indicated immunity while an excessive, i.e. hypersensitive, reactivity did more harm than good. Nevertheless, the difference between a protective CMI reaction and a necrotic DH reaction could be one of degree. On the other hand there is considerable evidence that the differences are qualitative rather than merely quantitative.

Youmans (1979) has cited several studies in support of the concept that CMI and DH are distinct and dissociable phenomena and, in particular, has drawn attention to the fact that primary tuberculosis lesions are often resolving by the time that tuberculin conversion occurs, strongly suggesting that CMI is already active.

In a detailed study of mice infected with various mycobacteria, Rook and Stanford (1979) observed that a reaction to tuberculin peaking at about 20 hours after skin testing appeared about 10 days after infection, while a reaction peaking at about 40 hours appeared a month or so after infection. The former reaction resembles that demonstrable in mice infected with *Listeria monocytogenes* and represents a macrophage-activating CMI reaction. The latter reaction appeared to be the murine equivalent of the necrotic Koch phenomenon in guinea-pigs. These reactions were therefore termed the 'Listeria-type' and the 'Koch-type' reactions respectively. Interestingly, the less virulent strains of BCG and other non-pathogens preferentially elicited the Listeria-type reaction, while more virulent BCG strains and other mycobacteria that are pathogenic in the mouse elicited the Koch-type reaction. Furthermore, the induction of one type of reaction appeared to block the subsequent induction of the other. Thus the pattern of immune reactivity could be determined by the nature of the immunologically effective contact with mycobacteria early in life – a phenomenon termed 'Original Mycobacterial Sin' by Abrahams (1970). This concept has profound implications for the use and efficacy of BCG vaccination as discussed below (see page 78). Mystery still surrounds the nature of the Koch-type reactivity. It is probably due to mediators such as tumour necrosis factor (see page 65) and vasoactive substances released by activated macrophages, probably as a result of some additional signal which could be a lymphokine released from an as-yet unidentified subset of T cells or vitamin D metabolites, or both (Rook, 1986).

Tuberculin and tuberculin reactivity

Koch's Old Tuberculin was a filtrate of a broth culture which was then concentrated by evaporation in a heated water bath. This material contained various impurities derived from the medium and induced non-specific inflammatory reactions. To overcome this problem Siebert (1934) attempted to isolate the tuberculoproteins by precipitation with acetone and ammonium sulphate. The resulting preparation was termed purified protein derivative of tuberculin (PPD) and has been used widely since. Despite the name, PPD is not pure protein and it suffers from several disadvantages. Its mode of preparation involves prolonged cultivation of the tubercle bacillus to enable antigens to be released by autolysis, heating to 100°C to kill the bacilli, and then precipitation of the protein. All these steps cause denaturation of protein and the resulting preparation has been termed a 'rotted, boiled and pickled antigen' (Stanford and Rook, 1983). A much more rational approach to the preparation of skin testing reagents was adopted by Stanford and his colleagues, resulting in the production of the 'New Tuberculins'. The first of these was Burulin, prepared for studies on Buruli ulcer in Africa (Stanford *et al.*, 1975), but subsequently they were prepared from many mycobacterial species (Editorial, 1984). These reagents are prepared by harvesting mycobacteria during the growing phase from non-antigenic media, washing thoroughly, disrupting the cell mass in an ultrasonicator, sterilizing by repeated membrane filtration, and diluting to a suitable protein concentration. Thus the three denaturing processes are avoided. Immunodiffusion analysis has revealed that new tuberculins are relatively much richer in the species-specific antigens than PPD (Table 5.1).

Classically, a positive tuberculin reaction manifests as an area of induration which reaches a maximum after 48 or 72 hours. Erythema also occurs and may be much more extensive than the induration; but it is, by convention, ignored as it is difficult to see and measure in dark-skinned peoples. The dermal changes elicited by tuberculin are usually considered to be due to a single immunological phenomenon; namely, a Type IV or

Table 5.1 Relative amounts of group i (common), group ii (slow-grower-related) and group iv (species-specific) soluble antigens in purified protein derivative (PPD) and New Tuberculin

Reagent	Antigen group		
	i	ii	iv
PPD	++	+++	+
New Tuberculin	++	+	+++

delayed hyersensitivity reaction. There is now clear evidence that the reaction is the result of several quite distinct reactions. Pepys (1955) observed that tuberculin could induce atopic (Type I) and Arthus (Type III) as well as Type IV reactions. Subsequently, Kardjito and his colleagues confirmed that some patients developed a weal and flare reaction within minutes of skin testing and some developed a distinct area of erythema six to eight hours after testing. Interestingly, the latter reaction is seen much more frequently amongst patients with active tuberculosis than amongst healthy tuberculin positive individuals (Kardjito and Grange, 1982). It is also particularly prominent amongst healthy individuals, such as nurses, who are occupationally exposed to patients with tuberculosis (Table 5.2), suggesting that it plays an important protective role (El Ansary and Grange, 1984; Grange *et al.*, 1986). The exact nature of this early erythematous reaction is unknown but it does not appear to be a Type III reaction due to immune complex formation. It resembles the late phase response of Type I hypersensitivity and may be due to the delayed release of mediators from mast cells triggered by receptors derived from T-cells rather than IgE (Fig. 5.5). Thus it could be the human equivalent of a reaction described by Askanase and van Loverin (1983) in mice.

The late component of the tuberculin reaction may also result from qualitatively different reactions. Stanford and his colleagues have observed that some 48-hour tuberculin reactions in man are purple coloured, indurated, well demarcated and tender, while others are pink, soft, ill-defined and much less tender (Stanford and Lema, 1983). It has been postulated that these reactions correspond to the Listeria-type and Koch-type reactions described in the mouse (see page 69).

Table 5.2 Number of reactors, and mean size of reaction, to tuberculin at 6–8 hours and 48 hours amongst 416 patients with pulmonary tuberculosis, 123 healthy hospital staff and 126 healthy factory workers

Group	6–8 hour reaction		48 hour reaction	
	Reactors (%)	Size (mm)	Reactors (%)	Size (mm)
Tuberculosis (smear positive)	73	7.8	83	17.1
Tuberculosis (smear negative)	82	9.2	88	18.8
Healthy hospital staff	90	18.8	94	24.8
Healthy factory workers	7	5.8	82	14.8

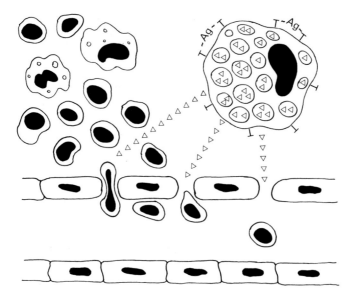

Fig. 5.5 Release of mediators from mast cells triggered by T cell-derived receptors. Cross linkage of T cell-derived receptors (T) on the mast cell by antigen (Ag) leads to the release of mediators (△) which cause vasodilation and accumulation of blood-derived white cells in the perivascular area (Askanase and van Loverin, 1983)

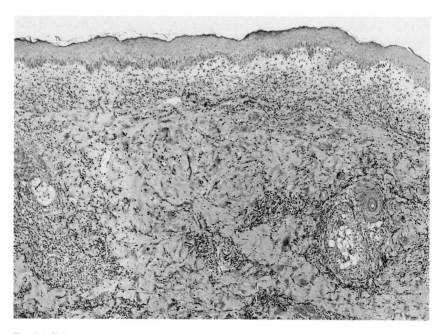

Fig. 5.6 Thin section of a biopsy of a tuberculin reaction at 48 hours, showing perivascular and periappendicular foci of inflammation. Courtesy Prof. J. Swanson Beck

The histological appearance of a positive tuberculin test at 48 hours has been described in detail by Beck and his colleagues (1986) and is shown in Fig. 5.6. There is a dense infiltrate of mononuclear cells (lymphocytes and monocyte/macrophages) around the capillaries and skin appendages (sweat glands and hair follicles). In addition, some of these mononuclear cells migrate out of the perivascular and periappendicular foci and migrate towards the epidermis. This migration, particularly that of the monocyte/macrophage cells, is greater in reactions to tuberculin than to leprosin in patients with tuberculosis, and vice versa in leprosy patients (Fig. 5.7). This suggests that this component of the reaction is affected by species-specific antigens. Although the tuberculin reaction is 'cell-mediated', the size of the reaction is not related to the intensity of the cellular infiltrate, measured as the area of the dermis occupied by the perivascular foci: some individuals with very large reactions have relatively few cells, while others who have no visible or palpable reaction have an intense cellular infiltrate. The clinically evident features of the reaction are almost certainly due to vasoactive and other mediators released by the cells. It is important to note that the reaction is 'cell-mediated' rather than 'cellular'. Clearly much more work is required to unravel the complexities of this reaction and the significance to protection of its various components.

Categories of tuberculin reactors

Multiple skin testing with a range of Stanford's new tuberculins has shown that individuals may be placed in three distinct categories according to their patterns of reactivity (Stanford *et al.*, 1981). Category 1 individuals react to any new tuberculin, indicating that they respond to the shared (group i) mycobacterial antigens. Category 2 individuals react to none of the reagents and this may be a genetic feature associated with HLA-D (class II) histocompatibility antigens (see page 85). There is no evidence that such non-responders are unable to develop immunity; indeed, as mentioned above, an intense cellular infiltrate may occur in the absence of clinically evident reactivity. Category 3 individuals react to some but not all reagents, indicating that they are responding to the species-specific (group iv) antigens.

Three important points arise from this categorization. First, an awareness of the category 2 non-responders is relevant to the use of BCG. Some unfortunate non-responders have been repeatedly and unnecessarily skin-tested and revaccinated in attempts to make them convert. The presence of a BCG scar is sufficient evidence that the vaccination has 'taken'. Second, surveys of the reactivity of the category 3 responders is a useful way of determining which mycobacteria are present in an environment without resorting to the time-consuming procedure of isolating and identifying strains from inanimate sources. Third, there is an interesting relation between categorization and disease. Studies on leprosy patients in Nepal (Stanford *et al.*, 1981) and on tuberculosis patients in Indonesia (Kardjito *et al.*, 1986) both show that most healthy individuals are category 1 reactors while the majority of patients with either disease are category 3 reactors

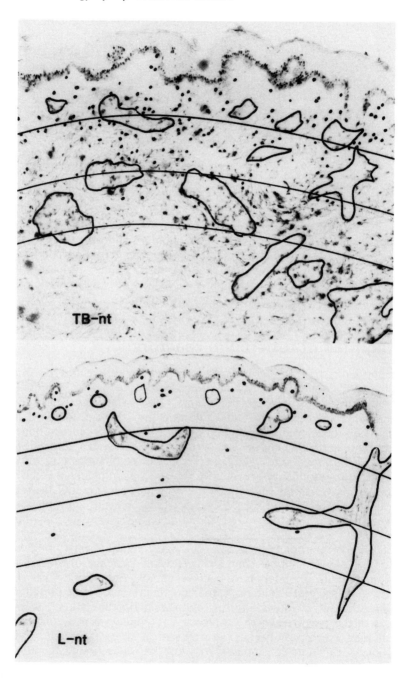

Fig. 5.7 Thin sections of a tuberculin reaction in a tuberculosis patient (TB-nt), a tuberculin reaction in a leprosy patient (L-nt), a leprosin-A reaction in a tuberculosis patient (TB-la), and a leprosin-A reaction in a leprosy patient (L-la). Courtesy Prof. J. Swanson Beck

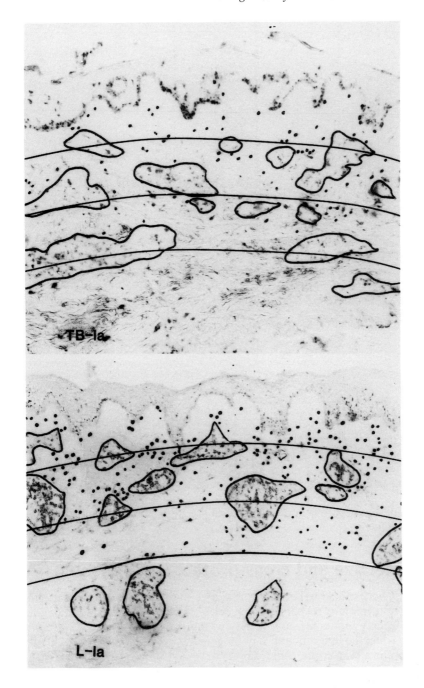

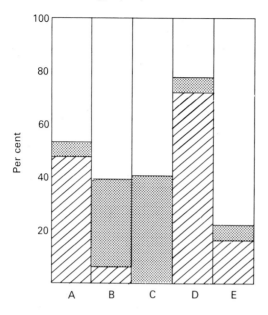

Fig. 5.8 Distribution of the three categories of tuberculin reactors amongst (A) healthy Nepalese individuals, (B) Nepalese patients with tuberculoid leprosy, (C) Nepalese patients with lepromatous leprosy, (D) healthy Indonesian individuals, and (E) Indonesian patients with pulmonary tuberculosis. Hatched: category 1; grey: category 2; white: category 3 reactors. Data from Stanford *et al.* (1981) and Kardjito *et al.* (1986)

(Fig. 5.8). It is not yet clear whether the category 3 reactors in the healthy population are those who are likely to develop disease if infected, or whether the disease process induces a lack of recognition of the common (group i) mycobacterial antigens. The relevance of this immunological 'blindness' to anergy in mycobacterial disease, particularly lepromatous leprosy, is discussed below (see page 80).

The practical aspects of skin testing in epidemiology and for the diagnosis of leprosy and tuberculosis are discussed in Chapters 6, 7 and 8 respectively.

Protective and cross-protective antigens

It is a widespread belief that 'protective antigens', i.e. those that elicit an effective immune response, are species- or strain-specific and that vaccines are usually ineffective unless they contain such antigens. This is certainly true when the determinant of pathogenicity is a strain-specific toxin or viral receptor site. On the other hand, there is no reason to assume that it is necessarily the case with intracellular pathogens such as the mycobacteria. Indeed, the available evidence strongly suggests that the shared mycobacterial antigens may be at least as protective as the species-specific ones and that BCG is as effective against leprosy as against tuberculosis (Stanford and Rook, 1983). There is also evidence that the protective antigens are those that are on the surface of, or are actively secreted by, living mycobacteria (Rook *et al.*, 1986). This may explain why it is necessary to use a live vaccine for the induction of protective immunity.

An understanding of the nature of the protective antigens in tuberculosis and leprosy is of crucial importance in attempts to develop better vaccines, particularly by 'genetic engineering'. Killed *M. leprae* has been added to BCG in order to enhance protection against leprosy by providing the appropriate species-specific antigens (Convit *et al.*, 1980). Although such antigens may be relevant, *M. leprae*, in common with *M. vaccae*, contains large amounts of common mycobacterial antigens relative to other species. It may well prove that supplementation of the common antigens increases protection and that the same effect may be obtained by adding the readily cultivable species *M. vaccae* (Bahr *et al.*, 1986).

The efficacy of BCG vaccination

The protective efficacy of BCG vaccine has been the subject of considerable controversy. The results of a number of major vaccine trials (Table 5.3) show that protection varies from 80 per cent to none at all. Many explanations for these discrepant results were advanced, including claims that the poor results seen in some studies were artefactual. In an attempt to reach a firm conclusion on the value of the vaccine, a major and well-controlled trial was organized in the Chingleput region of South India. Seven and a half years later it was reported that the incidence of disease was no lower among the vaccinated than among the non-vaccinated controls (Report: Tuberculosis Prevention Trial, 1979). Again, this led to a number of theories and explanations (World Health Organization, 1980). Of these, one of the most plausible is that the immune reactivity of the population had been conditioned by prior exposure to mycobacteria in the environment. In studies involving guinea-pigs, Palmer and Long (1969) found that such exposure afforded some protection against infection by *M. tuberculosis*. Vaccination with BCG increased this immunity but never to a

Table 5.3 Results of nine major BCG vaccine trials

Region	Year of commencement	Age range	Protection afforded (%)
North America*	1935	0–20 years	80
Chicago, USA	1937	3 months	75
Georgia, USA	1947	6–17 years	0
Illinois, USA	1948	Young adults	0
Puerto Rico	1949	1–18 years	31
Georgia, USA	1950	5 years	14
Great Britain	1950	14–15 years	78
South India	1950	All ages	31
South India	1968	All ages	0

*Amerindian population

level above that inducible by BCG alone. It was therefore concluded that BCG appeared to work well in populations not exposed to environmental mycobacteria, but that elsewhere the observed effect of the vaccine is diminished by the previously acquired natural immunity.

It may therefore be postulated that the Chingleput population had received sufficient 'natural vaccination' to induce maximum immunity. Indeed, the addition of BCG could even push individuals from a protective to a hypersensitive state, thereby reducing their resistance to disease. If this theory is correct, the vaccine should confer protection if given to neonates or young children before they experience a significant exposure to mycobacteria in the environment. Accordingly the World Health Organization is encouraging and sponsoring studies on neonatal vaccination in several countries and the results are encouraging (see, for example, Tidjani *et al.*, 1986).

An alternative explanation of the apparent failure of the Chingleput study was advanced by Stanford and his colleagues (1981) on the basis of their description of the Listeria-type and Koch-type reactions described above (page 69) and the finding that some forms of immunologically effective contact with mycobacteria induce the former reaction while others induce the latter. The effect of BCG is to induce a Listeria-type reaction or to boost the reactivity previously induced by environmental contact. Thus BCG may induce or boost protective immunity in some communities and will appear effective, or it will boost a necrotic Koch-type reaction in others and will not appear protective (Stanford *et al.*, 1981; Stanford and Rook, 1983). Again, it may be predicted that BCG will afford protection in the latter communities if given early enough in life to pre-empt the effect of environmental exposure.

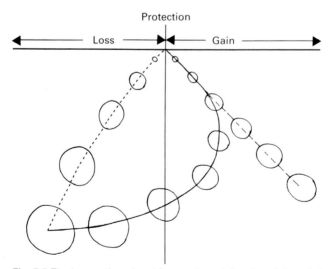

Fig. 5.9 The 'one pathway' and 'two pathway' theories of the relation between protective immunity and non-protective hypersensitivity. Solid line: the one pathway of protective immunity merging into hypersensitivity; broken line: the protective Listeria-type reactivity; dotted line: the tissue-damaging Koch-type reactivity. Circles denote reaction sizes

In summary, some workers claim that individuals in areas where BCG appears not to work acquire a high level of protection from the environment while others hold that such individuals are not protected owing to the induction of an inappropriate reaction that BCG cannot override. These concepts are summarized in Fig. 5.9. Proponents of the former ('one pathway') concept may recommend abandonment of BCG vaccination as the environment appears to provide the same service freely! Those supporting the second ('two pathway') concept would regard such a decision as being a very dangerous one and they strongly urge that neonates or infants should be vaccinated. These explanations are the subject of considerable controversy (briefly reviewed by Grange, 1986), but it is of very great importance for the future of the vaccine that the controversy be resolved.

Immunological spectra in mycobacterial disease

The clinical features and course of a mycobacterial disease is, in a very large measure, dependent upon the immunological reactivity of the patient. Consequently, the concept of a 'spectrum' of such reactivity, from highly active at one pole to absent at the other, has been developed. At first view this appears a reasonable and attractive idea but, as in most areas of mycobacterial immunology, the matter is not as straightforward as it seems to be.

The immune spectrum in leprosy

The great variation in the appearance and behaviour of the determinate forms of leprosy results from their position on an immunopathological spectrum described in detail by Ridley and Jopling (1966). For convenience, five points on the spectrum are recognized: the two polar forms, tuberculoid (TT) and lepromatous (LL), and three intermediate points, borderline tuberculoid (BT), mid-borderline (BB) and borderline lepromatous (BL). The clinical, immunological and histological features of these forms are shown in Table 5.4.

Table 5.4 Characteristics of the five points in the immunological spectrum of leprosy

Characteristic	TT	BT	BB	BL	LL
Bacilli in lesions	±	±/+	+	++	+++
Bacilli in nasal discharge	−	−	−	+	+++
Granuloma formation	+++	++	+	−	−
In vitro correlates of CMI	+++	++	+	±	−
Reaction to lepromin	+++	+	−	−	−
Anti *M. leprae* antibodies	±	±	+	++	+++
Macrophage maturity	mature				immature
Response to therapy	good				poor

At first view, the tuberculoid pole appears to be characterized by effective immunity which then decreases across the spectrum to the lepromatous pole where there is no apparent protective immune reactivity. Thus the lesions of tuberculoid leprosy are characterized by very few bacilli, many lymphocytes and granulomas containing mature epithelioid cells. In contrast, lesions of lepromatous leprosy contain few if any lymphocytes but numerous bacilli within immature macrophages. Thus, patients with tuberculoid leprosy are sometimes regarded as being near normal and those with lepromatous disease as being the most abnormal. For this reason, shifts in the position of the disease towards the tuberculoid and lepromatous poles of the spectrum are, respectively, termed 'upgrading' and 'downgrading'. An alternative view is that all forms of determinate leprosy are equally abnormal, but that the nature of the abnormality differs. The very fact that the patient with tuberculoid leprosy has overt disease indicates that the strong cell-mediated immune response is inappropriate or ineffective. The mechanism of effective immunity to leprosy must be sought in those contacts who either never develop the disease or display self-limiting indeterminate lesions.

The factors that determine whether an individual will develop disease and, if so, what form are poorly understood. It has been suggested that the outcome of infection is related to the time taken for cell-mediated immunity (CMI) to develop (Godal *et al.*, 1974). Thus in healthy contacts, a rapid onset of CMI would eliminate the bacilli before lesions were evident. In tuberculoid leprosy a slight delay would permit enough multiplication of the bacilli to render their removal more difficult, while an indefinite delay in the onset of CMI would result in lepromatous leprosy. The alternative, and more widely accepted, view is that the type of disease is 'predestined' by various factors, including genotypes and prior exposure to environmental mycobacteria (see page 70), and is the result of balances between the various T-cell subsets that modulate and control immune reactivity, particularly the T-helper and T-suppressor cells.

There is little doubt that leprosy at or near the lepromatous pole results from some form of specific anergy towards the leprosy bacillus. This anergy is poorly understood and could be due to one or more of several mechanisms.

One of the problems in understanding anergy in lepromatous leprosy is that the patients usually react normally to tuberculin and to skin testing reagents prepared from other mycobacterial species which clearly share many antigens with *M. leprae*. It has already been mentioned (see page 73) that patients with leprosy or tuberculosis only appear to respond to the species-specific mycobacterial antigens, i.e. they are category 3 responders. A further lack of recognition of the species-specific (group iv) antigens of *M. leprae* would eliminate immune reactivity to this species while permitting reactivity to the group iv antigens of other species. (As noted in Chapter 2, *M. leprae* is one of a small group of mycobacteria that possess only group i and iv antigens: they lack the group ii and iii antigens that characterize the slow and rapid growers respectively.)

The mechanism of the immunological 'blindness' to the common mycobacterial antigens in leprosy (and tuberculosis) is unknown. The

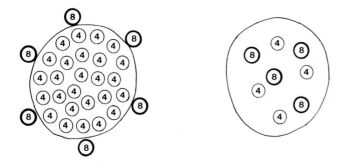

Tuberculoid   Lepromatous

Fig. 5.10 The distribution of CD4 (putative helper phenotype) and CD8 (putative suppressor phenotype) T-cells in lesions of tuberculoid and lepromatous leprosy. Data from Modlin *et al.* (1983)

superimposed anergy to the specific antigens of *M. leprae* is probably due to suppressor T-cells. Immunocytological studies on the lesions of tuberculoid and lepromatous leprosy (Fig. 5.10) have shown that there are many T-cells in tuberculoid lesions but that CD8 T-cells (putative suppressor phenotype) are in a minority and occur on the outside of the granuloma. In lepromatous lesions there are relatively few T-cells, but a much higher proportion are of the CD8 type and these are found within the lesion (Modlin *et al.*, 1983). Clones of CD8 T-cells from lepromatous lesions are able to suppress proliferation of CD4 T-cells (putative helper phenotype) specific for *M. leprae* (Bloom, 1986).

The anergy in lepromatous leprosy could also be the result of defects in antigen presentation and recognition, in T-cell proliferation due to defects in interleukin 2 or other mediators, or in production of gamma interferon or other macrophage activating factors by the helper T-cells. Patients with lepromatous leprosy have impaired production of gamma interferon (Nogueira *et al.*, 1983) and there is evidence that this may be the cause or the result of a failure of Schwann cells to present antigen to the T-cells (Samuel *et al.*, 1987b).

The immune phenomena underlying the spectrum in leprosy are complex and still poorly understood, but it is most important for the development of immunotherapy that the mechanisms of the various reactions, and lack thereof, are elucidated. An inappropriate therapeutic interference with the immune response could be disastrous.

The immune spectrum in tuberculosis

In view of the description of the immune spectrum in leprosy, the existence of a similar spectrum in tuberculosis has been postulated (Lenzini *et al.*, 1977). Undoubtedly, there is a gradation of immune reactivity in tuberculosis, but this gradation has little in common with the defined spectrum of leprosy. In particular, the lack of immune reactivity seen in some cases of tuberculosis is non-specific and, unless due to some underlying cause such as AIDS, renal failure or malignant disease, it reverts to normal on successful treatment of the infection. It should be noted that miliary disease is, contrary to some claims, not necessarily indicative of reduced immunity. Proudfoot (1971) drew the important distinction between cryptic disseminated tuberculosis with numerous minute necrotic foci teeming with bacilli but with little or no cellular response, and classical miliary disease with typical granulomas and few bacilli. The former is rapidly fatal while the latter may be surprisingly chronic. There is far too much toxicity and tissue destruction in unreactive tuberculosis to permit a chronic anergic form of the disease akin to lepromatous leprosy to develop.

The immune spectrum in *Mycobacterium ulcerans* infection

Immunological reactivity in this relatively uncommon disease is bizarre and fascinating. The disease, more fully described in Chapter 9, commences as a skin nodule that may resolve or progress to overt ulceration. During the ulcerative stage the lesions resemble those of lepromatous leprosy: there are many bacilli but very little evidence of a cellular response and the patient fails to react to Burulin, a skin test reagent prepared from the causative organism (Stanford *et al.*, 1975). A stage is reached when an effective immune response ensues, a lymphocytic infiltrate and granulomatous reaction is seen, the bacilli decline in number and then disappear, the patient reacts to Burulin and the lesion eventually heals. Thus the patients appear to convert spontaneously from an anergic state to one of protective cell-mediated immunity and the spectrum of immunological reactivity is therefore related to time (Fig. 5.11). An elucidation of the underlying

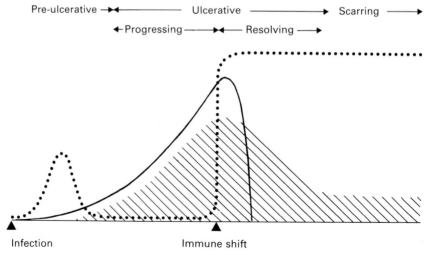

Fig. 5.11 The time course of *Mycobacterium ulcerans* infection (Buruli ulcer). Solid line: bacillary load; dotted line: reactivity to Burulin and correlates of cell-mediated immunity. The shaded area indicates the extent of clinically evident lesion

mechanism of this remarkable event could shed important light on the immune responses in the more prevalent mycobacterial diseases and might suggest new approaches to immunotherapy.

Immunotherapy

Many attempts have been made to treat tuberculosis and leprosy by stimulating the patient's immune reactivity. Koch's use of Old Tuberculin was the first such attempt and it appeared to be of value in the case of skin tuberculosis. The efficacy of vitamin D therapy in some cases probably also results from its role in macrophage activation.

Most of the difficulties experienced in the therapy of tuberculosis and leprosy stem from the long duration of treatment, even in the case of modern short-course therapy. As outlined in Chapter 10, the great majority of bacilli are killed within days of commencing therapy, but prolonged treatment is essential to destroy the few remaining parsisters. If the latter can be eliminated by a stimulated immune reactivity, it might be possible to reduce the duration of chemotherapy to a few weeks. In addition to vitamin D, agents that offer promise are levamisole (Sher *et al.*, 1981) and recombinant gamma interferon (INF-γ). Results of preliminary studies on the use of INF-γ in patients with lepromatous leprosy are very promising. Injection of small amounts into skin lesions leads rapidly to granuloma formation and destruction of bacilli (Nathan *et al.*, 1986; Samuel *et al.*, 1987a).

An alternative approach to immunotherapy is to modulate the immune response. On the hypothesis that infection by mycobacterial pathogens leads to the production of necrotic Koch-type reactions in some individuals rather than protective reactions, attempts are being made to switch reactivity from the former to the latter pathway (Bahr *et al.*, 1986).

Although no immunotherapeutic procedure is in regular clinical use, this is a very important field of mycobacterial research and modern technology should soon lead to exciting advances.

Humoral factors and serodiagnosis

Although it is unlikely that antibody plays a significant part in the immune response to mycobacteria, there have been numerous attempts to utilize the humoral immune response for diagnostic purposes (reviewed by Grange, 1984). Despite the huge amount of effort, no serodiagnostic test has yet proved suitable for routine clinical use. The reason for this is that, although antibodies are undoubtedly produced in mycobacterial disease, the overlap between levels in patients and either healthy individuals or those with other diseases is unacceptably large (Fig. 5.12). 'Natural' antibodies are probably due to contact with mycobacteria and related genera in the environment, but the use of purified specific antigens in diagnostic tests has proved disappointing: most of the humoral immune response is directed towards shared antigens. A greater, but still incomplete, discrimination between patients and controls has been

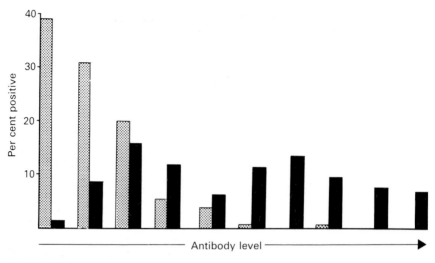

Fig. 5.12 Distribution of levels of IgG antibody to an ultrasonicate of *M. tuberculosis* in in Indonesian patients with pulmonary tuberculosis and comparable healthy subjects. Grey: controls; black: patients.

obtained by the use of a set of monoclonal antibodies in a solid-phase competition assay (Ivanyi *et al.*, 1983). This enables the amount of antibody against a single shared or specific epitope to be determined. Assay of antibody in the four subclasses of IgG also improves the discriminatory power of serodiagnosis (Gibson *et al.*, 1987). At present, these sophisticated assays are too complex and expensive for routine clinical use but they may form the basis for simpler procedures.

Serodiagnosis of leprosy has been attempted by using the PGL-1 antigen but, as explained in Chapter 7, a good distinction between patients and healthy contacts is achieved in the case of multibacillary forms of the disease but not with paucibacillary forms.

The use of serological tests to detect antigen in sputum, tissues and fluids is an attractive concept which has not yet been extensively investigated. The results of preliminary studies on a latex agglutination test for detection of tubercle bacilli in cerebrospinal fluid were promising (Krambovitis *et al.*, 1984).

Non-antibody humoral factors may have some relevance in mycobacterial disease. These include acute-phase reactants, some of which may have an immunoregulatory function. Haptoglobin has been shown to affect lymphocyte function in tuberculosis (Grange *et al.*, 1985) and also in cancer, possibly by modulating synthesis of prostaglandins and other arachidonic acid metabolites. It is often stated that corticosteroids suppress cell-mediated immune reactions in man and that steroid therapy suppresses tuberculin reactivity and is a cause of reactivation of quiescent tuberculosis. In fact, there is little firm evidence for the latter two claims; indeed, high-dose therapy has little effect on the established tuberculin reactivity (Lowe *et al.*, 1987). Other 'humoral' factors that possibly affect immune reactions include Vitamin D (see page 65), autoantibodies and circulating immune complexes.

Mycobacterial adjuvants

An adjuvant is a substance that enhances the antibody response to unrelated antigens administered with it. Adjuvants are therefore used extensively for the production of antibody. One of the best known is Freund's complete adjuvant which consists of a water-in-oil emulsion containing killed tubercle bacilli. The adjuvant activity of mycobacteria resides principally in the peptidoglycan (murein) of the cell wall, although trehalose dimycolate (cord factor), RNA and certain peptidoglycolipids also appear to possess this activity. In the case of peptidoglycan, the minimum structure required for adjuvant activity is muramyl dipeptide (see page 15), a water soluble molecule that has been synthesized.

Despite many studies, the mode of action and natural function of adjuvants remain shrouded in mystery. They may represent a primitive immune recognition system which could respond to certain common microbial components. In addition to enhancing antibody production, adjuvants also affect T-cells and can thereby cause some degree of macrophage activation. Indeed this property might be of some relevance in the containment of mycobacterial infections prior to the onset of specific cell-mediated immunity. Theories as to the mode of action of adjuvants include assistance in antigen presentation to T-cells, effects on lymphocyte traffic, derepression of immune reactions and non-specific mitogenic effects on B-cells. The latter property may explain the high levels of autoantibodies that are found in lepromatous leprosy and, to a lesser extent, in advanced tuberculosis. The subject of mycobacterial adjuvants has been reviewed in detail by Stewart-Tull (1983).

Genetic factors in mycobacterial immunity

As only a minority of those infected with mycobacteria develop overt disease, many attempts have been made to establish some genetic marker of susceptibility or resistance. In the last century, the fact that tuberculosis was spread very much within families led to the widespread notion that the disease was an hereditary one and this belief was one of the major stumbling blocks to the acceptance of Koch's discovery.

Many attempts have been made to link susceptibility of tuberculosis and leprosy to the class I antigens (HLA-A and HLA-B) of the major histocompatibility complex. Although some studies have claimed to show a low but significant association of a particular HLA type to overt disease, the results vary from region to region and no clear-cut pattern has yet emerged. More recently, attention has turned to the class II (HLA-D determined) antigens which determine the Ia antigens involved in presentation of foreign antigen to the lymphocytes. Although HLA-D phenotypes have not yet been shown to affect the susceptibility to disease,

there is evidence that they determine skin test responses to mycobacterial antigens. Thus individuals who do not respond to testing with such antigens (category 2 non-responders) do not express the HLA-DR3 specificity (van Eden *et al.*, 1983) and significantly larger reactions to tuberculin, but not to antigens prepared from *M. leprae* or other mycobacteria, were elicited in leprosy patients with HLA-DR4 specificity than in patients lacking this specificity (Ottenhoff *et al.*, 1986). The significance of these findings to the genetic control of immune responses to mycobacteria clearly merit further investigation. For a review of immunogenetic factors in leprosy see Ottenhoff and de Vries (1987).

In fine, the host's mechanism for defence against mycobacterial disease resembles an orchestra with some lead performers such as the antigen-specific lymphocytes and the macrophages and others with minor roles. The immunological orchestra is conducted by genetic determinants and also by phenetic factors, particularly exposure to non-pathogenic mycobacteria in the environment. The way in which all the parts perform determines whether an infected individual will enjoy the harmony of health or the cacophony of disease.

References

Abrahams, E.W. (1970) Original mycobacterial sin. *Tubercle* **51**: 316–21.

Askanase, P.W. and van Loveren, H. (1983) Delayed-type hypersensitivity: activation of mast cells by antigen-specific T-cell factors initiates the cascade of cellular interactions. *Immunology Today* **4**: 259–64.

Bahr, G.M., Stanford, J.L., Rook, G.A.W., Rees, R.J.W., Abdelnoor, A.M. and Frayha, G.J. (1986) Two potential improvements to BCG and their effect on skin test reactivity in the Lebanon. *Tubercle* **67**: 205–18.

Beck, J.S., Morley, S.M., Gibbs, J.H., Potts, R.C., Ilias, M.I., Kardjito, T., Grange, J.M., Stanford, J.L. and Brown, R.A. (1986) The cellular responses of tuberculosis and leprosy patients and of healthy controls in skin tests to new tuberculin and leprosin A. *Clinical and Experimental Immunology* **64**: 484–9.

Beutler, B., Greenwald, D., Hulmes, J.D., Chang, M., Pan, Y.C.E., Mathison, J., Ulevitch, R. and Cerami, A. (1985) Identity of tumour necrosis factor and the macrophage-secreted factor cachectin. *Nature* **316**: 552–3.

Bloom, B.R. (1986) Learning from leprosy: a perspective on immunology and the third world (American Association of Immunologists Presidential Address). *Journal of Immunology* **137**: i–x.

Convit, J., Ulrich, M. and Aranzazu, N. (1980) Vaccination in leprosy – observations and interpretations. *International Journal of Leprosy* **48**: 62–5.

Draper, P. (1981) Mycobacterial inhibition of intracellular killing. In: *Microbial Perturbation of Host Defences*, pp. 143–64. Edited by F. O'Grady and H. Smith. London: Academic Press.

Editorial (1984) New tuberculins. *Lancet* **i**: 199–200.

El Ansary, E.H. and Grange, J.M. (1984) Qualitative differences in tuberculin reactivity in patients with tuberculosis, occupational contacts and non-contacts. *Tubercle* **65**: 191–4.

Gibson, J.A., Grange, J.M., Beck, J.S. and Kardjito, T. (1987) Specific antibody in the subclasses of immunoglobulin G in patients with smear-positive pulmonary tuberculosis. *European Journal of Respiratory Diseases* **70**: 29–34.

Godal, M.B., Myrvang, B., Stanford, J.L. and Samuel, D.R. (1974) Recent advances in the immunology of leprosy with special reference to new approaches in immunoprophylaxis. *Bulletin de l'Institut Pasteur* **72**: 273–310.

Grange, J.M. (1984) The humoral immune response in tuberculosis: its nature, biological role and diagnostic usefulness. *Advances in Tuberculosis Research* **21**: 1–78.

Grange, J.M. (1986) Environmental mycobacteria and BCG vaccination. *Tubercle* **67**: 1–4.

Grange, J.M., Beck, J.S., Harper, E.I., Kardjito, T. and Stanford, J.L. (1986) The effect of exposure of hospital employees to patients with tuberculosis on dermal reactivity to four new tuberculins. *Tubercle* **67**: 109–18.

Grange, J.M., Kardjito, T., Beck, J.S., Ebeid, O., Kohler, W. and Prokop, O. (1985) Haptoglobin: an immunoregulatory role in tuberculosis? *Tubercle* **66**: 41–7.

Ivanyi, J., Krambovitis, E. and Keen, M. (1983) Evaluation of a monoclonal antibody (TB72) based serological test for tuberculosis. *Clinical and Experimental Immunology* **54**: 337–45.

Kardjito, T., Beck, J.S., Grange, J.M. and Stanford, J.L. (1986) A comparison of the responsivenes to four new tuberculins among Indonesian patients with pulmonary tuberculosis and healthy subjects. *European Journal of Respiratory Diseases* **69**: 142–5.

Kardjito, T. and Grange, J.M. (1982) Diagnosis of active tuberculosis by immunological methods. 2: Qualitative differences in the dermal response to tuberculin in patients with active pulmonary disease and healthy tuberculin-positive individuals. *Tubercle* **63**: 275–8.

Koch, R. (1891) Weitere Mitteilungen über ein Heilmittel gegen Tuberkulose. *Deutsche Medizinische Wochenschrift* **17**: 101–2.

Krambovitis, E., McIllmurray, M.B., Lock, P.E., Hendrickse, W. and Holzel, H. (1984) The rapid diagnosis of tuberculous meningitis by latex particle agglutination. *Lancet* **ii**: 1229–31.

Ladefoged, A., Bunch-Christensen, K. and Guld, J. (1976) Tuberculin sensitivity of guinea pigs after vaccination wth varying doses of BCG of twelve different strains. *Bulletin of the World Health Organization* **53**: 435–43.

Lagrange, P.H. (1984) Cell mediated immunity and delayed-type hypersensitivity. In: *The Mycobacteria: A Sourcebook*, Part B, Edited by G.P. Kubica and L.G. Wayne. New York: Marcel Dekker. pp. 681–720.

Lenzini, L., Rottoli, P. and Rottoli, L. (1977) The spectrum of human tuberculosis. *Clinical and Experimental Immunology* **27**: 230–7.

Lowe, J.G., Beck, J.S., Gibbs, J.H., Brown, R.A., Potts, R.C., Grange, J.M. and Stanford, J.L. (1987) Skin test responsiveness to four new tuberculins in patients with chronic obstructive airways disease receiving short term high doses of, or long term maintenance treatment with, prednisolone: clinical appearances and histometric studies. *Journal of Clinical Pathology* **40**: 42–9.

Lowrie, D.B. (1983) The macrophage and mycobacterial infections. *Transactions of the Royal Society of Tropical Medicine* **77**: 646–55.

Modlin, R.L., Hofman, F.M., Taylor, C.R. and Rea, T.H. (1983) T-lymphocyte subsets in the skin lesions of patients with leprosy. *Journal of the American Academy of Dermatology* **8**: 182–9.

Nathan, C.F., Kaplan, G., Levis, W.R., Nusvat, A., Witmer, M.D., Sherwin, A.S., Job, C.K., Horowitz, C.R., Steinman, R.M. and Cohn, Z.A. (1986) Local and systemic effects of intradermal recombinant interferon-γ in patients with lepromatous leprosy. *New England Journal of Medicine* **313**: 6–15.

Nogueira, N., Kaplan, G., Levy, E., Sarno, E.N., Kushner, P., Granelli-Piperno, A., Vieira, L., Colomer Gould, V., Levis, W., Steinman, R., Yip, Y.K. and Cohn, Z.A. (1983) Defective interferon production in leprosy. *Journal of*

Experimental Medicine **158**: 2165–70.

Ottenhoff, T.H.M. and de Vries, R. (1987) *Recognition of M. leprae antigens.* Dordrecht: Martinus Nijhoff.

Ottenhoff, T.H.M., Torres, P., de las Aguas, J.T., Fernandez, R., van Eden, W., de Vries, R.R.P. and Stanford, J.L. (1986) Evidence for an HLA-DR4-associated immune-response gene for *Mycobacterium tuberculosis.* A clue to the pathogenesis of rheumatoid arthritis. *Lancet* **ii**: 310–12.

Palmer, C.E. and Long, M.W. (1969) Effects of infection with atypical mycobacteria on BCG vaccination and tuberculosis. *American Review of Respiratory Disease* **94**: 553–68.

Pepys, J. (1955) The relationship of nonspecific and specific factors in the tuberculin reaction. *American Review of Tuberculosis* **71**: 49–73.

Proudfoot, A.T. (1971) Cryptic disseminated tuberculosis. *British Journal of Hospital Medicine* **5**: 773–80.

Report: Tuberculosis Prevention Trial (1979) Trial of BCG vaccines in South India for tuberculosis prevention. *Bulletin of the World Health Organization* **57**: 819–27.

Rich, A.R. and McCordock, H.A. (1929) An enquiry concerning the role of allergy, immunity and other factors of importance in the pathogenesis of human tuberculosis. *Bulletin of the Johns Hopkins Hospital* **44**: 273–422.

Ridley, D.S. and Jopling, W.W. (1966) Classification of leprosy according to immunity: a five group system. *International Journal of Leprosy* **34**: 255–73.

Rook, G.A.W. (1983) An integrated view of the immunology of the mycobacterioses in guinea pigs, mice and men. In: *The Biology of the Mycobacteria*, Vol. 2, pp. 279–319. Edited by C. Ratledge and J.L. Stanford. New York: Academic Press.

Rook, G.A.W. (1986) The immunopathology of tuberculosis. In: *Mycobacteria of Clinical Interest*, pp. 3–13. Edited by M. Casal. Holland: Elsevier.

Rook, G.A.W. and Stanford, J.L. (1979) The relevance to protection of three forms of delayed skin test response evoked by *M. leprae* and other mycobacteria in mice: correlation with the classical work in the guinea-pig. *Parasite Immunology* **1**: 111–23.

Rook, G.A.W., Steele, J., Barnass, S., Mace, J. and Stanford, J.L. (1986) Responses to live *M. tuberculosis*, and common antigens, of sonicate-stimulated T cell lines from normal donors. *Clinical and Experimental Immunology* **63**: 105–10.

Samuel, N.M., Grange, J.M., Samuel, S., Lucas, S., Owilli, O.M., Adalla, S., Leigh, I.M. and Navarette, C. (1987a) A study of the effects of intradermal administration of recombinant gamma interferon in lepromatous leprosy patients. *Leprosy Review* **58**: 389–400.

Samuel, N.M., Mirsky, R., Grange, J.M. and Jessen, K.R. (1987b) Expression of major histocompatibility complex class I and class II antigens in human Schwann cell cultures and effects of infection with *Mycobacterium leprae*. *Clinical and Experimental Immunology* **65**: 500–509.

Seibert, F.B. (1934) The isolation and properties of the purified protein derivative of tuberculin. *American Review of Tuberculosis* **30**: 713–20.

Sher, R., Wadee, A.A., Joffe, M., Kok, S.H., Inkamp, F.M.J.H. and Simson, I.W. (1981) The *in vivo* and *in vitro* effects of levamisole in patients with lepromatous leprosy. *International Journal of Leprosy* **49**: 159–66.

Stanford, J.L. and Lema, E. (1983) The use of a sonicate preparation of *Mycobacterium tuberculosis* (new tuberculin) in the assessment of BCG vaccination. *Tubercle* **64**: 275–82.

Stanford, J.L., Nye, P.M., Rook, G.A.W., Samuel, N.M. and Fairbank, A. (1981) A preliminary investigation of the responsiveness or otherwise of patients and staff of a leprosy hospital to groups of shared or species specific antigens of mycobacteria. *Leprosy Review* **52**: 321–7.

Stanford, J.L., Revill, W.D.L., Gunthorpe, W.J. and Grange, J.M. (1975) The production and preliminary investigation of Burulin, a new skin test reagent for *Mycobacterium ulcerans* infection. *Journal of Hygiene* **74**: 7–16.

Stanford, J.L. and Rook, G.A.W. (1983) Environmental mycobacteria and immunization with BCG. In: *Medicial Microbiology*, Vol. 2, pp. 43–69. Edited by C.S.F. Easmon and J. Jeljaszewic. New York: Academic Press.

Stanford, J.L., Shield, M.J. and Rook, G.A.W. (1981) Hypothesis: How environmental mycobacteria may predetermine the protective efficacy of BCG. *Tubercle* **62**: 55–62.

Stewart-Tull, D.E.S. (1983) Immunology important constituents of mycobacteria: adjuvants. In: *The Biology of the Mycobacteria*, Vol. 2, pp. 3–84. Edited by C. Ratledge and J.L. Stanford. New York: Academic Press.

Tidjani, O., Amedome, A. and ten Dam, H.G. (1986) The protective effect of BCG vaccination of the newborn against childhood tuberculosis in an African community. *Tubercle* **67**: 269–81.

Turner, H.M. (1953) Correlation of quantitative Mantoux reactions with clinical progress in pulmonary tuberculosis. *Tubercle* **34**: 155–65.

van Eden, W., DeVries, R.R.P., Stanford, J.L. and Rook, G.A.W. (1983) HLA-DR3 associated genetic control of response to multiple skin tests with new tuberculins. *Clinical and Experimental Immunology* **52**: 287–92.

Wilson, G.S., Schwabacher, H. and Maier, I. (1940) The effect of desensitization of tuberculous guinea-pigs. *Journal of Pathology and Bacteriology* **50**: 89–109.

World Health Organization (1980) *BCG Vaccination Policies*. Technical Report Series No. 652. Geneva: WHO.

Youmans, G.P. (1979) Relation between delayed hypersensitivity and immunity in tuberculosis. In: *Tuberculosis*, pp. 302–16. Edited by G.P. Youmans. Philadelphia: W.B. Saunders.

6 Epidemiology and control of mycobacterial disease

Epidemiological investigations are undertaken to detect changing trends in the incidence and prevalence of mycobacterial disease in a community. The main aims of such investigations are to determine 'natural' trends in the infections and to predict their future behaviour, to assist in the design of control measures and to assess the effectiveness of such measures (Styblo, 1978).

The epidemiology of tuberculosis

The very nature of tuberculosis renders studies of its behaviour in the community very difficult. As outlined in Chapter 8, not all those infected with *Mycobacterium tuberculosis* develop overt disease and not all those with such disease are infectious. Moreover, disease may develop months, years or even decades after infection and overt disease may last for over a year. Thus the number of cases of tuberculosis in a community at a given time – the *'point prevalence'* – is not the same as the number of new cases registered in a given year, i.e. the *annual incidence*. Neither of these is the same as the *annual infection rate*.

In practice, determinations of the incidence or prevalence of tuberculosis are notoriously unreliable and are dependent on the efficacy of case-finding programmes and notification schemes. Even in countries with well-organized medical services and mandatory notification schemes, many cases go unreported.

The three major tools for the study of the epidemiology of tuberculosis are sputum examination (microscopy and culture), mass radiology and tuberculin testing. In principle, the prevalence of pulmonary tuberculosis could be determined by a radiological screening of the entire population and a careful bacteriological examination of the sputum of those with pulmonary lesions. Such extensive screening programmes are rarely undertaken. Although mass miniature radiology (MMR) was widely used for the detection of pulmonary tuberculosis, its contribution to case-finding was, as shown in Table 6.1, remarkably small (Toman, 1976). Even when MMR was extensively used, most cases of tuberculosis were diagnosed as a result of patients presenting with symptoms. The reason for this is that a patient can develop radiological lesions and symptoms, and become infectious, within a relatively short period of time and that the usual interval between MMR examinations is too long. Even when individuals were X-rayed at 4-monthly intervals (an impractical interval in terms of

Table 6.1 Mode of detection of smear-positive tuberculosis by mass radiography (MMR) and by patients presenting with symptoms (Toman, 1976)

Study	Study period	Mode of detection		
		MMR (%)	Symptoms (%)	Other (%)
Canada				
Saskatchewan	1960–69	12	66	22
Ontario	1967–68	13	66	21
Czechoslovakia	1967–69	13	65	22
Netherlands				
Entire country	1951–67	13–15	54–58	27–30
Rotterdam	1961–65	19	55	26

organization, cost and radiation exposure), 21 per cent of detected cases had extensive disease. Indeed, MMR may even be detrimental to control as an individual with symptoms may wait for his next annual appointment rather than seek medical attention more promptly. For these reasons MMR for routine screening has, on the advice of the World Health Organization, been abandoned. It is nevertheless of value for the screening of certain high-risk groups, such as the occupants of common lodging houses (Capewell *et al.*, 1986).

The most reliable and informative epidemiological approach is the estimation of the *annual infection rate* or annual risk of infection. In this sense, 'infection' is defined as a contact with *M. tuberculosis* that leads to the acquisition of tuberculin positivity. This excludes tuberculin sensitivity that results from BCG vaccination or contact with mycobacteria in the environment. In some regions the latter is no problem as such contact induces only small reactions, but in other regions reactivity due to exposure to environmental mycobacteria is not so easily differentiated from that due to tubercle bacilli (Fig. 6.1). The method for determining the annual infection rate has been described in detail by Styblo (1984) and is based on mass tuberculin testing surveys of representative groups of unvaccinated schoolchildren or young adults. The estimation of the risk of infection may be direct or indirect. In the direct method, a group of individuals is tested at intervals and the number of tuberculin converters in a given time is determined. In the indirect method the number of reactors in a given age group is determined annually: this does not indicate in which year the individuals become infected but it shows how many of the particular age group have escaped infection. Changes in the rate of infection are easily calculated from changes in the proportion of reactors of a given age in successive years.

It is important to appreciate that the annual infection rate is not equivalent to the annual incidence of disease. Only a minority of those infected develop overt disease and in many cases the disease does not develop in the same year as the initial infection occurred. The proportion of those infected who develop overt disease is expressed as the *disease*

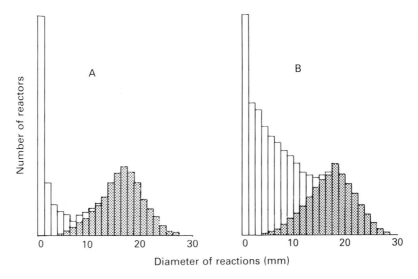

Fig. 6.1 Size distribution of tuberculin reactions in regions with (A) low and (B) high levels of sensitization by environmental mycobacteria. The shaded area indicates sensitivity due to immunologically effective contact with *M. tuberculosis*

ratio: this varies from region to region. Overall, it is estimated that between 10 and 20 per cent of infected individuals will subsequently develop overt disease; three-quarters of them within five years and the others at any time during the rest of their lives (Sbarbaro, 1975). Thus, worldwide, it is estimated that 100 million individuals become infected annually and at least 10 million develop overt disease. About 5 million progress to infectious disease and around 3 million die (WHO, 1982).

The calculation of the annual infection rate does not depend on the effectiveness of case-finding programmes and it gives up-to-date information of the extent of transmission of tubercle bacilli within the community. It is therefore a particularly useful means of assessing the efficiency of control measures designed to reduce such transmission.

Transmissibility depends on the prevalence of open or infectious patients in the community and their opportunities for contact with uninfected individuals. This is expressed as the *transmission or contagious parameter*, i.e. the average number of individuals infected by one source case. It has been shown that the number of infectious individuals in a community may be extrapolated with a fair degree of accuracy from the annual infection rate. In general, a 1 per cent annual infection rate indicates that there are 50 infectious cases for every 100,000 members of the population. This implies that, on average, each patient with open disease infects 20 other individuals each year. The actual number of individuals infected by one source case depends on the length of time that the individual is infectious. About 65 per cent of patients with untreated open pulmonary tuberculosis die within 4 years of becoming infectious, with an average survival of 14 months (Springett, 1971). From 26 to 33 per cent heal spontaneously and a minority become chronic excretors.

Transmission of tuberculosis occurs principally within households or other groups of people living in close proximity. By far the most important transmitters of disease are those patients with enough bacilli in the sputum to be evident on standard microscopical examinations. It has been estimated that such smear-positive patients must have at least 5000 bacilli in one mililitre of sputum. In a study in Holland (Geuns *et al.*, 1975) it was found that children with household exposure to a smear-positive case had a 50 per cent chance of being infected, compared with only a 6 per cent chance if the source case was smear-negative. Furthermore, the chance of a casual contact being infected by a smear-negative patient was exceedingly small. Hence for practical purposes, only smear-positive patients are infectious.

Surveys conducted in developed countries show that annual infection rates vary from 0.3 per cent to less than 0.1 per cent and that the rate is declining by about 12 per cent annually (Bleiker and Styblo, 1978). In the developing countries the rate is from 20 to 50 times higher (2 to 5 per cent) and shows little or no tendency to decline. Indeed, in absolute terms, the number of cases of tuberculosis is increasing annually in such countries in proportion to the population growth. Given a population of 3000 million in the developing world and an average annual infection rate of 3 per cent, it is estimated that 100 million individuals become infected each year.

Tuberculosis is divided into primary and post-primary, or secondary, disease (see page 119). For practical epidemiological purposes these are defined as disease developing within or after five years of infection respectively (Styblo, 1978). There has been much debate as to what extent post-primary tuberculosis is due to endogenous reactivation of the primary infection or to exogenous reinfection. The endogenous reactivation theory is based on the unproven assumptions that bacilli are able to survive for years or even decades within lesions and that the primary infection induces a life-long immunity to superinfection. Evidence for exogenous reinfection has been reviewed by Styblo (1978). In particular, intensive control measures amongst the Eskimo populations not only caused a large reduction in the tuberculosis infection rates, but also caused a reduction in the incidence of disease amongst older, previously infected, individuals. This strongly suggests that the high incidence of overt disease amongst the elderly was due, at least in part, to exogenous reinfection.

Whether or not exogenous reinfection occurs is a question of practical relevance to the planning of tuberculosis control programmes. Styblo (1984) has argued that if exogenous reinfection is rare, tuberculosis control in high-prevalence regions should be based mainly on BCG vaccination to prevent or limit primary disease. If, on the other hand, previously infected individuals are subject to reinfection, then priority should be given to the detection and treatment of the infectious cases. From the available evidence it appears likely that post-primary disease may have either an endogenous or exogenous origin and that the relative incidence of the two types of infection in a community at a given time is determined in the prevalence of infectious cases.

ESTIMATED ENDEMICITY OF LEPROSY IN THE WORLD 1983

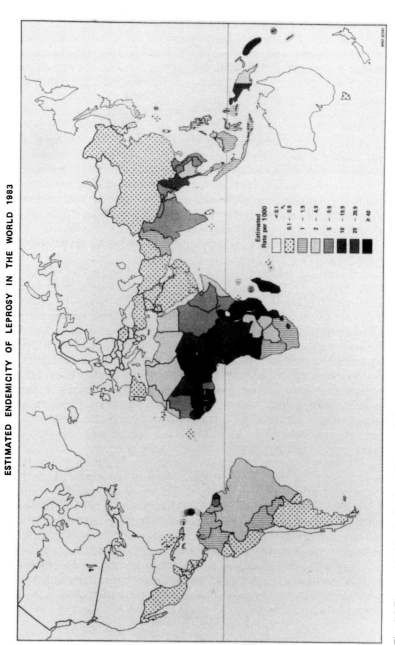

Fig. 6.2 The worldwide distribution of leprosy. This map was established by the Leprosy Unit of the World Health Organization, based on information received in 1983, and is reproduced by permission of the WHO.

The epidemiology of leprosy

If the study of the epidemiology of tuberculosis is fraught with difficulties, these pale into insignificance in comparison with the problems posed by leprosy. As *Mycobacterium leprae* has never been cultured *in vitro*, the possibility that this organism can live and replicate in, and infect man from, some natural environmental niche cannot be excluded. Furthermore, leprosy may not be entirely restricted to man: there have been a few reports of the natural occurrence of the disease in armadillos, a chimpanzee and a mangabey monkey (see page 102). Despite these claims, and reports of transmission of the disease from infected armadillos to their handlers, leprosy is a disease that, for practical purposes, is only transmitted from person to person. As in the case of tuberculosis, not all those with disease appear to be infectious. It is likely that the major sources of infection are those patients who are at or near the lepromatous end of the immunological spectrum and who consequently have a high bacterial load. Accordingly, it is important to determine the proportion of patients who have the multi-bacillary forms of the disease. This proportion appears to vary considerably from country to country (Sansarriq *et al.*, 1979) but it is uncertain how reliable these reported differences are.

The worldwide distribution of leprosy is shown in Fig. 6.2. Despite this distribution, leprosy cannot be regarded as a tropical disease: until relatively recently the disease occurred in Scandinavia and Siberia and a few cases are still encountered in southern Europe and in the southern parts of the USA (Texas and Louisiana). In absolute terms, the estimated number of victims of leprosy has not changed significantly over the last two decades. As the population of the world has increased markedly during that time the actual prevalence of leprosy, in contrast to tuberculosis, has declined. Although there were about 5.3 million registered cases of leprosy in 1980, the actual number is conservatively estimated by the World Health Organization to be between 10 and 11 million (WHO, 1977; Sansarricq, 1983). As leprosy is a disease that frequently afflicts its victims for many years or decades the prevalence of the disease at a given time is much higher than the annual incidence of new cases.

Contrary to a long-held opinion, leprosy bacilli are not shed in appreciable quantities from the skin. Instead, transmission appears to be principally from the nose. Indeed a patient with lepromatous leprosy may shed over 10^7 organisms daily, equivalent to the number of tubercle bacilli expectorated by a patient with open tuberculosis (Davey and Rees, 1974). The portal of entry of the leprosy bacillus is unknown. It is possible that the bacilli are inhaled and reach the skin and nerves by blood-borne dissemination from a primary focus in the upper respiratory tract or the lung. The leprosy bacillus remains viable in nasal secretions for up to one week and immunocompromised mice have been infected with aerosols (Rees and McDougall, 1977). A further but unproven possibility is that the disease is transmitted by biting insects: patients with lepromatous leprosy have enough bacilli in their circulation to be taken up by blood-sucking insects.

The view that leprosy is one of the least infectious of the communicable diseases has been rather overstressed in recent years, probably to counteract the old and popular notion that the disease is highly contagious. In fact, infection with the leprosy bacillus occurs fairly readily but, as with tuberculosis, only a minority of those infected develop overt disease. Hence the annual infection rate is much higher than the incidence of new cases of disease.

Skin testing surveys to determine the annual infection rate have not been widely applied to the study of leprosy owing to difficulties in preparing suitable skin test reagents in large enough quantities. Indeed, in a scholarly comparison of the epidemiology of tuberculosis and leprosy, Fine (1984) noted that while studies on the former disease concentrate on *infection*, as determined by tuberculin testing, those on the latter concentrate on overt disease.

The interval between infection and the appearance of signs is lengthy, usually 3–5 years, although it may be shorter or considerably longer. Children appear much more likely than adults to develop leprosy after infection (Jopling, 1984), but the disease is rare in those less than five years of age, possibly owing to the long incubation period. In many regions, new cases appear most often amongst young adults although, as with tuberculosis, the mean age of those affected increases in regions where the disease is in decline.

The epidemiology of opportunist mycobacterial disease

This topic is discussed in Chapter 9.

Control of tuberculosis and leprosy

The aim of control measures is the prevention of transmission of disease from person to person. This may, in principle, be achieved by detecting infectious cases and rendering them non-infectious, by raising the immunity of the uninfected population by vaccination or, in some circumstances, by administering antimycobacterial drugs prophylactically to individuals with a high risk of infection. Although these measures appear straightforward in principle, their application in practice has met with enormous difficulties. Indeed there are few parts of the world where public health measures have had an appreciable impact on the behaviour of mycobacterial diseases. Yanez (1982) has remarked that control programmes designed by international agencies and authorities are usually conceptually irreproachable but almost unworkable in practice. Unfortunately, dogmatic assertions often prevail over imaginative, realistic and practical proposals (Chaulet, 1983).

Patients with open tuberculosis and infectious forms of leprosy can be rendered uninfectious by modern chemotherapy. Although excellent in principle, control by chemotherapy has, in fact, proved disappointing. The success of chemotherapy as a control measure depends on the percentage of infectious cases in the community that are detected, the interval between onset of infectivity and diagnosis, the effectiveness of the regimen, and the compliance of the patient (Rouillon *et al.*, 1976).

Control by reduction of infectious cases requires motivation, education, organization and, inevitably, financial backing. Correct motivation must exist at governmental and local levels: administrators, health workers and patients must be united by a genuine desire to rid the community of the afflictions.

The first step in a control programme is case-finding, which may be passive or active. Passive case-finding relies on individuals with symptoms taking the initiative to attend the health clinic. In practice, this is notoriously unreliable. Even with widespread public education, there are often many reasons why patients lack motivation to seek help before they have developed irreversible physical damage and, in the case of leprosy, stigmatizing signs, and will have infected many other individuals. These reasons include a fear of prolonged hospitalization, loss of earnings (especially if the patient has a family to support), loss of social status or even complete social ostracism. Education must therefore be aimed at the entire community and should instil the view that tuberculosis and, more especially, leprosy are no more fearsome than other diseases. Prejudices are often deeply ingrained and such education is no easy matter. Fear of loss of the ability to support a family can be overcome by educating the public that, if diagnosed early, therapy of mycobacterial disease does not interfere with normal employment.

As patients will often have infected many of their contacts before presentation, dapsone therapy was very good at reducing the prevalence of active leprosy, but its impact on the number of new cases was less impressive (Sansarricq, 1982). For this reason, it is doubtful whether the incidence of leprosy will be subtantially affected by the replacement of dapsone monotherapy by more effective multi-drug regimens. The same argument applies to tuberculosis: one of the reasons for the abandonment of mass miniature radiography was its failure to diagnose cases early (see page 91).

Active case-finding for tuberculosis suspects, usually defined as individuals with a cough of more than one month duration, involves a deliberate search. Many techniques have been evaluated, particularly by Aluoch and his colleagues in Kenya (Aluoch *et al.*, 1985; Nsanzumuhire *et al.*, 1981). The most reliable, though expensive and time-consuming, method is a direct approach to the head of every household in a community. Approaches to community leaders were less effective. Indirect methods, such as asking women attending antenatal or child-care clinics to deliver letters to suspects within their families were relatively unsuccessful. In general, the success of case-finding, whether active or passive, is related to the distance that the suspects live from the hospital or clinic. Thus there is an important need to establish health care facilities in rural areas. Similar principles apply to leprosy case-finding, although great care must be paid to confidentiality.

The provision of free, fully supervised, effective chemotherapy is essential for the reduction of infectious cases. Most patients with tuberculosis or leprosy are unable to bear the cost of therapy, particularly

Table 6.2 Patient compliance in a tuberculosis programme in a Somali refugee camp (Shears, 1984)

Patient compliance	Numbers	%
Total number of patients starting chemotherapy	600	100
Number completing 12 months' treatment	55	9
Number completely lost during therapy	301	50
Number attending at time of assessment	244	41
Number defaulting a few weeks at time of assessment (of 244)	70	28
Number defaulting in intensive phase of treatment (of 70)	14	20

now that multi-drug regimens are considered essential for the latter disease (see page 163). Experience has shown that curative chemotherapy is essential for the control of tuberculosis, and this means the use of supervised short-course regimens (see page 156). Inadequate regimens or irregularly administered drugs will merely result in a high prevalence of chronic excretors of bacilli.

Even with very good organization, the provision of fully supervised chemotherapy is beset with many problems. In a tuberculosis control programme in Somali refugee camps (Shears, 1984), half the patients absconded before completion of therapy and the great majority of the remainder defaulted intermittently. The most frequent excuses for default were that the patients felt better and saw no need for further therapy or that they had social engagements such as weddings or religious festivals. Similar difficulties have been encountered in other programmes for both tuberculosis and leprosy, stressing the need for much shorter therapeutic regimens, possibly involving some form of immunotherapy. With both diseases case-holding is as important as case-finding and success of the latter depends on education, the amount of social and welfare support provided and the efficacy of 'defaulter-retrieval' procedures.

Control by enhancement of herd immunity is achieved non-specifically by improvement of the general health and nutrition of the community, and specifically by vaccination with Bacille Calmette Guerin (BCG). The contribution of vaccination to the decline in tuberculosis appears small, even in countries where BCG has been shown to be highly effective, but several points must be considered. When the vaccine is first introduced, it is offered to young tuberculin-negative individuals: infected individuals of all age groups are excluded. Thus vaccination programmes will not prevent the development of reactivation disease in adults. In countries where tuberculosis is in decline, most reported cases are of this type and these are the major source of infection. In such situations, vaccination will have little impact on the incidence of tuberculosis in general and particularly not on the prevalence of infectious cases. Farga (1978) has stressed that a case of tuberculosis prevented by vaccination is not equivalent to a prevented source of infection. This is true at the present time but clearly primary (non-infectious) disease prevented now will lessen the incidence of post-primary disease (often infectious) in the future, particularly if such disease results from endogenous reactivation rather than exogenous infection (see page 93). Thus, at the start of a BCG immunization programme, BCG has a good direct effect but has no indirect effect, i.e. it does not prevent the disease in the unvaccinated population. As the programme continues the indirect effect will become more apparent. Nevertheless, the main value of BCG lies in its well-proven ability to reduce the incidence of the serious forms of childhood disease such as tuberculous meningitis. The place for BCG vaccination in the prevention of leprosy requires further investigation. From the present evidence (see page 76) it appears likely that the vaccine is of value in those regions where it also gives good protection against tuberculosis.

When studying the changing trends of the disease, it is particularly necessary to consider 'natural' trends as well as those due to the benefits of medical science, otherwise the contribution of the latter may be overestimated. It is salutory to recall that tuberculosis was in decline in Great Britain and in other industrially developed countries long before the advent of BCG vaccination and effective chemotherapy. It is even more relevant that indigenous leprosy died out in Great Britain, Scandinavia and other developed nations in the absence of any 'scientific' intervention. 'Natural' changes may, in a large part, really be sociological changes. Probably the most important factor leading to the decline in tuberculosis during the last century in the developed world was the reduction in overcrowding in the home and in workplaces. Reduction of overcrowding reduces the contagion parameter, and in a relatively wealthy country such as the USA, a source case may only infect two or three individuals annually (Johnston and Wildrick, 1974) as opposed to 20 or so in the poorer nations. Accordingly the control of tuberculosis and leprosy requires an 'holistic' approach, involving all aspects and levels of society and requiring changes in sociological and political attitudes over which the medical profession has a limited influence.

Although 'natural' changes in the behaviour of mycobacterial disease may be underestimated in the evaluation of control measures, they may be overemphasized when considering the eradication of such disease in low-prevalence areas. It is often assumed that the downward trend of tuberculosis observed in the industrially developed nations will continue until the disease is extinct. Unfortunately this attitude has encouraged a premature loss of interest amongst clinicians and public health authorities, with the result that source cases infect many individuals before they are detected. Complacency and backsliding by the medical profession towards tuberculosis were evident decades before the introduction of chemotherapy and BCG vaccination (Williams, 1908) but have been accentuated by these innovations. Indeed, Fine (1984) remarked that BCG vaccination has done more to eradicate the study of tuberculosis than to eradicate the disease itself. It is most important for health authorities to remember that tuberculosis and leprosy are remarkably tenaceous afflictions, both in the individual patient and in the community, and that 'no one is safe until all are safe'.

The principles of tuberculosis surveillance and control have been reviewed in detail by Styblo (1986). Useful practical guides to the organization of control programmes for tuberculosis and leprosy in developing countries have, respectively, been prepared by the Oxfam Health Unit (1985) and the World Health Organization (1980).

Control of bovine tuberculosis

In contrast to leprosy, there are animal reservoirs of tuberculosis that pose a serious threat to man. Cattle form by far the most important reservoir and no tuberculosis programme is complete without serious consideration of bovine disease. Tuberculosis in cattle is principally pulmonary and is spread from animal to animal by the aerogenous route. Although milk is the usual vector of transmission to man, the udder is involved in only about 1 per cent of infected cows. In contrast to man, infection in cattle almost invariably results in a progressive and ultimately infectious disease. Control is based on detecting infected animals and slaughtering them. This is highly effective when organized at a national level, pursued on a region-by-region basis and when the farmers are adequately compensated. Such schemes, when rigorously applied, cause a large and rapid drop in the incidence of bovine tuberculosis. Unfortunately complete eradication has not proved possible owing to the carriage of the bovine tubercle bacillus in wild animals, such as the badger in Great Britain and the opossum in New Zealand. A further threat to cattle comes from human beings infected with bovine tubercle bacilli. For further details on bovine tuberculosis, see Collins and Grange (1983).

Although control measures for bovine tuberculosis are clearly not applicable to human disease, lessons can be learnt from the former. In particular, veterinary practitioners are very aware of the ever-present danger of a resurgence in the incidence of disease and have refused to permit a very low incidence of the disease to diminish their vigilence. Indeed, it was a veterinary surgeon (Torning, 1965) rather than a physician who remarked that 'if tuberculosis is eradicated from the minds of the doctors, if only for a short time, its eradication from the bodies of their patients will be deplorably postponed!'

References

Aluoch, J.A., Swai, O.B., Edwards, E.A., Stott, H., Darbyshire, J.H., Fox, W., Stephens, R.J. and Sutherland, I. (1985) Studies on case-finding for pulmonary tuberculosis in outpatients at 4 district hospitals in Kenya. *Tubercle* **66**: 237–49.

Bleiker, M.A. and Styblo, K. (1978) The annual tuberculosis infection rate and its trend in developing countries. *Bulletin of the International Union Against Tuberculosis* **52**: 295–9.

Capewell, S., France, A.J., Anderson, M. and Leitch, A.G. (1986) The diagnosis and management of tuberculosis in common hostel dwellers. *Tubercle* **67**: 125–32.

Chaulet, P. (1983) But the emperor has no clothes on! A critique on the report of the joint IUAT/WHO study group. *Bulletin of the International Union Against Tuberculosis* **58**: 153–6.

Collins, C.H. and Grange, J.M. (1983) The bovine tubercle bacillus. *Journal of Applied Bacteriology* **55**: 13–29.

Davey, T.F. and Rees, R.J.W. (1974) The nasal discharge in leprosy: clinical and bacteriological aspects. *Leprosy Review* **45**: 121–34.

Farga, V. (1978) A turning point in the fight against tuberculosis. *Bulletin of the International Union Against Tuberculosis* **53**: 228–9.

Fine, P. (1984) Leprosy and tuberculosis – an epidemiological comparison. *Tubercle* **65**: 137–53.

Geuns, H.A. van, Meijer, J. and Styblo, K. (1975) Results of contact examination in Rotterdam. *Bulletin of the International Union Against Tuberculosis* **50**: 107–21.

Johnston, R.F. and Wildrick, H.K. (1974) State of the art review. The impact of chemotherapy on the care of patients with tuberculosis. *American Review of Respiratory Disease* **109**: 636–64.

Jopling, W.H. (1984) Handbook of Leprosy, 3rd edn. London: Heinemann.

Nsanzumuhire, H., Aluoch, J.A., Karuga, W.K., Edwards, E.A., Stott, H., Fox, W. and Sutherland, I. (1981) A third study of case-finding methods for pulmonary tuberculosis in Kenya, including the use of community leaders. *Tubercle* **62**: 79–94.

Oxfam Health Unit (1985) *Guidelines for Tuberculosis Control Programmes in Developing Countries*. Oxfam Practical Guide No. 4.

Rees, R.J.W. and McDougall, A.C. (1977) Airborne infection with *Mycobacterium leprae* in mice. *Journal of Medical Microbiology* **10**: 63–8.

Rouillon, A., Perdrizet, S. and Parrot, R. (1976) Transmission of tubercle bacilli: the effects of chemotherapy. *Tubercle* **57**: 275–99.

Sansarricq, H. (1982) The general situation of leprosy in the world (Kellersberg Memorial Lecture 1981). *Ethiopian Medical Journal* **20**: 89–105.

Sansarricq, H. (1983) Recent changes in leprosy control. *Leprosy Review*, special issue, 7s–16s.

Sansarricq, H., Seal, K. and Walter, J. (1979) Distribution of leprosy throughout the world. *Bulletin of the International Union Against Tuberculosis* **54**: 354–8.

Sbarbaro, J.A. (1975) Tuberculosis: the new challenge to the practising clinician. *Chest* **68** (Suppl.): 354–8.

Shears, P. (1984) Tuberculosis control in Somali refugee camps. *Tubercle* **65**: 111–16.

Springett, V.H. (1971) Ten-year results during the introduction of chemotherapy for tuberculosis. *Tubercle* **52**: 73–87.

Styblo, K. (1978) State of the art. I: Epidemiology of tuberculosis. *Bulletin of the International Union Against Tuberculosis* **53**: 141–52.

Styblo, K. (1984) Epidemiology of tuberculosis. In: *Infectionskrankheiten und ihre Erreger*, Vol. 4/VI, pp. 77–161. Jena: Gustav Fischer.

Styblo, K. (1986) Tuberculosis control and surveillance. In: *Recent Advances in Respiratory Medicine*, No. 4, pp. 77–108. Edited by D.C. Flenley and T.L. Petty. Edinburgh: Churchill Livingstone.

Toman, K. (1976) Mass radiology in tuberculosis control. *WHO Chronicle* **30**: 51–7.

Torning, K. (1965) Bovine tuberculosis. *Diseases of the Chest* **47**: 241–6.

Williams, L. (1908) The worship of Moloch. *British Journal of Tuberculosis* **1**: 56–63.

World Health Organization (1977) *WHO Expert Committee on Leprosy: Fifth Report*, WHO Technical Report Series No. 607. Geneva: WHO.

World Health Organization (1980) *A Guide to Leprosy Control*. Geneva: WHO

World Health Organization (1982) *Tuberculosis control*, WHO Technical Report Series No. 671. Geneva: WHO

Yanez, A. (1982) Restrictive factors in the application of a tuberculosis control programme. *Bulletin of the International Union Against Tuberculosis* **57**: 252–7.

7 Leprosy

Leprosy is a very chronic disease that was, and unfortunately often still is, regarded with a particular revulsion and horror. The disease is unique among the bacterial diseases in that the nerves are the principal sites of the lesions. The nerve involvement is largely responsible for the crippling deformities and blindness that have earned the disease its dread reputation. An additional stigmatizing feature is the involvement of the nasal bones which may lead to collapse of the nose (Fig. 7.1). It is, perhaps, understandable that the sight of a crippled and blind outcast in a state of abject misery and poverty should invoke a fear in others that they could endure the same fate. The management of the disease does not therefore merely consist of administering antileprosy drugs: it involves the physical and psychological rehabilitation of the patients and removal of the stigma and prejudice by education of the community. This is of particular importance as a fear of ostracism causes many sufferers to avoid seeking help until irreversible physical damage has occurred, by which time they may have infected many others.

Early accounts of a disease that could well have been leprosy are found in an Indian text, the *Charaka Samhita*, written between 600 and 400 BC. It is thought that the disease was brought from India to Greece by the army of Alexander the Great in the fourth century BC. The Greeks called the disease *elephantiasis graecorum*, probably on account of the ichthyosis and wooden-hard oedema of the legs of some of the sufferers. (Filarial elephantiasis was also recognized at that time and termed *elephantiasis arabum*.) Other Greek names were *leontiasis* and *satyriasis* – in allusion to the facial deformities. The term *lepra* or *lepros*, used to translate the Hebrew word *Tsarat*, referred to a scaling skin disease, possibly psoriasis or a fungal infection. Thus Biblical leprosy was almost certainly not the disease that bears the name today.

Although Koch's postulates have never been fulfilled with respect to leprosy, the evidence that it is caused by *Mycobacterium leprae* is overwhelming. With few exceptions, naturally occurring disease is restricted to man. The exceptions include the description of a leprosy-like disease in a chimpanzee (*Chimpansee troglodytes*) from Sierra Leone (Donham *et al.*, 1977) and in a sooty mangabey monkey (*Cercocebus atys*) from Nigeria (Gerone, 1982). More interestingly, about 10 per cent of free-living armadillos caught in various parts of Louisiana, USA, were found to have a disease caused by organisms indistinguishable from *M. leprae* (Meyers *et al.*, 1977). Five patients from this region developed leprosy after receiving cuts whilst wrestling with armadillos (Lumpkin *et al.*, 1983)!

Fig. 7.1 Stigmatizing features of leprosy: loss of eyebrows and nasal collapse

Experimentally, disease may be established in the armadillo, immunocompromised mice and the sooty mangabey monkey.

Owing to the nature of the disease, there was a widely held assumption that leprosy is spread by skin-to-skin contact. Although this may occur, it is noteworthy that the epidermis is usually intact and remarkably free of bacilli, even in the multibacillary forms of the disease (see below). It now appears more likely that the bacilli are principally shed from the nose. Indeed, a patient with lepromatous leprosy may discharge up to 10^8 bacilli daily in the nasal secretions – a number roughly equivalent to the number of tubercle bacilli expectorated daily by a patient with open pulmonary tuberculosis. Patients towards the tuberculoid end of the spectrum discharge few or no bacilli and are much less infectious than those at or near the lepromatous pole. Nevertheless, cases have arisen as a result of contact with such paucibacillary patients.

The portal of entry of *M. leprae* is unknown. It has been suggested that the bacilli enter the skin through cuts and abrasions. There are a few recorded cases of leprosy following dog bites, abrasions and tattooing (Sehgal, 1986). It has also been suggested that bacilli may be transmitted by insect bites and that they contaminate food and are ingested. It is also both possible and likely that they are inhaled and reach the skin by blood-borne dissemination from a primary focus in the respiratory tract.

Clinical features of leprosy

A remarkable feature of leprosy is the diversity of forms in which the disease manifests itself. This great variety is, as outlined in Chapter 5, the result of the immune responses of the patients who may be classified according to their position on a 'spectrum' of clinical, immunological and histological features. This spectrum (described in detail in Chapter 5), is divided into five points as follows:

tuberculoid	TT
borderline tuberculoid	BT
mid-borderline	BB
borderline lepromatous	BL
lepromatous	LL

Early lesions with characteristics that do not enable them to be classified are termed indeterminate (I or Idt). Some lepromatous patients show anergy from the onset (polar cases: LLp) while others represent an extreme downgraded borderline form of the disease (subpolar leproma-

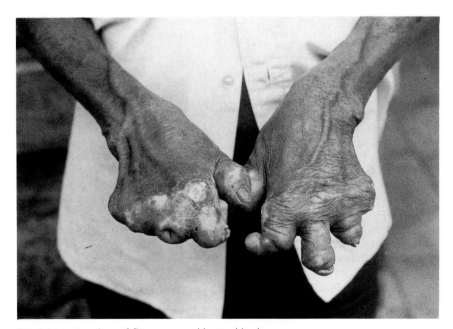

Fig. 7.2 Leprosy: loss of fingers caused by trophic changes

tous: LLs). Cases at the extremes of the spectrum (TT and LLp) are usually immunologically stable but the intermediate forms can shift on the spectrum, often with the development of severe reactions (see below). Patients may either undergo 'upgrading' or 'reversal' towards the tuberculoid pole or 'downgrading' towards the lepromatous pole. It is uncertain whether the spectrum is a continuous one. The distribution of patients in the various forms suggests that there are two populations, one towards the tuberculoid end and the other towards the lepromatous end of the spectrum. Whatever the form of leprosy, the cardinal features are the skin lesions and nerve damage leading to local anaesthesia and paralysis. In lepromatous leprosy nerve damage is generalized but fairly slowly progressive, while in the tuberculoid form of the disease only a few nerves are involved but progression is often rapid owing to granuloma formation. Leprosy reactions (see below) may cause severe and irreversable nerve damage very rapidly. Anaesthesia and paralysis cause the trophic changes and deformities that are characteristic of the disease (Figs. 7.2 and 7.3).

The eye is particularly vulnerable in leprosy: indeed, of all chronic infections, leprosy has the highest incidence of eye complications and, in the absence of adequate care, blindness occurs in about 5 per cent of patients (ffytche, 1981; Murray *et al.*, 1986). The eye may be damaged by corneal ulceration, iridocyclitis or, in the more anergic forms of the disease, by direct bacillary invasion. Corneal ulceration may result from corneal anaesthesia or from paralysis of the eyelids (Fig. 7.4). Iridocyclitis is most frequent in lepromatous leprosy and two forms are encountered: a chronic form thought to be secondary to nerve involvement and an acute

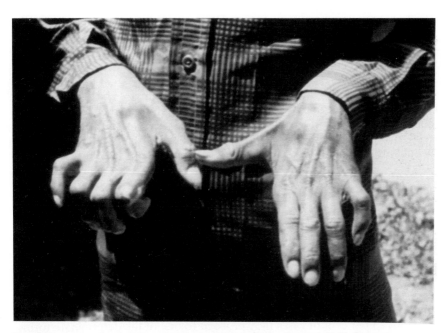

Fig. 7.3 Leprosy: claw hands caused by muscle paralysis. Courtesy N.M. Samuel

Fig. 7.4 Leprosy: blindness caused by corneal ulceration secondary to facial paralysis

form which is an ocular manifestation of erythema nodosum leprosum (see page 112). Recurrent attacks of iridocyclitis may cause glaucoma or cataracts.

The presenting features of leprosy are very complex and there is great variation from case to case, even within a given point on the spectrum. The clinical features of leprosy have been reviewed by Bryceson and Pfaltzgraff (1979), Jopling (1984) and Thangaraj and Yawalkar (1986). The following is only a brief summary.

Indeterminate leprosy

This is the earliest clinically evident form of the disease. The patient presents with one or more skin lesions of variable appearance. Microscopical examination reveals a mononuclear cell infiltration around capillaries and skin appendages (hair follicles and sweat glands). There is often a proliferation of the Schwann cells of the dermal nerves and a careful search may reveal occasional acid-fast bacilli in nerves and erector pili muscles. In the majority of patients the indeterminate lesions heal spontaneously, but in about 25 per cent of cases they proceed to one of the determinate forms described above.

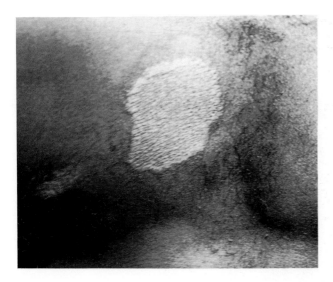

Fig. 7.5 Tuberculoid leprosy: distinct lesion with an elevated edge and hypopigmentation

Tuberculoid (TT) leprosy

There are a few skin lesions, usually one to four in number, which have distinct elevated edges (Fig. 7.5) and show a loss of tactile sensation. Hence they are sometimes termed 'maculoanaesthetic' lesions. There are also usually changes in pigmentation, impairment of sweating and reduced hair growth, and lesions may show central areas of healing (Fig. 7.6).

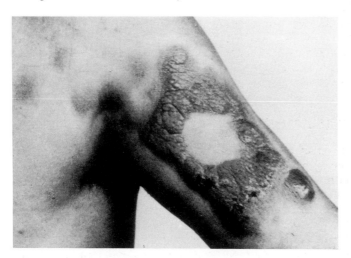

Fig. 7.6 Borderline tuberculoid leprosy: lesion showing central healing

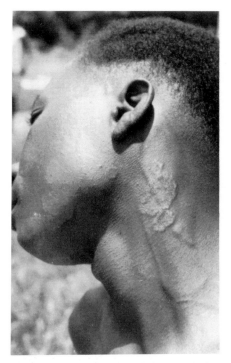

Fig. 7.7 Borderline tuberculoid leprosy: lesions on the neck and thickening of underlying greater auricular nerve

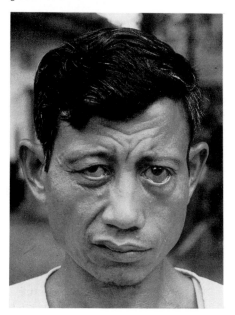

Fig. 7.8 Paralysis of the facial nerve (Bell's palsy) causing loss of naso-labial fold and lagophthalmos – a condition predisposing to corneal ulceration (cf. Fig. 7.4)

Biopsies of lesions show a heavy granulomatous infiltrate of lymphocytes, epithelioid cells and non-vacuolated giant cells, but bacilli are virtually never seen. The granulomas resemble those of non-caseating tuberculosis, hence the name tuberculoid leprosy. Lesions may also occur in nerves: these include cutaneous nerves near the lesions and major nerve trunks such as the ulnar, posterior tibial and common peroneal. Affected nerves are greatly thickened by an intense cellular infiltrate and permanent damage occurs rapidly (Figs. 7.7 and 7.8).

Borderline forms

These forms have features intermediate between the two polar forms. As the immune responses wane the lesions become more numerous and less discrete. The number of acid-fast bacilli increases as the histological evidence of cell-mediated immune responsiveness decreases. Furthermore, the tactile sensory loss diminishes towards the anergic pole.

Borderline tuberculoid (BT) leprosy

The skin lesions resemble those of TT leprosy but are smaller and more numerous. Large nerves may be affected but palpable lesions in the cutaneous nerves are seldom found. Biopsies show fewer lymphocytes than TT lesions but non-vacuolated giant cells are present. Bacilli are rarely found.

Borderline (BB) leprosy

There are numerous skin lesions with a greater tendency towards a symmetrical distribution than in the above forms. Biopsies show no vascular cuffing, few lymphocytes and no giant cells. Acid-fast bacilli are regularly seen.

Borderline lepromatous (BL) leprosy

Lesions are smaller and more numerous and biopsies show scanty lymphocytes and numerous acid-fast bacilli. Macrophages, histiocytes and foamy cells are seen.

Lepromatous (LL) leprosy

This is the anergic form of leprosy. The disease is not localized: there is widespread involvement of the skin and nerves. Many small and ill-defined macules appear and coalesce to form extensive lesions. Numerous small cutaneous nerves become progressively involved and anaesthesia gradually develops in a symmetrical fashion. In some cases the dermal infiltration is

Fig. 7.9 Lepromatous leprosy: marked nodulation of the skin.

diffuse while in others large nodules develop (Fig. 7.9). Ear lobes, in particular, may show gross thickening (Fig. 7.10).

Acid-fast bacilli are present in large numbers in the skin, blood stream and internal organs. Other organs and structures directly involved include the bone marrow, lymph nodes, eyes, larynx, muscles, liver and spleen. A patient with LL leprosy may harbour as many as 10^{11} leprosy bacilli. Direct invasion of the nasal bones leads to collapse of the nose, while involvement of the phalangeal bones exacerbates the destruction of the fingers and toes. Testicular involvement leads to infertility and to enlargement of the breasts. The latter may be so embarrassing to the patient that simple mastectomy is required.

Other forms: histoid, primary polyneuritic and Lucio's leprosy

Histoid leprosy is a hyperactive form of lepromatous leprosy manifesting as firm nodules containing numerous spindle-shaped macrophages. These lesions are particularly associated with relapses of lepromatous leprosy due to cessation of therapy or the development of drug resistance. The exact pathogenesis of histoid leprosy is unknown.

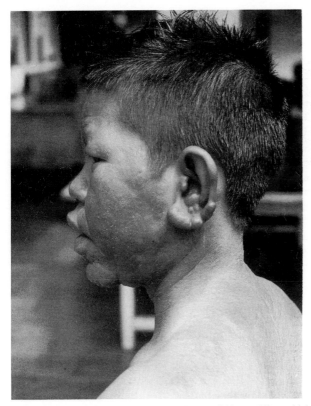

Fig. 7.10 Lepromatous leprosy in a 15-year-old boy: diffuse thickening of the skin and marked enlargement of the ear lobes

Primary (or pure) polyneuritic leprosy is seen in India but rarely elsewhere. There is an asymmetric involvement of peripheral nerve trunks in the absence of visible skin lesions.

Lucio's leprosy occurs principally, but not exclusively, in Mexico (Latapi and Chevez Zamora, 1948). There is a uniform thickening of the skin, a loss of body hair, including eyebrows, and a heavy bacillary load, but there are no nodules or other discrete lesions. There is a widespread sensory loss but no paralysis or eye damage. Such patients are prone to an unusual type of reaction termed Lucio's phenomenon (see below).

Reactions in leprosy

Hypersensitivity reactions often occur in leprosy and, if untreated, may lead to serious disability. Two main types of reactions are recognized: Jopling type 1 and 2 (Jopling, 1978). Type 1 reactions are cell-mediated and type 2 reactions are associated with the formation of immune complexes. Thus they are classified as types IV and III hypersensitivity reactions respectively in the scheme of Gell and Coombs.

Type 1 reactions occur in patients in the middle part of the immunological spectrum who shift either towards the lepromatous pole (downgrading reaction) or towards the tuberculoid pole (reversal reaction). (The term 'reversal' reflects the concept, no longer held, that early leprosy is of a tuberculoid type and that the other forms develop from it by a loss of immune reactivity.) The clinical features of the reactions vary according to the patient's position on the spectrum, but there is no evident difference between the downgrading and reversal reactions. Reversal reactions tend to occur after the commencement of therapy, while downgrading reactions occur if therapy is inadequate or has lapsed. Skin lesions become swollen, erythematous and oedematous and may even ulcerate. Nerves become swollen and painful and irreversible damage may occur. Surgical decompression of nerves may be required as an emergency procedure. The patient may also develop general malaise, dactylitis, orchitis or laryngitis and the latter may prove fatal by causing respiratory obstruction. The reaction is due to a flare up of delayed hypersensitivity, although the exact mechanism and cause remain poorly understood. There are usually no significant haematological changes.

Type 2 reactions occur in patients at, or very near to, the lepromatous end of the spectrum. They are due to the formation of complexes between circulating antibody and antigen released from dead bacilli. Accordingly, such reactions occur principally in treated cases, usually one to two years after the commencement of chemotherapy. They are also precipitated by vaccinations, other infections such as malaria, various forms of physical or mental stress and pregnancy. The cardinal feature of a type 2 reaction is *erythema nodosum leprosum* (ENL). The ENL lesions consist of erythematous plaques or nodules which appear in crops and fade within a few days. In severe reactions, the lesions may ulcerate and discharge liquefied necrotic tissue (*erythema necroticans*). In addition to ENL, reactions may cause arthritis, periosteitis, orchitis and iridocyclitis. Nerves may be involved although with less serious consequences than with type 1 reactions. In addition there may be a severe constitutional disturbance. Circulating immune complexes may be detected and may cause renal damage leading to proteinuria. The serum levels of the third component of complement and immunoglobulin in the IgG and IgM classes are elevated.

A third and rare type of reaction is Lucio's phenomenon which occurs in the variant of lepromatous leprosy known as Lucio's leprosy (see above). Unlike ENL, this reaction is almost always seen in untreated patients. Indeed, the reaction responds well to therapy, particularly to rifampicin, and typical ENL reactions may then occur as in other forms of lepromatous leprosy. The reaction manifests as small, painful and red patches, mainly on the limbs. The centres of the patches become necrotic and ulcerate and then heal, leaving a cicatrical scar. The lesions thus somewhat resemble the papulonecrotic tuberculides which are very occasionally seen in tuberculosis (see page 135).

The histopathology of leprosy

The histopathological features of leprosy reflect the immunological features of the disease. At the tuberculoid pole the histological features are those of a granulomatous response and the absence of bacilli, while at the lepromatous pole there are numerous bacilli, mostly within macrophages, but no evidence of an immune reaction. The various borderline forms have, as expected, characteristics intermediate between the two polar forms.

At the TT pole there are many mature epithelioid cells and Langhans giant cells in the granuloma. There may be a destructive granulomatous infiltration of nerves and the epidermis. In the BT and BB forms, the epithelioid cells become less in number and maturity and there is less nerve involvement by the granulomas.

In the BL and LL forms there are macrophages which have not matured into epithelioid cells: these are laden with bacilli. Nerve involvement is much less obvious, although in BL leprosy a cellular infiltration gives the perineurium a laminated or 'onion skin' appearance. Lymphocytes are numerous at or near to the TT pole and surround the granulomas. There are relatively few lymphocytes in BB or LL lesions, but in BL lesions there are many lymphocytes that infiltrate diffusely through the granuloma. The lymphocyte phenotypes vary across the spectrum: there are relatively more CD8 cells (putative suppressor cells) towards the LL pole (see page 81).

The diagnosis of leprosy

The diagnosis is made by careful history-taking and clinical examination, slit-skin smears and, if possible, skin biopsies. The latter have been compared to X-rays in tuberculosis: they aid the diagnosis and are a means of following the progress of the disease and the response to therapy. They also enable the patient to be accurately graded according to the immunological spectrum. At clinical examination the nature, location and extent of skin lesions are recorded. Areas of anaesthesia and thickened nerves are sought. The extent of physical handicap is assessed by a full neurological and orthopaedic examination. The eye should be carefully examined to exclude ocular complications such as iridocyclitis and corneal ulceration. Nasal signs and symptoms such as partial obstruction, discharge and bleeding may be helpful. Indeed nasal stuffiness may be the first noticeable feature of the disease, although the patient may not mention it unless specifically asked (Jopling, 1984).

Slit-skin smears are taken from lesions and from standard sites such as the ear, chin and elbows. The skin is pinched between the thumb and forefinger and a small cut is made. This cut should reach the dermis but without causing bleeding as blood causes acid-fast artefacts. Tissue fluid is scraped out with the tip of the scalpel blade held at 90 degrees to the line of the incision and smeared in a standard way on a glass slide. Nasal scrapings are taken from the lower turbinate or septum with a small curette. After Ziehl–Neelsen staining, the slides are examined with a ×100 magnification objective and the number of bacilli are expressed as the Bacterial Index (BI), as follows:

6+   numerous clumps of bacilli in each microscopic field
5+   100 to 1000 bacilli in each field
4+   10 to 100 bacilli in each field
3+   1 to 10 bacilli in each field
2+   1 to 10 bacilli in 10 fields
1+   1 to 10 bacilli in 100 fields

It is generally assumed that solid-staining bacilli are viable but that those that stain in a weak or irregular fashion are dead. Although this may not be strictly true, the percentage of solid-staining bacilli, the morphological index (MI), gives a good clinical guide to the viability of the bacilli and the effect of therapy, as well as indicating non-compliance or the emergence of drug resistance.

Skin biopsies are used to confirm doubtful cases and to classify the disease according to the immunological spectrum. The examination of biopsies requires skill and experience and is discussed in detail by Ridley (1977).

It is of the greatest importance that leprosy is diagnosed correctly. A wrong diagnosis may have the direst consequences for the patient and his or her family.

Skin testing

Skin testing is of no diagnostic value in leprosy; but it is useful epidemiologically as a guide to the extent of exposure of a community to the leprosy bacillus, and it assists in determining the position of the patient on the immunological spectrum. There are two types of reagents: lepromins (Mitsuda or integral lepromins) and leprosins. The former are autoclaved suspensions of bacilli-laden tissues and the latter are filtered soluble antigens derived from bacilli separated from the host's tissues. The suffix H or A indicates that the reagent is derived from human or armadillo material respectively. Dharmendra lepromin is a widely used tissue-free bacillary extract of human origin, but many workers are now using the armadillo-derived equivalent.

Two different reactions are elicited by the skin testing reagents: the early or Fernandez reaction and the late or Mitsuda reaction. The former is seen particularly with the leprosin type reagents. It is analogous in its timing, gross appearance and histological features to the tuberculin reaction, except that tissue necrosis does not occur. In common with the tuberculin reaction, a positive reaction indicates previous exposure to the relevant antigens. Thus the test may be positive in healthy contacts as well as in immunologically reactive patients (Godal and Negassi, 1973). Furthermore, uninfected individuals responding to the common mycobacterial antigens (category 1 reactors, see page 73) may react to leprosin.

The cruder tissue-containing lepromins elicit the Mitsuda reaction which is seen 3–5 weeks after testing. It manifests as an erythematous papule or nodule which may ulcerate. The immunological mechanism is poorly understood, but it appears to be related to the ability of the patient to mount a granulomatous immune response to the leprosy bacillus. Thus reactivity is strongest in TT cases, weaker in BT cases but negative in BB, BL and LL cases.

Serological tests

Attempts are being made to develop a reliable serological test for leprosy based on the detection of antibody to a unique antigenic determinant on phenolic glycolipid I (PGL-1) of *M. leprae* (see page 16). This determinant has been synthesized and should therefore become available in amounts large enough for widespread use. Clinical studies have shown that significant levels of antibody are present in a large majority of sera from multibacillary cases (LL and BL) but in only a minority of those from patients with paucibacillary disease (BT and TT) (Qinxue *et al.*, 1986). In addition, a monoclonal antibody to PGL-1 may be used to detect the specific antigen in the blood and provides a useful and more specific alternative to Ziehl–Neelson staining for the demonstration of *M. leprae* in slit-skin smears and biopsies.

The epidemiology of leprosy

This topic is discussed in Chapter 6.

Prevention of leprosy

As in the case of tuberculosis, the spread of leprosy is preventable by detecting infectious patients as early as possible and treating them, and by increasing the resistance of the community. Chemoprophylaxis has also been used in certain rather special situations (see page 163). The general socio-economic measures that are so important in the control of tuberculosis also affect the spread of leprosy.

Early detection of leprosy benefits both the patient, by preventing or minimizing deformity, and the community by reducing the risk of transmission. Such early detection requires careful education of the community, especially in view of the prejudices associated with this disease. Patients are much more likely to attend the clinic if they have no need to fear banishment, ostracism and loss of employment and if they know that minimal disease can be completely cured with no residual effects.

The prevention of leprosy by BCG vaccination is discussed in Chapter 5.

Management of the leprosy patient

Effective chemotherapy (described in Chapter 10) is, as in tuberculosis, of the greatest importance but it is only one aspect of the treatment of the leprosy patient. Other aspects of management include the control of immunological reactions, care of anaesthetic limbs, prevention and treatment of ocular damage, surgery to remove stigmatizing signs and to restore function to deformed limbs, and physical, occupational and social rehabilitation. If diagnosed early, leprosy is treatable on an outpatient basis. Segregation or banishment is unnecessary, inhumane and ineffective. Indeed the fear of segregation may deter the patient from seeking medical help. Much can be achieved to alleviate the misery of leprosy in the individual patient and in the community by the application of a little medical care and a lot of common sense.

Control of reactions

Reactions must be treated as emergencies in order to avert serious and irreversible nerve damage, deformity or blindness. Anti-inflammatory drugs are used to reduce neuritis, iridocyclitis and pain and to prevent skin ulceration. Mild reactions may respond to aspirin, chloroquine or antimonial drugs, but steroids are required for more serious episodes. Clofazimine is anti-inflammatory as well as being anti-leprotic, and thalidomide, although notorious for its teratogenic properties, is very useful for treating type 2 reactions. Oral zinc sulphate is also useful for treatment of type 2 reactions and may enable the dose of other anti-inflammatory agents to be reduced.

Gross swelling in major nerve trunks may rapidly cause serious and permanent loss of function. Surgical decompression of the nerve by opening the sheath may partially or totally avert such damage.

Care of anaesthetic limbs

Damage and deformity occur mainly because, being painless, injuries, splinters and infections are neglected. Burns may occur because a patient is not aware that he is handling a hot object. If the patient often picks up heavy objects, callosities and ulcers may develop at pressure points. Thorough education of the patient and family as to the cause of damage and the ways of preventing it are essential. Training must be strict enough to instil a strong sense of self-discipline in the patient.

Anaesthetic feet require particular care: pressure points and chafing must be prevented. In some leprosaria, the patients make special shoes as part of their occupational therapy.

Prevention of blindness

Loss of vision is a very tragic complication of leprosy, particularly as it is often preventable. Both types of hypersensitivity may damage the eye. Thus, in addition to the general management of such reactions, the local administration of steroids by drops or by subconjunctival injection may be required. In cases of iridocyclitis mydriatics are used to prevent the formation of adhesions between the iris and the lens. Post-iridocyclitic adhesions, cataracts or glaucoma require the appropriate surgical procedures.

Anaesthesia of the cornea is a very difficult complication to manage as it readily predisposes to ulceration. Regular and careful inspection of the affected eye is essential. Paralysis of the eyelids predisposes to corneal ulceration and tarsorrhaphy may be required. In some cases, surgical transposition of a segment of the temporalis muscle may restore function to the eyelids.

Surgical management

Surgery is required for the correction of stigmatizing deformities and for improvement of the function of limbs. It is also, as outlined above, required for the prevention or treatment of blindness and for the decompression of swollen nerves.

Facial operations include replacement of eyebrows and reconstruction of the collapsed nose. Foot drop or claw hand may be amenable to treatment by tendon transposition. Other orthopaedic operations include wedge resections and arthrodeses. In the case of a severely deformed limb, amputation and the provision of a prosthesis may give a better functional result than attempts to correct the deformity.

Physical rehabilitation

This involves management of the disability, prevention of further disability and training for an occupation which is within the patient's capability. Physiotherapy and remedial exercises are required to avoid stiffness of joints and to strengthen muscle groups. Simple pieces of locally made equipment may considerably enhance the quality of life of patients with deformities. Individuals who have lost their fingers or hands may be provided with leather wrist straps to which cutlery and other utensils may be attached. The pharmaceutical company Ciba–Geigy have sponsored a programme for the local manufacture of 'Grip-aids' (Thangaraj and Yawalker, 1986). These are made from epoxy putty and are moulded on to tools, pens and cutlery to fit the patient's hands.

Psychological and social rehabilitation

This difficult aspect of management involves a reorientation of the patient and society towards the disease. Education at all levels of society should instil the idea that leprosy is just another disease. This is not easy in regions where there are deeply entrenched prejudices and superstitions. If, however, the stigma of leprosy can be removed and patients encouraged to seek treatment at an early stage of the disease, a very great part of the suffering and misery associated with this misunderstood affliction will be averted.

The self-esteem of the patients will be maintained if they can be treated as outpatients in a general medical unit and if they lose neither their social status nor occupation. If, sadly, deformity, physical handicap or prejudice render this impossible, long-term care may be necessary. This may take the form of sheltered workshops or craft centres, or leprosy villages. In some countries such care is provided by voluntary agencies – examples are *Amigos de los Enfermos de Lepra* (friends of leprosy patients) in Spain and the *Dharma Wanita* (a womens' association) in Indonesia.

It is to be hoped that the time will soon come when enlightened attitudes and comprehensive medical services will obviate the need for such long-term care.

References

Bryceson, A. Pfaltzgraff, R.E. (1979) *Leprosy*. Edinburgh: Churchill Livingstone.

Donham, K.H. and Leininger, J.R. (1977) Spontaneous leprosy-like disease in a chimpanzee. *Journal of Infectious Diseases* **136**: 132–6.

ffytche, T.J. (1981) Iritis in leprosy. *Transactions of the Ophthalmological Society of the United Kingdom* **101**: 325–7.

Gerone, P.J. (1982) Hansen's disease in a sooty mangabey monkey. *The Star* **41**: 1.

Godal, T. and Negassi, K. (1973) Subclinical infection in leprosy. *British Medical Journal* **3**: 557–9.

Jopling, W.H. (1978) *Handbook of Leprosy*, 2nd edn. London: William Heinemann.

Jopling, W.H. (1984) *Handbook of Leprosy*, 3rd edn. London: William Heinemann.

Latapi, F. and Chevez Zamora, A. (1948) 'Spotted' leprosy of Lucio (la lepra 'manchada' de Lucio); introduction to its clinical and histological study. *International Journal of Leprosy* **16**: 421–30.

Lumpkin, L.R., Cox, G.F. and Wolf, J.E. (1983) Leprosy in five armadillo handlers. *Journal of the American Academy of Dermatology* **9**: 899–903.

Meyers, W.M., Walsh, G.P., Brown, H.L., Rees, R.J.W. and Convit, J. (1977) Naturally acquired leprosy-like disease in the nine-banded armadillo (*Dasypus novemcinctus*): reactions in leprosy patients to lepromins prepared from naturally infected armadillos. *Journal of the Reticuloendothelial Society* **22**: 369–75.

Murray, P.I., Kerr Muir, M.G. and Rahi, A.H.S. (1986) Immunopathogenesis of acute lepromatous uveitis: a case report. *Leprosy Review* **57**: 163–8.

Qinxue, W., Ganyun, Y., Xinyu, L., Qi, L. and Lilin, Z. (1986) A preliminary study on serological activity of a phenolic glycolipid from *Mycobacterium leprae* in sera from patients with leprosy, tuberculosis and normal controls. *Leprosy Review* **57**: 129–36.

Ridley, D.S. (1977) *Skin Biopsy in Leprosy*. Basel: Ciba-Geigy.

Sehgal, V.N. (1986) Leprosy following mechanical trauma. *Leprosy Review* **57**: 74–6.

Thangaraj, R.H. and Yawalkar, S.J. (1986) *Leprosy for Medical Practitioners and Paramedical Workers*. Basle: Ciba–Geigy.

8 Tuberculosis

Tuberculosis in man is caused by the human tubercle bacillus (*Mycobacterium tuberculosis*) and, much less often, by *M. bovis* or *M. africanum*. The lung is the most frequently affected organ but the disease has been termed *Morbus Percorpus*, stressing that it may involve virtually any organ or system of the body. Tuberculosis may therefore mimic many other diseases and often presents a serious diagnostic challenge, especially in countries where the disease is now rare and often overlooked.

The name tuberculosis was probably first used by Shönlein in 1939. Older epithets include *phthisis* and *consumption*, both alluding to the marked wasting characteristic of advanced disease. Non-pulmonary manifestations, particularly cervical lymphadenitis, were known as *scrofula* (see page 127).

The terminology and classification of tuberculosis

Many of the problems in understanding the pathogenesis and clinical types of tuberculosis stem from the terminology and classification, most of which is a legacy of the extensive clinical and pathological descriptions of the disease published in the early decades of this century. At that time, tuberculosis was rife and primary infection usually occurred during childhood. Consequently primary disease, i.e. that following the initial infection, was also termed childhood-type disease, while secondary disease, which usually occurs after an interval of several years, became known as adult-type disease. It was widely assumed that secondary disease was always due to reactivation of a primary infection, rather than to exogenous reinfection, and it was thus termed post-primary tuberculosis. Despite much controversy in the past, it is now realized that this type of tuberculosis may be the result of exogenous reinfection as well as to endogenous reactivation, particularly in high-prevalence areas. Thus the term 'reinfection tuberculosis' is preferable to post-primary tuberculosis, although the latter term is well established and its usage will therefore probably continue.

The clinical manifestations of tuberculosis are so variable in type, extent and site that any classification, except into the obvious categories of pulmonary, non-pulmonary and widespread, is of little value. Likewise, attempts to place patients on an immunological spectrum with a prognostic usefulness, as in leprosy, have generally failed. Radiological classifications of lesions are highly subjective and their prognostic values have been rendered obsolete by the introduction of effective chemotherapy.

Diagnosis of tuberculosis

The diagnosis of tuberculosis is not easy. Radiological appearances, though helpful, are notoriously non-specific and culture of the causative organism, though specific, is time-consuming and fraught with difficulties. Many cases, particularly those of non-pulmonary disease, are still diagnosed in the absence of bacteriological proof. The tuberculin test is of limited usefulness in individual patients; positive tests are not conclusive evidence of active disease, while negative tests do not exclude it.

A positive tuberculin test indicates that the patient has experienced immunologically effective contact with mycobacteria. This may be due to active tuberculosis, past infection, past BCG vaccination or sensitization by environmental mycobacteria. Reactions due to the latter cause can usually be differentiated from those due to infection by tubercle bacilli or BCG vaccination on the basis of size. In some countries BCG is no longer used routinely and tuberculin reactivity is thus more likely to indicate active or quiescent infection with *M. tuberculosis*.

Skin testing is performed in several different ways. In the Mantoux technique, a measured quantity (usually 0.1 ml) is given by intradermal injection. By giving an exact dose, reactions may be compared by size, but the method requires some technical skill. In the Heaf method a spring-loaded gun is used to fire six needles into the skin, to a depth of 2 mm, through a drop of concentrated tuberculin. Although technically easy, the amount of tuberculin entering the skin is variable. Thus Heaf testing is unsuitable for quantitative studies. The tine test is similar in principle to the Heaf test except that the tuberculin is freeze-dried on to the spikes (tines) of a disposable applicator. The tines are pressed into the skin and the tuberculin dissolves in the tissue fluids. The tine test is suitable for single testings but opinions vary as to its reliability. An important technical point is that the tines should remain in the skin for several seconds so that sufficient tuberculin dissolves off.

For full details of tuberculin testing in clinical practice, see Caplin (1980).

Features in the history and clinical examination may be diagnostically helpful but are rarely conclusive. Likewise, haematological features such as a high erythrocyte sedimentation rate (ESR), a mild anaemia and elevated lymphocyte count are no more than suggestive. More sophisticated haematological tests, such as the assay of acute-phase reactants, are no more informative than the ESR (Grange *et al.*, 1984). Assay of lymphocyte-derived adenosine deaminase in pleural, pericardial and cerebrospinal fluids appears promising (Ocana *et al.*, 1986). Biopsies may be submitted to histological, as well as microbiological, examination. The introduction of the fibreoptic bronchoscope has rendered lung biopsy as well as bronchial lavage and brushing (see page 51) a simple and safe procedure (Burk *et al.*, 1978), although unfortunately the equipment and expertise are not universally available.

Expense ?

In view of the great diagnostic difficulties posed by tuberculosis, there have been thousands of attempts to develop immunological and biochemical tests for this disease but none has achieved clinical usefulness. It is to be hoped that the introduction of monoclonal antibodies and DNA probes will facilitate the development of highly sensitive and specific tests for the presence of mycobacteria in sputum, fluids and tissues.

The natural history of tuberculosis

As outlined in Chapter 6, infection by the tubercle bacillus does not always lead to clinically evident disease and not all those with disease develop open or infectious forms.

The first event in the pathogenesis of tuberculosis, whether inapparent or overt, is the implantation of bacilli in tissue. The most frequent portal of entry is the lung, usually resulting from the inhalation of airborne droplets containing a few bacilli, expectorated by individuals with 'open' pulmonary disease (see page 93). Less frequently, bacilli may be ingested and lodge in the tonsil or in the wall of the intestine. Such infection is particularly associated with the consumption of contaminated milk or milk products. A third but rare mode of infection is direct implantation of bacilli into the skin. This is a health hazard faced by those working with infected material or cultures of tubercle bacilli. Such skin lesions were termed 'prosector's warts' and it was in this manner that Laennec, the inventor of the stethoscope, acquired tuberculosis which led to his death.

Dissemination of bacilli from the site of implantation occurs via the lymphatics to the regional lymph nodes. Lesions developing at the implantation site and in the nodes form the *primary complex* (Fig. 8.1). In

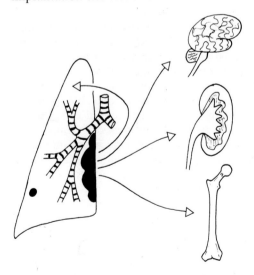

Fig. 8.1 Primary tuberculosis: the Ghon focus, primary complex of Ranke (Ghon focus and lymphatic component), and haematogenous dissemination to other organs including brain, kidney, bone and other parts of the lung

the lung, the implantation lesion is termed the Ghon focus and this, together with the involved lymph nodes, form the primary complex of Ranke. The lymph node component is often much larger than the Ghon focus, especially in children. In the case of tonsillar infection, the initial focus is usually inapparent: cervical lymphadenopathy is the only clinical manifestation of the disease.

Haematogenous dissemination of bacilli to many organs, including other parts of the lung and the pleurae, occurs during the early stage of the disease: within days or even hours. Primary tuberculosis is thus a widely disseminated infection – a fact not usually realized. The subsequent events vary considerably from patient to patient: the primary complex may resolve without becoming clinically evident, or a severe and rapidly fatal disease may ensue. Despite the differences, a general pattern or 'timetable' of the disease is evident, particularly in children, and has been described in detail by Wallgren (1948):

Stage 1:   (duration 3–8 weeks) The primary complex develops and conversion to tuberculin positivity occurs.

Stage 2:   (duration about 3 months) Serious forms of disease due to haematogenous dissemination occur. These include meningitis and miliary tuberculosis.

Stage 3:   (duration 3–4 months) Tuberculous pleurisy may develop, owing either to haematogenous spread or to direct spread from an enlarging primary focus.

Stage 4:   This stage lasts until the primary complex resolves which, in the case of untreated but non-progressive disease, may take up to three years. During this stage the more slowly developing extrapulmonary lesions, such as those in bones, joints and the kidneys, appear.

The evolution of the primary focus and its histological appearance reflects the host's immune response. In the case of pulmonary infection, the inhaled bacilli are probably first engulfed by the alveolar macrophages in which they multiply. Other alveolar macrophages, together with polymorphonuclear cells, engulf these bacilli and form a focus of acute but non-specific inflammation. After a few days tissue and blood-borne monocytes surround the bacilli, and mature into epithelioid cells. Some epithelioid cells fuse to form multinucleate Langhans giant cells. This is the first stage in the formation of the granuloma, which may be defined as a chronic compact aggregate of activated macrophages around a granuloma-genic agent. Initially, this macrophage activation is probably due to adjuvants present in the bacilli, but after about two weeks specific lymphocyte-mediated activation occurs (see page 64). The lymphocytes are seen as a zone around the epithelioid cells.

Unless the infection is rapidly contained, many macrophages within the centre of the granuloma are killed and replaced by new ones attracted to the site by lymphokines. The central dead cells and bacilli form a cheese-like, or caseous, material. Such 'caseation' is characteristic of the tuberculous granuloma or 'tubercle' (Fig. 8.2). At the same time, fibroblasts appear in the outer layers of the granuloma and produce a layer

Fig. 8.2 Thin section of a tuberculous granuloma showing caseation, epithelioid cells, lymphocytes and multinucleate giant cells

of collagen which contains the infection. In most individuals the initial infection subsides and the collagen capsule contracts to form a cicatrical scar which often becomes calcified. This scar may persist but in a substantial number of cases it is completely absorbed. In some cases, repeated periods of extension of infection followed by healing and fibrosis occur and lead to the formation of an 'onion skinned' or 'coin' lesion. The fact that tissue necrosis and cavity formation is not usually characteristic of primary lesions indicates that the disease is usually controlled by cell-mediated immunity before the development of necrotic delayed hypersensitivity reactions.

In a minority of individuals, particularly children, the primary complex causes serious complications. These include pleurisy or pericarditis due to erosion of an enlarging Ghon focus into the pleural cavity or of an affected lymph node into the pericardium. Alternatively, but rarely, an enlarging focus may erode into a bronchus, causing cavity formation or tuberculous

primary complex

bronchopneumonia. A grossly enlarged lymph node may compress one of the major bronchi, leading to collapse of a lobe or segment of the lung – a condition termed *epituberculosis*. For full details of such complications, see Miller (1982).

In the absence of such local complications or of an extrapulmonary manifestation of the disease, the infection becomes quiescent and may remain so for years, decades or for the remainder of the individual's life. The state of 'suspended animation' in which the bacilli persist for such a long period of time is not understood, but presumably they are kept in check by the host's immune response. If, for any reason, this response is weakened, the bacilli can escape their bondage and multiply, thereby causing secondary, or post-primary, tuberculosis.

Conditions predisposing to reactivation include old age, malnutrition, intercurrent infection, malignant disease, administration of steroids or other immunosuppressive drugs, and the acquired immune deficiency syndrome (AIDS). In the latter case, tuberculosis is often diagnosed before the underlying disorder, while the reverse is usually the case with infection by the less pathogenic 'atypical' mycobacteria (Duncanson *et al.*, 1986). In principle, such reactivation could occur in any site in which bacilli had been seeded by the initial haematogenous spread. In practice, the most usual site of reactivation is, for reasons that are far from clear, the apex of the lung.

Post-primary pulmonary lesions differ from primary ones in several important respects. These differences are almost certainly related to the patient's immune reactivity: patients developing post-primary disease are usually tuberculin-positive, implying that they are able to express necrotic Koch-type responses to the bacillus. Local progression and central caseous necrosis is much more marked in post-primary lesions. Caseous material and surrounding tissue often liquefy (probably owing to proteoloytic enzymes released by macrophage); and if the lesion ruptures into a bronchus, discharge of the liquified material leads to cavity formation, and aspiration of bacilli into other bronchi leads to the development of many secondary lesions. Thus, the typical radiological or post-mortem appearance of advanced pulmonary tuberculosis is of one or more cavities in the apices and mottled areas of caseation with patchy fibrosis and emphysema in the lower lobes (Figs 8.3 and 8.4). Less frequently, the focus ruptures into the pleural cavity, causing tuberculous empyema.

The tuberculous cavity is the ideal site for the growth of the tubercle bacillus. The temperature is optimum, there is an abundance of oxygen, and various nutrients are derived from the necrotic cavity wall. Accordingly, the cavity wall is teeming with bacilli and these readily gain access to the sputum and are expectorated. Such patients are said to have 'open' tuberculosis and are infectious. Surprisingly, cavities can heal and become obliterated. Indeed, about 20 per cent of cases of open tuberculosis heal spontaneously. In the pre-chemotherapeutic era the only measures of proven usefulness in the treatment of the disease were those operative procedures that induced the closure of cavities by collapsing the lung. These included artificial pneumothorax, or thoracoplasty whereby a more permanent collapse was achieved by the excision of large parts of the chest

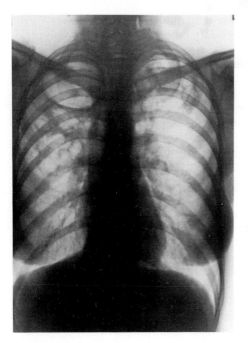

Fig. 8.3 Post-primary pulmonary tuberculosis: chest radiograph showing widespread disease with large cavities in both upper lobes

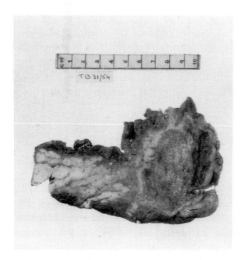

Fig. 8.4 Post-primary pulmonary tuberculosis: resected upper lobe from the patient in Fig. 8.3, showing numerous caseous lesions and a large cavity lined with granulation tissue

wall. Such mutilating surgery is now, mercifully, obsolete.

A further characteristic of post-primary tuberculosis is that lymphatic or haematogenous spread of disease is rare. (This is probably due to the obliteration of draining lymphatics by the necrotic reaction.) On the other hand, as mentioned above, secondary lesions in the lung follow the release of liquified caseous material into the bronchus. Secondary non-pulmonary lesions, usually small ulcers, are due to the direct inoculation of bacilli from sputum. These ulcers may occur in the larynx, mouth, small and large intestines and the anus.

Symptoms of pulmonary tuberculosis

Uncomplicated primary disease is usually symptomless. Older accounts refer to the development of erythema nodosum, phlyctenular conjunctivitis and fever at the time of tuberculin conversion, but these features are rarely encountered nowadays and hardly ever raise a suspicion of tuberculosis. Post-primary disease causes local and generalized symptoms. Localized symptoms include coughing, sputum, haemoptysis and pleuritic pain. General symptoms include fatigue, anorexia, weight loss, fever and night sweats. None of these are characteristic for tuberculosis and many other causes must be considered, particularly in low-prevalence regions.

Neonatal and congenital tuberculosis

Neonatal tuberculosis is probably more prevalent than realized as very young children are often not included in case-finding programmes. Infection is almost always from a parent or other member of the household. Neonates and children less than three years of age are particularly likely to develop serious complications of primary disease, such as miliary disease or meningitis. Congenital tuberculosis is very rare but, if undiagnosed, it is rapidly fatal (Snider and Bloch, 1984). It probably results from haematogenous spread via the umbilical artery or ingestion of infected amniotic fluid. Many organs, including the lung, are involved.

Types of tuberculosis according to organ involvement

The descriptions above refer principally to pulmonary tuberculosis. The pathogenesis, evolution and histology of non-pulmonary forms of the disease are essentially similar, but the clinical features and management are substantially different. Whereas virtually all cases of pulmonary tuberculosis are treated by antituberculous drugs alone, additional procedures are often required for non-pulmonary disease. Indeed, such disease is sometimes termed 'surgical tuberculosis'. Some cases result from the direct implantation of the bacilli in a non-pulmonary site, but most are

Table 8.1 Nature of bacteriologically confirmed cases of non-pulmonary tuberculosis in South-East England, 1977–84, with percentages in parentheses (Grange *et al.*, 1985)

Type of lesion	Ethnic origin of patient	
	European	Asian
Lymph node	446 (30.3)	1065 (55.6)
Orthopaedic	287 (19.5)	473 (24.7)
Urinary tract	527 (35.9)	122 (6.4)
Male genital	25 (1.7)	9 (0.5)
Female genital	36 (2.5)	20 (1.0)
Central nervous	75 (5.1)	82 (4.3)
Abdominal	62 (4.2)	130 (6.8)
Other	12 (0.8)	13 (0.7)
Totals	1470	1914

due to haematogenous dissemination from a primary pulmonary focus.

Non-pulmonary manifestations of tuberculosis are not uncommon: the actual frequency and type of disease varies from region to region and between ethnic groups within a given country. A survey of bacteriologically confirmed tuberculosis in South-East England from 1977 to 1983 (Grange *et al.*, 1985) showed that 21 and 50 per cent of isolates from ethnic European and Asian patients respectively were from non-pulmonary sites. As shown in Table 8.1, the site of the lesions varied between the two ethnic groups.

For convenience, non-pulmonary tuberculosis may be considered under the following headings:

> lymph nodes (other than intrathoracic nodes)
> bones and joints
> central nervous system
> urinary and male genital tracts
> female genital tract
> abdominal: intestinal and peritoneal
> skin
> less common sites
> disseminated disease.

Lymph nodes

Tuberculous lymphadenitis of the neck is also known as scrofula. The origin of this name is obscure: it is thought to be a diminutive of the latin word *Scrofa* – a brood sow. Some authors claim that the name is derived from the likeness of the cluster of nodes to piglets feeding from the sow, while others believe it reflects the somewhat sow-like features imparted to the victims by the swelling of the face and neck and the associated anaemia. The disease was also termed the King's evil in the belief that the condition could be cured by the touch of a reigning monarch. A good clinical description of the disease by Richard Wiseman (1696) also contains

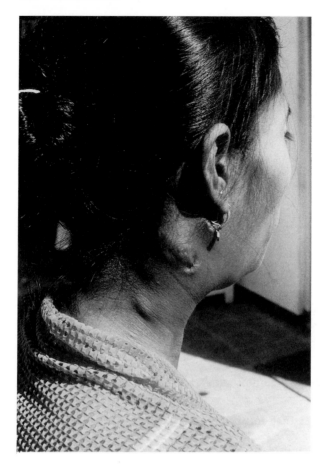

Fig. 8.5 Tuberculous cervical lymphadenitis (scrofula) with involvement of the tonsillar node

a theological discussion of the nature of this therapeutic procedure.

The common aetiology of scrofula and pulmonary tuberculosis was firmly established by Jean Antoine Villemin (1868) during his classical studies on the transmissibility of tuberculosis.

During the time when milk was frequently contaminated by bovine tubercle bacilli, lymphadenitis often resulted from a primary tonsillar lesion and the cervical glands were usually involved (Fig. 8.5). Many lesions proceeded to sinus formation (see below), and it has been suggested that the high winged collar of Victorian times was introduced to cover the resultant scarring.

In regions where milk-borne tuberculosis is uncommon, cervical lymphadenitis is more usually due to the human tubercle bacillus (or to other species such as *M. avium* – see page 145) and the supraclavicular nodes are often involved. In some cases, such node involvement may represent an upward extension of an intrathoracic primary complex.

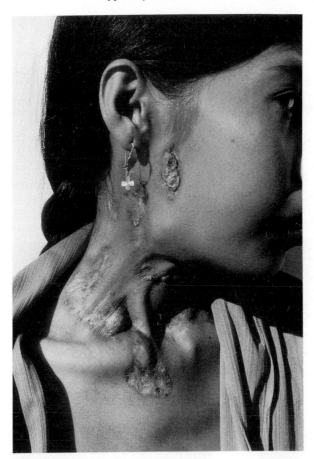

Fig. 8.6 Tuberculous cervical lymphadenitis with secondary lupus vulgaris of the skin (scrofuloderma)

Tuberculous lymphadenitis is the most prevalent form of extrapulmonary tuberculosis seen in Great Britain and is particularly common in patients of Asian ethnic origin (Table 8.1). It also, for unknown reasons, tends to occur more frequently in females in all ethnic groups.

If untreated, affected glands may undergo necrosis and the disease may spread into the surrounding tissues and ultimately to the skin with the formation of sinuses. The skin around the sinuses may thus become infected and lupus-like lesions develop – a condition termed scrofuloderma (Fig. 8.6).

Tuberculous lymphadenitis responds to standard chemotherapy, although relapse appears to occur more frequently than with pulmonary disease. Accordingly some physicians continue therapy for longer than for pulmonary disease. Occasionally glands undergo massive enlargement during the course of therapy. This is almost certainly the result of a hypersensitivity reaction and responds well to steroid therapy.

Bone and joint tuberculosis

This form of tuberculosis is almost always the result of haematogenous dissemination of bacilli from a primary focus elsewhere, usually the lung. Most cases present six months to three years after the initial infection. Any bone or joint may be affected but the most frequent site is the spine. This site is followed in frequency by the large joints of the lower limb (hip, knee and ankle) and then by the large joints of the upper limb (shoulder, elbow and wrist). Mulitple cystic lesions may occur in disseminated tuberculosis (Fig. 8.7).

Spinal tuberculosis was well decribed by Hippocrates who considered it to be closely related to pulmonary tuberculosis (Major, 1945). It is often termed Pott's disease, after Sir Percival Pott (1713–88), a surgeon at St Bartholomew's Hospital, London. Untreated, it leads to severe deformity and handicap. Any part of the spine may be affected but the disease is most

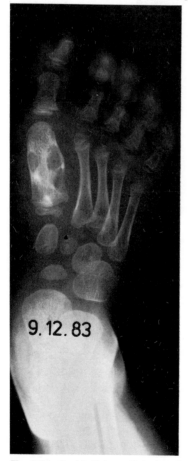 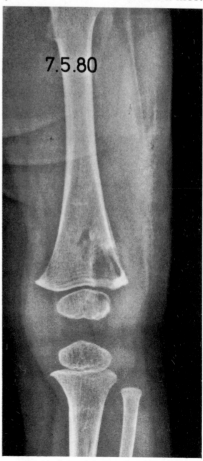

Fig. 8.7 *Left* – Multiple cystic lesions in the first metatarsal of the right foot in a 30-month-old girl with disseminated BCG infection. *Right* – BCG osteitis: osteolytic lesion in the distal metaphysis of the left femur in a 6-month-old boy. Both photographs courtesy Kurt Schopfer

frequent at or near the 10th thoracic vertebra. Erosion of the bone causes vertebral collapse leading to the characteristic angular deformity or 'gibbus'. Abscess formation is common, either around the spinal cord leading to paralysis or in the surrounding soft tissue. Abscesses track along fascial planes and may come to the surface well away from the site of disease. Psoas abscesses resulting from infection of the lumbar vertebrae may reach a point in the thigh below the inguinal ligament. Diagnosis is not easy and is often delayed, particularly when, as often happens, radiological signs are minimal. The presentation may be bizarre: cases have presented as acute abdominal emergencies leading to exploratory laparotomies (Humphries *et al.*, 1986). Specimens are not easily obtained and the number of bacilli present is remarkably low. Histological as well as bacteriological examinations should be performed on all biopsy material.

Chemotherapy has had a particularly beneficial impact on this relatively common form of tuberculosis which previously caused so much pain and immobility. Indeed, there have been more clinical trials of the chemotherapy and surgical management of spinal tuberculosis than of any other form of non-pulmonary tuberculosis, including meningitis. The aim of these studies, conducted by the British Medical Research Council Working Party on Tuberculosis of the Spine (1986), has been to devise the most suitable chemotherapeutic regimens and to compare the results of conservative management, with or without bed rest or a plaster jacket, with simple debridement or more radical excision and anterior spinal fusion (the 'Hong Kong operation'). It has been shown that 6-month or 9-month regimens (see Chapter 10) are highly effective and that, in the absence of complications, bed rest, plaster jackets and surgical debridement are unnecessary. There is some evidence that the radical Hong Kong operation reduces deformity and hastens healing, but this procedure requires surgical skill and resources that are not widely available. Accordingly, non-surgical management is advocated for all except those with complications such as severe kyphosis or paralysis.

Tuberculosis of other bones and joints may mimic a wide range of other conditions, especially the various arthritic conditions and diagnosis is not easy. Chemotherapy is effective and joint mobility should be maintained during treatment unless prevented by pain or muscle spasm. Disorganized joints may require subsequent arthroplasty or other orthopaedic procedures.

Two orthopaedic conditions occasionally occur in patients with tuberculosis but are not a direct result of bacillary invasion. These are Poncet's disease and hypertrophic osteoarthropathy. Poncet's disease is a form of polyarthritis similar to that inducible in certain strains of rats by injection of Freund's complete adjuvant: it may therefore have an immunological basis. The condition resolves on treatment of the underlying disease (Wilkinson and Roy, 1984). Hypertrophic osteoarthropathy is characterized by periosteal inflammation and subperiosteal new bone formation. Although usually associated with lung cancer, it has also occasionally been associated with pulmonary tuberculosis in both man and animals (Webb and Thomas, 1986). The pathogenesis of this condition, and its relationship to Poncet's disease, is unknown.

Osteitis is a rare complication of BCG vaccination in neonates and young children (Schopfer *et al.*, 1982; Lotte *et al.*, 1984) (Fig. 8.7). An unusually high incidence of osteitis following neonatal BCG vaccination occurred in Sweden between 1972 and the cessation of routine BCG immunization in 1975 (Böttiger, 1982). It is thought that the introduction of a more virulent vaccine strain was the cause.

Central nervous system

Three types of tuberculosis are encountered in the central nervous system: solitary space-occupying tuberculomas, disseminated miliary lesions and meningeal involvement. The latter is the commonest and it is the most serious form of tuberculosis: its occurrence constitutes a medical emergency. Tuberculous meningitis occurs in all age groups and at any time after initial infection. It is, fortunately, one of the less common of the extra-pulmonary forms of tuberculosis but about 100 cases occur in Great Britain annually.

The disease results from the rupture of a meningeal or subcortical lesion (Rich's focus) with liberation of tubercle bacilli into the cerebrospinal fluid. This leads to the development of many tubercles on the meninges, causing meningeal inflammation and the secretion of a thick exudate, and on the cerebral vessels with a resulting endarteritis. Organizing exudates at the base of the brain may lead to strangulation of the nerves, especially the optic and auditory nerves, and hydrocephalus due to obstruction to the flow of cerebrospinal fluid. Endarteritis may proceed to thrombosis, causing cerebral infarction leading to convulsions or paralysis.

Clinically, cases are classified into three stages:

1. The patient is fully conscious and rational with non-specific symptoms, no focal signs and little or no evidence of meningitis.
2. The patient is mentally confused and/or has focal neurological signs.
3. The patient is deeply stuporose or comatose and/or has complete hemiplegia, paraplegia or quadriplegia.

Patients may present with more insidious features, including apathy, irritability, personality changes or depression. Occasionally a convulsion is the first indication of disease, particularly in a child. For further details of the clinical aspects of tuberculous meningitis, see Parsons (1979).

Classically the cerebrospinal fluid shows an increase in lymphocyte content and protein level and a decrease in the glucose level, but in many cases these parameters are normal. In other cases there is a high polymorph count, suggestive of a non-acid-fast bacterial infection. The definitive diagnosis is made by the microscopical detection of acid-fast bacilli in centrifuged deposits of the fluid.

Tuberculomas manifest as space-occupying lesions of the brain or spinal cord. Where available, computerized tomography is of great diagnostic value as lesions often have characteristic appearances (Tandon and Bhargava, 1985). Small lesions often resolve with medical treatment alone, but surgery is required for larger lesions, if sight is threatened or, in the case of tuberculomata of the spinal cord, if there is paralysis.

Urinary and male genital tracts

Renal lesions result from haematogenous spread from a primary lesion and usually appear several years after the initial infection. Lesions, which may be unilateral or bilateral, are often found in the renal cortex, possibly owing to the relatively high oxygen tension in this region. The disease progresses towards the medulla and lesions may eventually rupture into the renal pelvis. Tubercle bacilli thus entering the urine may cause secondary foci in the ureters, bladder, epididymis and testis.

Symptoms include frequency, dysuria, renal colic and haematuria. Pyonephrosis, with renal pain and fever, may develop, or the patient may present with a swollen testis. In many cases, symptoms are of a vague 'orthopaedic' nature and diagnosis is often delayed. If undiagnosed and untreated the disease may progress to ureteric obstruction, shrinkage and fibrosis of the bladder and even to renal failure.

Treatment of uncomplicated cases consists of standard modern chemotherapy. Surgery may be required for the removal of destroyed kidneys, relief of ureteric obstruction or augmentation of shrunken bladders. For further details, see Gow and Barbos (1984).

Female genital tract

In contrast to tuberculosis of the male genital organs, which is usually due to spread from the renal tract, lesions in the female genital tract are virtually always the direct result of blood-borne dissemination. Sexually transmitted tuberculosis is extremely rare.

The disease usually commences in the epithelium of the fallopian tubes and spreads to the endometrium. Patients usually present with infertility, pelvic pain and either excessive menstrual bleeding or amenorrhoea. The infection may spread from the fallopian tubes and cause tuberculous peritonitis.

Treatment is by standard modern triple chemotherapy (see page 156), although continuation of therapy for one year has been recommended (Sutherland, 1981). Chemotherapy has greatly reduced the need for surgical intervention, but this may still be required for patients with persistent pelvic pain, pyosalpinx or excessive uterine bleeding. Surgical procedures for infertility are usually unsuccessful but pregnancy may follow *in vitro* fertilization.

For further details of this form of tuberculosis and its management, see Sutherland (1985).

Abdominal tuberculosis

This is divisible into intestinal and peritoneal disease. The former is often due to primary infection and was prevalent in Europe in the days when milk-borne bovine tuberculosis was common. Not all abdominal tuberculosis is of bovine origin, and the disease is still encountered in Great Britain and other regions where milk-borne infection has been virtually eradicated (Palmer *et al.*, 1985). Peritonitis is due to lymphatic or haematogenous dissemination from a primary site elsewhere, or to local spread from an infected viscus, such as a fallopian tube.

Intestinal tuberculosis most frequently occurs in the ileocaecal region and results in mucosal hypertrophy. This, together with enlarged lymph nodes, often presents as a tender mass in the right iliac fossa. Patients may present with complications including malabsorption, intestinal obstruction, fistulae or peritonitis. Massive rectal bleeding is a rare but life-threatening presentation, usually requiring surgery for diagnosis and treatment (Pozniak *et al.*, 1985).

Lesions in the stomach and small intestine are often due to bacilli in sputum being swallowed by patients with post-primary pulmonary tuberculosis. Ulceration rather than hypertrophy is usual, and intestinal perforation leading to peritonitis may occur.

Tuberculous peritonitis occurs principally in two groups of individuals: young women and older alcoholics, usually male. A study in Lesotho (Menzies *et al.*, 1986) showed that tuberculous peritonitis accounted for 42 per cent of all cases of ascites but that the condition was very difficult to diagnose, especially in the elderly alcoholic. The symptoms, signs and laboratory findings were non-specific and the 'doughy abdomen', often cited as being characteristic of this condition, was never encountered. Laparoscopy with peritoneal biopsy may provide a diagnosis, although special facilities are required and the procedure may fail owing to adhesions. Open peritoneal biopsy through a 3–4 cm midline incision under local anaesthetic is recommended as a safe and useful alternative (Falkner *et al.*, 1985).

Even with chemotherapy the prognosis is poor, particularly in elderly alcoholics: the mortality rate in the Lesotho series was 26 per cent.

Skin tuberculosis

This form of tuberculosis has a long and fascinating history and the various lesions have acquired a plethora of quaint, picturesque but outmoded names. Skin lesions may result from primary inoculation of bacilli into the skin, autoinoculation with contaminated sputum (tuberculosis orificialis cutis), underlying scrofula with sinus formation (scrofuloderma, page 129), or from haematogenous dissemination.

Primary cutaneous inoculation tuberculosis is an occupational hazard of medical laboratory workers, and it also occurs in children following wounds to the skin and exposure to infectious adults (Miller, 1982). In general, primary skin tuberculosis remains localized and is a relatively benign disease: Laennec (see page 121) was an unfortunate exception.

Lupus vulgaris is a chronic form of skin tuberculosis that usually occurs on the nose, cheeks or neck. It is characterized by elevated red–brown semi-translucent nodules which may subsequently coalesce and ulcerate. If compressed with a glass slide the nodules have an opalescent 'apple jelly' appearance. In the pre-chemotherapeutic era lupus vulgaris was the cause of severe facial deformities, and secondary skin cancer was not uncommon. Treatment with large doses of vitamin D caused dramatic improvements in many cases (Prosser Thomas, 1950; see page 65).

The tuberculides are rarely encountered and poorly understood skin lesions occurring in patients with tuberculosis. In the most frequently seen form, papulonecrotic tuberculide, the lesions usually manifest as multiple pink papules which then develop black necrotic centres and subsequently heal. It has been suggested (Morrison and Fourie, 1974) that the lesions are due to the trapping of blood-borne antibody-coated bacilli in the skin, with the development of an Arthus (Type III) reaction followed by a delayed hypersensitivity (Type IV) reaction that eliminates the bacilli, as in Robert Koch's original experiments (see page 68). Occasionally the bacilli survive and lupus vulgaris develops. It is likely that erythema induratum (Bazin's disease) and tuberculosis-associated idiopathic gangrene of the extremities (Morrison and Fourie, 1974) are due to similar reactions occurring in larger blood vessels. For a further discussion of tuberculides, see Grange (1982).

Other sites

Any other site or organ may be involved, but those worthy of special mention include the eye, the heart and the pericardium.

The eye may be involved by haematogenous spread from a primary focus elsewhere or by an extension of skin tuberculosis. Choroidal lesions frequently occur in miliary disease and the optic nerve may be affected in tuberculous meningitis. Ocular tuberculosis has been reviewed by Dinning and Marston (1985).

Tuberculous pericarditis may occur as a manifestation of miliary disease or it may follow the rupture of an affected mediastinal lymph node into the pericardial sac. Exudates cause distension of the pericardium, and the resulting pressure on the heart (cardiac tamponade) may cause heart failure. Healing with fibrosis and calcification may result in constrictive pericarditis, requiring surgical relief. Although generally uncommon, there is, for unknown reasons, a high incidence of tuberculous pericarditis in the Transkei (Strang, 1984). Very rarely, the heart itself is involved, leading to heart block which may prove fatal.

Disseminated tuberculosis

Two major types of disseminated tuberculosis are encountered: miliary and cryptic disseminated. The former usually occurs as a manifestation of primary tuberculosis and is characterized by multiple discrete granulomas macroscopically resembling millet seeds (Latin: *milium*, a millet seed). These occur throughout the body and, on chest X-ray, may produce a characteristic 'snow storm' appearance. Occasionally they are seen on the retina by ophthalmoscopy. Lesions also occur in the kidney, and tubercle bacilli are found in the urine of about 25 per cent of patients with miliary disease. The disease may be acute and rapidly progressive, but in other cases it is surprisingly chronic and insidious (Hoyle and Vaizey, 1937). Meningitis or pleural involvement may occur. Acid-fast bacilli are demonstrable in sputum in about half the cases and in liver biopsies in about a quarter of cases. Probably the best diagnostic procedure, if facilities are available, is transbronchial biopsy through a fibreoptic bronchoscope.

Cryptic disseminated tuberculosis occurs in elderly and immunosuppressed individuals. In contrast to miliary tuberculosis, the lesions show very little cellular infiltration but consist of minute necrotic foci teeming with acid-fast bacilli (Proudfoot, 1971). The lesions are often too small to be visible on chest X-ray and the tuberculin test is almost always negative. Patients usually present with non-specific features such as fever, weight loss and anaemia. Biopsy of the lung, liver or bone marrow may be required for the diagnosis. In the absence of therapy the disease is often rapidly fatal.

References

Böttiger, M., Romanus, V., der Verdier, C. and Boman, G. (1982) Osteitis and other complications caused by generalized BCG-itis: experiences in Sweden. *Acta Paediatrica Scandinavica* **71**: 471–8.

Burk, J.R., Viroslav, J. and Bynum, L.J. (1978) Miliary tuberculosis diagnosed by fibreoptic bronchoscopy and transbronchial biopsy. *Tubercle* **59**: 107–10.

Caplin, M. (1980) *The Tuberculin Test in Clinical Practice: An Illustrated Guide.* London: Baillière Tindall.

Dinning, W.J. and Marston, S. (1985) Cutaneous and ocular tuberculosis: a review. *Journal of the Royal Society of Medicine* **78**: 576–81.

Duncanson, F.P., Hewlett, D., Maayan, S., Estapan, H., Perla, E.N., McLean, T. Rodriguez, A., Miller, S.N., Lenox, T. and Wormser, G.P. (1986) *Mycobacterium tuberculosis* infection in the acquired immunodeficiency syndrome: a review of 14 patients. *Tubercle* **67**: 295–302.

Falkner, M.J., Reeve, P.A. and Locket, S. (1985) The diagnosis of tuberculous ascites in a rural African community. *Tubercle* **66**: 55–9.

Gow, J.G. and Barbos, A.S. (1984) Genitourinary tuberculosis: a study of 1117 cases over a period of 34 years. *British Journal of Urology* **56**: 449–55.

Grange, J.M. (1982) Mycobacteria and the skin. *International Journal of Dermatology* **21**: 497–503.

Grange, J.M., Kardjito, T. and Setiobudi, I. (1984) A study of acute-phase reactant proteins in Indonesian patients with pulmonary tuberculosis. *Tubercle* **65**: 23–39.

Grange, J.M., Yates, M.D. and Collins, C.H. (1985) Subdivision of *Mycobacterium tuberculosis* for epidemiological purposes: a seven year study of the 'Classical' and 'Asian' types of the human tubercle bacillus in South-East England. *Journal of Hygiene* **94**: 9–21.

Hoyle, C. and Vaizey, M. (1937) *Chronic Miliary Tuberculosis.* Oxford: Oxford University Press.

Humphries, M.J. and Gabriel, M. (1986) Spinal tuberculosis presenting with abdominal symptoms – a report of two cases. *Tubercle* **67**: 303–307.

Lotte, A., Wasz-Hockert, O., Poisson, N., Dumitrescu, N., Verron, M. and Couvet, E. (1984) BCG complications. *Advances in Tuberculosis Research* **21**: 107–93.

Major, R.H. (1945) *Classic Descriptions of Disease*, 3rd edn. pp. 52–3. Springfield: Charles C. Thomas.

Medical Research Council Working Party on Tuberculosis of the Spine (1986) Tenth Report: A controlled trial of six-month and nine-month regimens of chemotherapy in patients undergoing radical surgery for tuberculosis of the spine. *Tubercle* **67**: 243–59.

Menzies, R.I., Alsen, H., Fitzgerald, J.M. and Mohapeloa, R.G. (1986) Tuberculous peritonitis in Lesotho. *Tubercle* **67**: 47–54.

Miller, F.J.W. (1982) *Tuberculosis in Children*. Edinburgh: Churchill Livingstone.

Morrison, J.G.L. and Fourie, E.D. (1974) The papulonecrotic tuberculide from Arthus reaction to lupus vulgaris. *British Journal of Dermatology* **91**: 273–70.

Ocana, I., Martinez-Vazquez, J.M., Ribera, E., Segura, R.M. and Pascual, C. (1986) Adenosine deaminase activity in the diagnosis of lymphocytic pleural effusions of tuberculous, neoplastic and lymphomatous origin. *Tubercle* **67**: 141–5.

Palmer, K.R., Patil, D.H., Basran, G.S., Riordan, J.F. and Silk, D.B. (1985) Abdominal tuberculosis in urban Britain – a common disease. *Gut* **26**: 1296–305.

Parsons, M. (1979) *Tuberculous Meningitis: A Handbook for Clinicians*. Oxford: Oxford University Press.

Pozniak, A.L., Dalton-Clarke, H.J. and Ralphs, D.N.L. Colonic tuberculosis presenting with massive rectal bleeding. *Tubercle* **66**: 295–300.

Prosser Thomas, E.W. (1950) Tuberculosis and sarcoidosis of the skin. In: *Modern Practice of Dermatology*, pp. 303–24. Edited by G.B. Mitchell Heggs. London: Butterworth.

Proudfoot, A.T. (1971) Cryptic disseminated tuberculosis. *British Journal of Hospital Medicine* **5**: 773–80.

Schopfer, K., Matter, L., Brunner, C., Pagon, S., Stanisic, M. and Baerlocher, K. (1982) BCG osteomyelitis: case report and review. *Helvetica Paediatrica Acta* **37**: 73–81.

Snider, D.E. and Bloch, A.B. (1984) Congenital tuberculosis. *Tubercle* **65**: 81–2.

Strang, J.I. (1984) Tuberculous pericarditis in Transkei. *Clinical Cardiology* **7**: 667–70.

Sutherland, A.M. (1981) The treatment of tuberculosis of the female genital tract with rifampicin, ethambutol and isoniazid. *Archives of Gynaeology* **230**: 315-19.

Sutherland, A.M. (1985) Tuberculosis of the female genital tract. *Tubercle* **66**: 79–83.

Tandon, P.N. and Bhargava, S. (1985) Effect of medical treatment on intracranial tuberculoma – a CT study. *Tubercle* **66**: 85–9.

Villemin, J.A. (1868) *Etudes Experimentales et Cliniqes sur Tuberculose*. Paris: Baillière et Fils.

Wallgren, A. (1948) The time table of tuberculosis. *Tubercle* **29**: 245–51.

Webb, J.G. and Thomas, P. (1986) Hypertrophic osteoarthropathy and pulmonary tuberculosis. *Tubercle* **67**: 225–8.

Wilkinson, A.G. and Roy, S. (1984) Two cases of Poncet's disease. *Tubercle* **65**: 301–3.

Wiseman, R. (1696) A Treatise of the King's Evill. In: *Eight Chirurgical Treatises*, 3rd edn. London: Tooke and Meredith.

9 Other mycobacterial diseases

The other mycobacterial diseases are caused by species that normally exist as harmless environmental saprophytes. Worldwide, the diseases are, relative to tuberculosis and leprosy, uncommon. Nevertheless, unlike tuberculosis, their incidence is not in decline in the West. Indeed, there appears to be an absolute increase in the numbers of reported cases in some regions. Furthermore, as tuberculosis declines, the incidence of the other diseases shows a relative increase so that, in some regions, they have now become a major cause of mycobacterial disease. There is no ideal collective epithet for such disease: in the absence of a better term, the expression 'opportunist' mycobacterial disease will be used here.

These opportunist diseases are divisible into two major groups: those that result from traumatic inoculation of the causative bacilli into the skin or deeper tissues, and those which do not. The former group includes two named diseases, i.e. Buruli ulcer and swimming-pool granuloma, caused by *Mycobacterium ulcerans* and *M. marinum* respectively. Other mycobacterial species, notably the rapid growers *M. chelonei* and *M. fortuitum*, also cause lesions following various forms of trauma. Other infections, which presumably follow inhalation or ingestion of the bacilli, resemble the various pulmonary and non-pulmonary manifestations of tuberculosis. These forms are divisible into four main categories: localized lymphadenitis, localized pulmonary lesions, discrete non-pulmonary lesions and disseminated disease. For a detailed review of the clinical manifestations, see Wolinsky (1979).

Epidemiology

The epidemiology of opportunist mycobacterial disease differs from that of tuberculosis and leprosy in several important respects. The most fundamental difference is that person-to-person transmssion of the former is extremely rare. Disease almost certainly arises as a result of contact with bacilli in the environment. The probable vector is water and infection may result from drinking, washing or indulgence in aquatic sports. Accordingly, the incidence of opportunist mycobacterial disease is independent of that of tuberculosis and leprosy and is unaffected by control measures designed to interrupt the transmission of these diseases. This explains why the incidence of opportunist disease increases relative to tuberculosis in those regions where the latter is in decline.

The species causing opportunist disease in a given region reflect the

Table 9.1 Mycobacteria causing 'opportunist' disease in man

Widespread causes	Geographically restricted causes	Rare causes*
M. avium	M. marinum	M. asiaticum
M. intracellulare	M. simiae	M. gordonae
M. scrofulaceum	M. ulcerans	M. szulgai
M. kansasii	M. xenopi	M. flavescens
M. chelonei		M. thermoresistibile
M. fortuitum		M. haemophilum
		M. malmoense
		M. nonchromogenicum
		M. shimoidei
		M. terrae
		M. triviale

*For references to case reports, see Collins *et al.* (1986)

distribution of the various species in the environment. Worldwide, members of the *M. avium–intracellulare* group are the most frequent cause of such disease. In Great Britain the most frequent species thus encountered is *M. kansasii*, but in South London *M. xenopi* is the most prevalent species. *Mycobacterium simiae* is seen in Israel and Australia but is rare elsewhere, and infection due to *Mycobacterium ulcerans* (Buruli ulcer) is restricted to a number of localized regions in Australia and some tropical countries. The frequent and rare causes of opportunist disease are listed in Table 9.1. See also Grange and Yates (1986) and Collins *et al.* (1986).

Predisposing factors

It is well known that only a minority of individuals infected by one of the major mycobacterial pathogens, *M. tuberculosis* and *M. leprae*, develop overt and progressive disease. In the case of the other mycobacteria the ratio of disease to infection is even lower (Grange, 1987).

Being ubiquitous, contact between environmental mycobacteria and man is virtually unavoidable yet overt disease is uncommon. It is, of course, possible that the frequently observed skin test reactivity to environmental mycobacteria is the result of small, 'silent', self-limiting lesions. In many, but certainly not all, patients with overt disease there is an obvious predisposing factor. These factors include traumatic breaches of the skin, pre-existing pulmonary disease or damage and generalized immunosuppressive disorders.

Predisposing pulmonary conditions include industrial dust disease (pneumoconiosis and silicosis), chronic obstructive airway disease, emphysema, cystic fibrosis, tumours and old, incompletely closed tuberculous cavities. Systemic causes include lymphoproliferative disorders (particularly Hodgkin's disease and hairy cell leukaemia), disseminated solid tumours, autoimmune disease, renal failure, immunosuppressive drug regimens, congenital immunodeficiencies and the acquired immune deficiency syndrome (AIDS). Indeed, mycobacterial disease may be the presenting feature of the predisposing disorder. Accordingly, patients with opportunist mycobacterial disease must be thoroughly investigated for any underlying condition.

Diagnosis

Diagnosis is almost always based on the isolation of the causative organism, thereby permitting identification of the isolate and its differentiation from *M. tuberculosis*. In the case of *Mycobacterium ulcerans* infection (Buruli ulcer), the clinical characteristics are usually pathognomonic.

Being an obligate pathogen, a single isolate of *M. tuberculosis* from sputum or urine is of diagnostic significance (provided that there has been no cross-contamination in the laboratory). By contrast, environmental mycobacteria may occur as pathogens, as secondary saprophytes of diseased tissues, as transient commensals or as contaminants of specimen containers. Several 'epidemics' of opportunist mycobacterial disease in hospitals have been due to the practice of washing sputum pots or urinals with contaminated tap water (Collins *et al.*, 1984). Single isolates of a light growth from sputum or urine are usually indicative of contamination or transient colonization, while repeated heavy growths are obtained in the case of disease. As a general rule, pulmonary disease should be considered if a heavy growth of the same strain is obtained on a minimum of two occasions, at least one week apart, from a patient with compatible clinical and radiological features and in whom other causes, including tuberculosis, have been carefully excluded.

Isolates of mycobacteria from biopsies are likely to be of significance, especially in the presence of a compatible histological appearance. In disseminated infections, bacilli may be cultured from many sites, including bone marrow. In AIDS patients, the bacilli may be cultured from blood (Macher *et al.*, 1983) and faeces (Stacey, 1986): indeed, this disease is probably the only one in which there is an indication for examining these materials for mycobacteria.

Attempts to develop immunological tests – skin testing, *in vitro* assays of cell-mediated immune responses and serology – have been limited and have not proved helpful.

M. ulcerans infection (Buruli ulcer)

This, the most bizarre of the mycobacterial diseases, was first described in the Bairnsdale district of Australia and the causative organism was named *M. ulcerans* (MacCallum *et al.*, 1948). Later, a similar disease was seen in the Buruli county of Uganda (hence the name Buruli ulcer), other parts of Africa, Mexico, Malaysia and New Guinea. Cases occur in 'hot spots', which tend to be low-lying marshy areas subject to periodic flooding. Epidemiological evidence suggests that *M. ulcerans* lives in the environ-

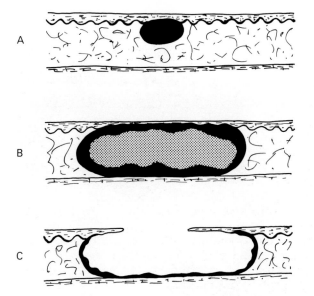

Fig. 9.1 *M. ulcerans* infection: evolution of the disease. A: Firm nodule attached to skin but not to deep tissue. B: Lesion extending to deep fascia and undergoing central necrosis. C: necrosis of overlying skin and discharge of liquefied necrotic tissue, leading to deeply undermined ulcer

ment (although it has never been isolated from inanimate sources) and is inoculated into the skin by spiky vegetation (Barker, 1973). In Uganda there is a close association between the distribution of the disease and the occurrence of a tall thorny grass, *Echinochloa pyrimidalis*.

The evolution of the disease is shown in Fig. 9.1. The earliest sign is a discrete firm nodule fixed to the skin but mobile over deep tissues. Although painless, it may itch and in parts of Zaire it is known as *mputa matadi* – the itching stone. In some individuals the lesion resolves, but in others it progresses to the ulcerative stage. At this stage the lesions contain many bacilli and there is little or no histological evidence of an immune reaction. Patients are negative on skin testing with Burulin, a reagent prepared from *M. ulcerans* (Stanford *et al.*, 1975). As the process spreads through the subdermal adipose tissue, the overlying skin becomes anoxic and necrotic. Ulceration then occurs with the escape of liquefied necrotic tissue and the formation of a deeply undermined ulcer (Fig. 9.2). The tissue necrosis may, at least in part, be due to a bacterial toxin (Hockmeyer *et al.*, 1978).

Lesions may occur on any part of the body but they are more common on exposed parts, especially the limbs. Lesions are single or, less frequently, multiple. In some cases lesions occur in the ends of the long bones and sinuses may develop. The disease is progressive for up to three years, but a stage is virtually always reached when, for unknown reasons, an effective immune response develops. There is a granulomatous

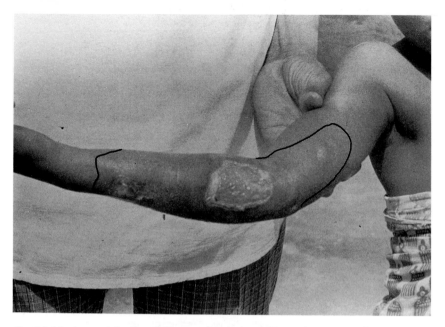

Fig. 9.2 *M. ulcerans* infection of the arm showing one ulcer and two areas of pre-ulcerative skin changes. The black line indicates the extent of the underlying disease

mononuclear cells, the bacilli diminish in number and then disappear, and the patient gives a positive dermal reaction to Burulin. Eventually the lesion heals (Fig. 9.3), but often with extensive fibrosis and contractures leading to crippling deformities (Fig. 9.4). Occasionally progression and healing may occur in different parts of the same lesion.

The treatment of Buruli ulcer differs according to the stage of the disease. In general, chemotherapy has not proved very successful, although the antileprosy drug clofazimine appears to have some effect (Lunn and Rees, 1964). Rifampicin is effective *in vitro* and in experimental infections in the mouse, but results in man have been less promising. In most cases, surgical excision is the treatment of choice. Simple excision of small early lesions under local anaesthetic is usually curative, and individuals living in endemic areas should be encouraged to present for such treatment at the early stage. Larger lesions require radical surgery to clear all the disease. This may involve extensive excision of skin and subdermal tissues and subsequent skin grafting. Resolving lesions require less radical surgery, but skin grafting may be necessary to prevent contractions and disfiguring scars. In all cases involving limbs, physiotherapy and splinting is required to prevent deformities.

M. marinum infection

This skin disease is known as swimming-pool granuloma, fish-tank granuloma and fish-fancier's finger. These names indicate that the disease

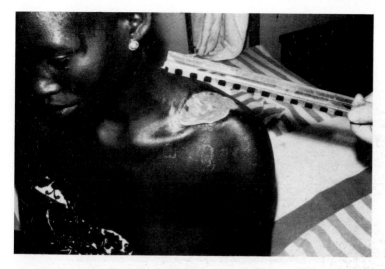

Fig. 9.3 *M. ulcerans* infection: a lesion in the healing stage showing an ulcer with a granulating base and no extension of the disease

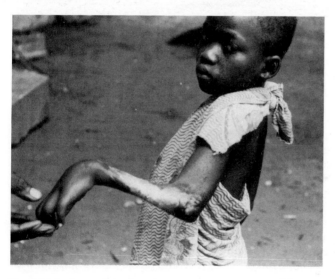

Fig. 9.4 *M. ulcerans* infection: the result of natural healing. The hand is oedematous and functionless and there are flexion contractures at the elbow and wrist

is usually acquired during leisure activities involving water. The disease and the causative organism were first described by Linell and Norden (1954) during the investigation of an outbreak of lesions amongst users of a swimming pool at Orebro in Sweden. The organism, which was isolated from the pool as well as from lesions, was termed *M. balnei* but it was later found to be identical to the fish tubercle bacillus, *M. marinum*, described by Aronson (1926). Several other epidemics associated with swimming

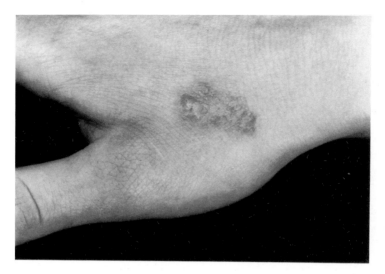

Fig. 9.5 *M. marinum* infection on the hand of a keeper of tropical fish. Courtesy Prof. W.C. Noble

pools have been reported, as well as isolated cases acquired while repairing boats, fishing and swimming in estuaries or the sea. A dolphin trainer developed a lesion after being bitten on the hand by one of his trainees. Other cases have principally occurred amongst keepers of tropical fish.

Lesions occur at sites of injury, which are usually the knees and elbows of swimming-pool users and the hands of fish fanciers (Fig. 9.5). The lesion commences as a solitary raised warty lesion, but secondary lesions sometimes develop along the draining lymphatics, as in sporotrichosis with which it may be confused. Occasionally deeper infections, such as tenosynovitis, have followed more penetrating injuries and a disseminated infection occurred in a renal transplant recipient. Multiple lesions developed in an infant whose bath was used by his father to clean his fish tank in. For a full review of these and other published cases, see Collins *et al.* (1985).

In the absence of treatment many *M. marinum* infections will eventually resolve, but surgical excision or antimicrobial therapy is usually resorted to. Effective drugs include minocycline, cotrimoxazole and rifampicin with ethambutol (see page 165).

Other inoculation-associated infections

These infections take three main forms: superficial warty lesions similar to those of *M. marinum* infection, localized post-injection abscesses, and joint infections and deeper lesions following penetrating injuries or surgical procedures.

Superficial warty lesions, sometimes with sporotrichoid spread, have occasionally been caused by *M. kansasii*, *M. szulgai*, *M. chelonei* and unidentified slow-growing scotochromogens (for references see Collins *et al.* (1985)).

Post-injection mycobacterial abscesses are usually caused by the rapid growers *M. chelonei* and *M. fortuitum*. These abscesses may occur sporadically, especially when dirty needles are used, and there have been several cases among diabetics. Small outbreaks have also occurred when several individuals have received injections from a batch of contaminated material. Examples include the use of a histamine solution (Inman *et al.*, 1969) and a batch of vaccine (Borghans and Stanford, 1973). These abscesses are usually localized and resolve after surgical drainage with curettage. In other cases, particularly in diabetics, the disease may spread locally and require antimicrobial therapy (Jackson *et al.*, 1981). Mycobacterial arthritis has resulted from the injection of contaminated steroid solutions into joint cavities (Lau, 1986).

Corneal lesions due to *M. fortuitum* and *M. chelonei* have followed abrasions or penetrating injuries (Gangadharam *et al.*, 1978).

Deeper lesions of the hand and foot have occurred in individuals occupationally exposed to injuries contaminated by soil, e.g. gardeners. A number of different species have been involved, including *M. terrae* which, as the name suggests, is associated with soil. There have been a few outbreaks of sternal infections due to *M. chelonei* or *M. fortuitum* following open heart surgery. These often required extensive debridement or even excision of the sternum. In some cases it seemed likely that the bone wax was contaminated, and in one instance the 'epidemic' ceased when the use of non-sterile ice to cool the cardioplegia solution was abandoned (Kuritsky *et al.*, 1983). Endocarditis, usually caused by *M. avium* or *M. chelonei*, has followed the insertion of contaminated pig heart valve xenografts and the need to sterilize and screen such materials has been emphasized (Tyras *et al.*, 1978). Infections due to *M. chelonei* or *M. fortuitum* have also occurred after the insertion of artificial hip joints and silicone implants for breast augmentation.

Disease not associated with traumatic inoculation

These diseases are divisible into four main groups: localized lymphadenitis, localized pulmonary lesions, localized non-pulmonary lesions, and disseminated diseases.

Localized lymphadenitis

The most frequent manifestation is a unilateral involvement of a cervical node in a child of less than five years of age. Very occasionally other nodes and age groups are affected. The infection is usually self-limiting, although in practice most nodes are excised for diagnostic purposes. The node probably represents the lymphatic component of a tonsillar lesion resulting from the ingestion of the causative agent. ('Silent' tonsillar infections may occur: several workers have isolated mycobacteria from tonsillectomy specimens.)

As with other opportunist mycobacterial disease, the causative agent varies from region to region. In the USA, a frequent cause is *M. scrofulaceum* (the 'scrofula scotochromogen'). A study in Great Britain showed that 56 of 67 cases were due to *M. avium–intracellulare*, five were due to *M. chelonei*, four to *M. scrofulaceum* and one each to *M. kansasii* and *M. xenopi* (Grange *et al.*, 1982).

Pulmonary disease

The lung is the organ most frequently involved in opportunist mycobacterial disease. It is likely that infection results from the inhalation of the bacilli in aerosols (see page 18). Disease usually occurs in individuals with predisposing lung damage or immune deficiencies. Nevertheless, a substantial proportion of cases, particularly those due to *M. avium–intracellulare* and *M. kansasii*, occur in those with no apparent predisposition.

The clinical features of such pulmonary disease are deceptively similar to those of tuberculosis. Radiology is of little specific diagnostic use, although one characteristic radiological feature is said to occur in some cases of opportunist disease (Cook *et al.*, 1971). This is described as a group of homogeneous shadows, one centimetre in diameter, round a translucent zone and with line shadows radiating from each lesion.

In view of the lack of pathognomonic clinical and radiological features, the diagnosis is made by cultivation and identification of the causative organism. As explained above, mere isolation is insufficient to establish a diagnosis as the strain may be a secondary saprophyte of diseased tissue or a contaminant. Occasionally diagnosis is established by examination of tissue obtained by bronchoscopic or open lung biopsy. Brushings or washings obtained through the fibreoptic bronchoscope are very useful specimens for diagnostic purposes for both tuberculosis and opportunist disease (Funahashi *et al.*, 1983). Diagnostic problems arise when there are repeated scanty isolates which may represent saprophytic colonization. Such colonization may, in some cases, be cleared by a course of 'bronchial hygiene', i.e. the inhalation of a nebulized bronchodilator and then a warm saline aerosol (Ahn *et al.*, 1982).

Non-pulmonary lesions

Solitary lesions occasionally occur in the genitourinary tract, bones and joints and, with extreme rarity, elsewhere. Although some have been associated with surgical or accidental trauma, others are presumably due to haematogenous dissemination from a primary focus elsewhere.

The diagnosis of genitourinary disease is complicated by the frequent colonization of the lower urethra by mycobacteria, usually rapidly growing scotochromogens. (Contrary to a widely publicized belief, *M. smegmatis* is

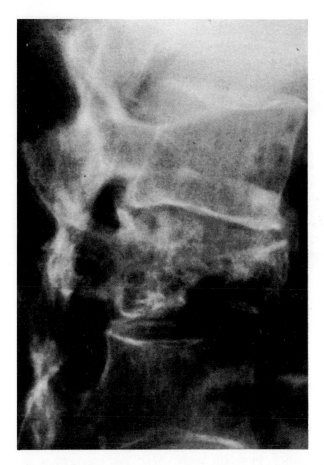

Fig. 9.6 Spinal collapse due to *M. xenopi* infection. Courtesy A.J. Prosser

very rarely found.) When possible, biopsy material should be examined. There are few descriptions of disease in this site: fourteen cases have been reviewed by Thomas *et al.* (1980).

Bone and joint disease, in the absence of a penetrating injury, is also rare. Prosser (1986) has described a case of spinal disease due to *M. xenopi* (Fig. 9.6) and has briefly reviewed six other cases, four due to *M. fortuitum*, one to *M. kansasii* and one to an unidentified strain. Some cases of bone or joint disease have occurred in patients receiving oral steroid therapy for systemic lupus erythematosus or rheumatoid arthritis (Zventina *et al.*, 1982). As the features of such infections resemble the underlying disorder, diagnosis may be delayed until considerable damage has occurred.

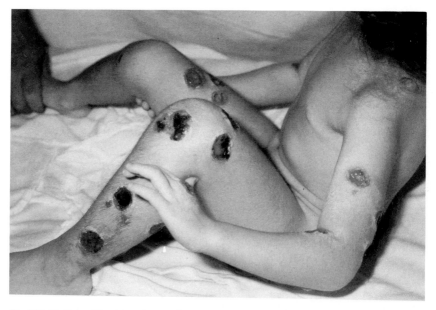

Fig. 9.7 Multiple skin ulcers due to disseminated *M. chelonei* infection in a 4-year-old girl. Courtesy Kurt Schopfer.

Disseminated disease

Two forms of disseminated disease are seen: multiple discrete lesions (Fig. 9.7) and a widespread infiltration of many organs. Such disease is particularly, but not exclusively, associated with congenital or acquired causes of immune deficiency. In recent years, AIDS has emerged as an increasingly important predisposing cause of such infections.

Most cases are due to the more pathogenic of the opportunists (namely, *M. avium–intracellulare*), but patients with more profound immunosuppression may be infected with any of the species listed in Table 9.1. The features of disseminated disease may be superimposed on the patient's underlying disorder and the presenting signs and symptoms may vary considerably. These include multiple skin lesions, generalized lymphadenopathy, hepatomegaly, fever and wasting. Haematological abnormalities are common, but it is difficult to ascertain whether they are cause or effect. The bone marrow is infiltrated by acid-fast bacilli in a high proportion of patients, but smears or biopsies may look deceptively normal unless stained by the Ziehl–Neelsen method (Cohen *et al.*, 1983).

The number of cases of disseminated disease has increased in recent years as a result of the more widespread use of immunosuppressive drugs in transplant surgery and the advent of AIDS. Both tuberculosis and opportunist mycobacterial disease are well recognized hazards of renal transplant surgery (Lloveras *et al.*, 1982). In some cases, reduction or cessation of the immunosuppressive drugs may lead to spontaneous resolution of an opportunist infection. In other cases, removal of the homograft and intensive antimicrobial therapy may be required. In one such case the causative organism, *M. chelonei*, was isolated from deionizer resin in the renal dialysis machine (Azadian *et al.*, 1981).

Mycobacterium haemophilum is a rare cause of disease and virtually all those affected have been immunosuppressed, either as a result of renal transplant or lymphoma. Such patients develop widespread granulomatous skin nodules which tend to develop into abscesses (Moulsdale *et al.*, 1983). An exception is a case of submandibular lymphadenitis in an otherwise healthy infant (Dawson *et al.*, 1981).

Mycobacteria are a major cause of secondary infection in AIDS victims. The situation is often complicated by the simultaneous occurrence of infections due to other pathogens, especially *Cryptosporidium* spp. and *Pneumocystis carinii*. The great majority of infections are due to *M. avium–intracellulare* (Kiehn *et al.*, 1985; Iseman *et al.*, 1985). Indeed, it has been suggested that the nature of the immune defect favours infection by this organism (Masur, 1982). On the other hand, the association may be explained by the geographical distribution of the bacilli and the patients and by the relatively high virulence of the *M. avium–intracellulare* group compared with other species. As a general rule, the diagnosis of opportunist mycobacterial disease is made after that of AIDS while tuberculosis tends to appear before the underlying immune defect is suspected (Duncanson *et al.*, 1986; Pinching, 1987).

Opportunist mycobacterial disease complicating AIDS is usually extensive and disseminated and acid-fast bacilli are present in the blood and faeces. Histologically, the lesions resemble those of lepromatous leprosy: there are many large, foamy macrophages containing clumps of acid-fast bacilli but no evidence of an immune response (Sohn *et al.*, 1983). It is, unfortunately, likely that there will be a substantial increase in the incidence of this type of disease in the future.

Other diseases of possible mycobacterial aetiology

It is possible that there are still undescribed mycobacterial species, and diseases of man and animals. There was, for example, a cluster of non-imunosuppressed patients living near the border of Canada and the USA who developed skin ulcers containing non-culturable acid-fast bacilli (Feldman and Hershfield, 1974). Likewise, a nodular skin disease termed *lepra bubalorum* occurring in water buffalo, and occasionally in bovines, in Indonesia has been described (Ressang, 1964). The skin nodules were teeming with acid-fast bacilli but many attempts to grow the organism *in vitro* and to transmit the disease to experimental animals failed. There have been no reports of this disease in recent years and it may well be extinct.

Despite many attempts, the aetiological agents of certain granulomatous diseases, notably sarcoidosis and Crohn's disease, have never been isolated. Several workers have postulated that mycobacteria, possibly lysogenic or cell wall-free forms, are the responsible agents (Burnham *et al.*, 1978; Mankiewicz and Beland, 1964; Mitchell *et al.*, 1976). At present there is no firm evidence for or against this concept. The development of specific DNA probes or monoclonal antibodies may soon enable this subject to be seriously reinvestigated.

References

Ahn, C.H., McLarty, J.W., Ahn, S.S., Ahn, S.I. and Hurst, G.A. (1982) Diagnostic criteria for pulmonary disease caused by *Mycobacterium kansasii* and *Mycobacterium intracellulare*. *American Review of Respiratory Disease* **125**: 388–91.

Aronson, J.D. (1926) Spontaneous tuberculosis in salt water fish. *Journal of Infectious Diseases* **39**: 315–20.

Azadian, B.S., Beck, A., Curtis, J.R., Cherrington, L.E., Gower, P.E., Phillips, M., Eastwood, J.B. and Nicholls, J. (1981) Disseminated infection with *Mycobacterium chelonei* in a haeodialysis patient. *Tubercle* **62**: 281–4.

Barker, D.J.P. (1973) Epidemiology of *Mycobacterium ulcerans* infection. *Transactions of the Royal Society of Tropical Medicine and Hygiene* **67**: 43–50.

Borghans, J.G. and Stanford, J.L. (1973) *Mycobacterium chelonei* in abscesses after injection of diptheria–pertussis–tetanus–polio vaccine. *American Review of Respiratory Disease* **107**: 1–8.

Burnham, W.R., Lennard-Jones, J.F., Stanford, J.L. and Bird, R.G. (1978) Mycobacteria as a possible cause of inflammatory bowel disease. *Lancet* **ii**: 693–6.

Cohen, R.J., Samoszuk, M.K., Busch, D. and Lagios, M. (1983) Occult infections with *M. intracellulare* in bone marrow biopsy specimens from patients with AIDS. *New England Journal of Medicine* **308**: 1475–6.

Collins, C.H., Grange, J.M., Noble, W.C. and Yates, M.D. (1985) *Mycobacterium marinum* infections in man. *Journal of Hygiene* **94**: 135–49.

Collins, C.H., Grange, J.M. and Yates, M.D. (1984) Mycobacteria in water. *Journal of Applied Bacteriology* **57**: 193–211.

Collins, C.H., Grange, J.M. and Yates, M.D. (1986) Unusual opportunist mycobacteria. *Medical Laboratory Sciences* **43**: 262–8.

Cook, P.L., Riddell, R.W. and Simon, G. (1971) Bacteriological and radiographic features of lung infection by opportunist mycobacteria. *Tubercle* **52**: 232–41.

Dawson, D.J., Blacklock, Z.M. and Kane, D.W. (1981) *Mycobacterium haemophilum* causing lymphadenitis in an otherwise healthy child. *Medical Journal of Australia* **2**: 289–90.

Duncanson, F.P., Hewlett, D., Maayan, S., Estepan, H., Perla, E.N., McLean, T., Rodriguez, A., Miller, S.N., Lenox, T. and Wormser, G.P. (1986) *Mycobacterium tuberculosis* infection in the acquired immunodeficiency syndrome: a review of fourteen patients. *Tubercle* **67**: 295–302.

Feldman, R.A. and Hershfield, E. (1974) Mycobacterial skin infections by an unidentified species: a report of 29 patients. *Annals of Internal Medicine* **80**: 445–52.

Funahashi, A., Lohaus, G.H., Poliotis, J. and Hranicka, L.J. (1983) Role of fibreoptic broncoscopy in the diagnosis of mycobacterial disease. *Thorax* **38**: 267–70.

Gangadharam, P.R.J., Lanier, J.D. and Jones, D.E. (1978) Keratitis due to *Mycobacterium chelonei*. *Tubercle* **59**: 55–60.

Grange, J.M. (1987) Infection and disease due to the environmental mycobacteria.

Transactions of the Royal Society of Tropical Medicine and Hygiene **81**: 179–82.

Grange, J.M., Collins, C.H. and Yates, M.D. (1982) Bacteriological survey of tuberculous lymphadenitis in South-East England: 1973–80. *Journal of Epidemiology and Community Health* **36**: 157–61.

Grange, J.M. and Yates, M.D. (1986) Infections caused by opportunist mycobacteria: a review. *Journal of the Royal Society of Medicine* **79**: 226–9; and *International Journal of Leprosy* **54**: 626–31.

Hockmeyer, W.T., Krieg, R.E., Reich, M. and Johnson, R.D. (1978) Further characterization of *Mycobacterium ulcerans* toxin. *Infection and Immunity* **21**: 124–8.

Inman, P.M., Beck, A., Brown, A.E. and Stanford, J.L. (1969) Outbreaks of injection abscesses due to *Mycobacterium abscessus*. *Archives of Dermatology* **100**: 141–7.

Iseman, M.D., Corpe, R.F., O'Brien, R.J., Rosenzweig, D.Y. and Wolinsky, E. (1985) Disease due to *Mycobacterium avium–intracellulare*. *Chest* **87** (Suppl. 2): 139s–149s.

Jackson, P.G., Keen, H., Noble, C.J. and Simmons, N.A. (1981) Injection abscesses due to *Mycobacterium chelonei* occurring in a diabetic patient. *Tubercle* **62**: 277–9.

Kiehn, T.E., Edwards, F.F., Brannan, P., Tsang, A.Y., Maio, M., Gold, J.W., Whimbey, E., Wong, B., McClatchy, J.K. and Armstrong, Y. (1985) Infections caused by *Mycobacterium avium* complex in immunocompromised patients: diagnosis, blood culture and faecal examination, antimicrobial susceptibility tests, morphological and sero-agglutination characteristics. *Journal of Clinical Microbiology* **21**: 168–73.

Kuritsky, J.M., Bullen, M.G., Broome, C.V., Silcox, V.A., Good, R.C. and Wallace, R.J. (1983) Sternal wound infection due to organism of the *Mycobacterium fortuitum* complex. *Annals of Internal Medicine* **98**: 938–9.

Lau, J.H.K. (1986) Hand infection with *Mycobacterium chelonei*. *British Medical Journal* **292**: 444–5.

Linell, F. and Norden, A. (1954) *Mycobacterium balnei*: a new acid-fast bacillus occurring in swimming pools and capable of producing skin lesions in humans. *Acta Tuberculosea Scandinavica* (Suppl. 33): 1–84.

Lloveras, J., Peterson, P.K., Simmons, R.L., and Najarian, J.S. (1982) Mycobacterial infections in renal transplant recipients: seven cases and a review of the literature. *Archives of Internal Medicine* **142**: 883–92.

Lunn, H.F. and Rees, R.J.W. (1964) Treatment of mycobacterial skin ulcers with a riminophenazine derivative (B663). *Lancet* **i**: 247.

MacCallum, P., Tolhurst, J.C., Buckle, C. and Simmons, H.A. (1948) A new mycobacterial infection in man. *Journal of Pathology and Bacteriology* **60**: 93–122.

Macher, A.M., Kovacs, J.A., Gill, V., Roberts, G.D., Ames, J., Park, C.H., Straus, S., Lane, H.C., Parvillo, J.E., Fauci, A.S. and Masur, H. (1983) Bacteraemia due to *Mycobacterium avium–intracellulare* in the acquired immunodeficiency syndrome. *Annals of Internal Medicine* **99**: 782–5.

Mankiewicz, E. and Beland, J. (1964) The role of mycobacteriophages and of cortisone in experimental tuberculosis and sarcoidosis. *American Review of Respiratory Disease* **89**: 707–20.

Masur, H. (1982) *Mycobacterium avium–intracellulare*: another scourge for individuals with the acquired immunodeficiency syndrome. *Journal of the American Medical Association* **248**: 3013.

Mitchell, D.N., Rees, R.J.W. and Goswami, K.K.A. (1976) Transmissible agents from human sarcoid and Crohn's disease tissue. *Lancet* **ii**: 761–5.

Moulsdale, M.T., Harper, J.M., Thatcher, G.N. and Dunn, B.L. (1983) Infection by *Mycobacterium haemophilum*, a metabolically fastidious acid-fast bacillus. *Tubercle* **64**: 29–36.

Pinching, A.J. (1987) The acquired immune deficiency syndrome: with special reference to tuberculosis. *Tubercle* **68**: 65–9.

Prosser, A. (1986) Spinal infection with *Mycobacterium xenopi*. *Tubercle* **67**: 229–32.

Ressang, A.A. (1964) Lepra bubalorum et bovinum. *Proceedings of the Royal Society of Medicine* **57**: 483.

Sohn, C.C., Schroff, R.W., Kliewer, K.E., Lebel, D.M. and Fligiels, S. (1983) Disseminated *Mycobacterium avium–intracellulare* infection in homosexual men with acquired cell-mediated immunodeficiency: a histologic and immunologic study of two cases. *American Journal of Clinial Pathology* **79**: 247–52.

Stacey, A.R. (1986) Isolation of *Mycobacterium avium–intracellulare* complex from faeces of patients with AIDS. *British Medical Journal* **293**: 1194.

Stanford, J.L., Revill, W.D.L., Gunthorpe, W.J. and Grange, J.M. (1975) The production and preliminary investigation of Burulin, a new skin test reagent for *Mycobacterium ulcerans* infection. *Journal of Hygiene* **74**: 7–16.

Thomas, E., Hillman, B.J. and Stanisic, T. (1980) Urinary tract infection with atypical mycobacteria. *Journal of Urology* **124**: 748–50.

Tyras, D.H., Kaiser, G.C., Barner, H.B., Laskowski, L.F. and Marr, J.J. (1978) Atypical mycobacteria and the xenograft valve. *Journal of Thoracic and Cardiovascular Surgery* **75**: 331–7.

Wolinsky, E. (1979) State of the art: non-tuberculous mycobacteria and associated disease. *American Review of Respiratory Disease* **119**: 107–59.

Zvetina, J.R., Demos, T.C. and Rubenstein, H. (1982) *Mycobacterium intracellulare* infection of the shoulder and spine in a patient with steroid-treated systemic lupus erythematosus. *Skeletal Radiology* **8**: 111–13.

10 Therapy of mycobacterial disease

Tuberculosis

From time immemorial, the search for a cure for tuberculosis has been a major preoccupation of the medical profession. Indeed, it is salutory to recall that effective drugs have been available for only a few decades. Prior to the epoch-making discovery of streptomycin by Waksman in 1948, huge numbers of remedies were advocated: many were bizarre, most were unpleasant and all were useless. Although Sir William Buchan's advice that the patient should suck woman's milk directly from the breast contrasts sharply in acceptability with John of Gaddeston's prescription of pigeon's dung and weasel's blood, it was probably equally ineffective. Perhaps the only remedy of some value was cod liver oil, first used by Percival in 1770. This, owing to its high vitamin D content, may have exerted a beneficial effect in skin tuberculosis (see page 65).

Nowadays, the most important factor in the treatment of tuberculosis is adequate chemotherapy. This far outweighs any dietary or psychological supportive measures. Modern chemotherapeutic regimens are curative in the great majority of cases provided that treatment is completed under full supervision. For this reason, much research has gone into establishing the shortest duration of therapy and smallest number of doses that are compatible with a high cure rate.

Sensitivity and resistance to antimicrobial agents

In the absence of mutational resistance, tubercle bacilli are remarkably uniform in their sensitivity to the antituberculous drugs, although bovine strains are naturally resistant to pyrazinamide. Unfortunately, mutational change to resistance occurs at a low and constant rate (Table 10.1), which

Table 10.1 Spontaneous mutation rate of *Mycobacterium tuberculosis* to rifampicin, isoniazid, ethambutol and streptomycin

Drug	Mutation rate
Rifampicin	2.3×10^{-10}
Isoniazid	2.6×10^{-8}
Ethambutol	1.0×10^{-7}
Streptomycin	3.0×10^{-9}

varies from drug to drug (David, 1970). Consequently, large populations of mycobacteria will inevitably contain a few bacilli that are resistant to a given drug. The purpose of drug sensitivity tests is not to detect these few mutants but to determine whether the great majority of the bacilli are sensitive. In the absence of the drug, the numbers of resistant mutants in a bacillary population remains low but in the presence of the drug the sensitive bacilli are killed and the resistant ones become dominant.

Principles of chemotherapy

The aim of therapy is to destroy all viable bacilli, rather than merely to reduce them to a very low level. This is not as straightforward as it might appear: drugs that are fully bactericidal *in vitro* may not achieve this effect *in vivo*. Mitchison (1985) has stressed the important difference between drugs that are bacteriostatic, those that are bactericidal under permissive conditions, and those that are capable of sterilizing lesions. Two further important properties of drugs are their ability to prevent the emergence of resistance to a second drug, and their efficacy at destroying a large part of the bacterial population rapidly, thereby greatly reducing the infectivity of the patient (Table 10.2).

The mycobacterial population in a patient is functionally divisible into four groups:

A   freely dividing extracellular bacilli
B   slowly dividing intracellular bacilli
C   intracellular bacilli showing only occasional short bursts of metabolic activity
D   dormant bacilli, within cells and in firm caseous material

The most powerful sterilizing drug is rifampicin, being active against groups A, B and C. Isoniazid is the most effective agent for destroying the freely multiplying extracellular bacilli that are found in large numbers in the walls of cavities, but it is not a good sterilizing drug as it has limited activity against the intracellular bacilli which become relatively more frequent as treatment progresses. By its bacteriocidal effect on replicating

Table 10.2 The ability of antituberculous drugs to sterilize lesions, reduce viable bacterial population rapidly and prevent the emergence of drug resistance

Drug	Sterilizing action	Early bactericidal	Prevention of drug resistance
Rifampicin	Good	Fair	Good
Pyrazinamide	Good	Poor	Poor
Isoniazid	Fair	Good	Good
Ethambutol	Poor	Fair	Fair
Streptomycin	Poor	Poor	Fair
Thiacetazone	Poor	Poor	Poor

bacilli, isoniazid is very good at preventing the emergence of drug resistance.

Streptomycin is not a sterilizing drug as, although it is bacteriocidal in alkaline areas of the cavity walls, it is totally ineffective at the low pH found within cells. By contrast, pyrazinamide is effective at a low pH, such as is found within macrophages and probably in anoxic areas of inflammatory lesions, but it is ineffective against organisms at a neutral or high pH. Thus neither streptomycin nor pyrazinamide are good at preventing emergence of drug resistance as they are able to destroy only part of the bacterial population.

Chemotherapy is divisible into three main phases. First, most of the freely multiplying extracellular organisms (group A) are destroyed. It is in this phase, which lasts a few days, that isoniazid has a major role. In the second phase, lasting perhaps a month or two, the remaining bacilli, mostly within macrophages, are killed by rifampicin and pyrazinamide. Finally, only dormant or near-dormant bacilli remain and are (it is hoped) eventually killed by rifampicin. Despite the fact that isoniazid is usually only bacteriocidal during the early phase of therapy, it is used throughout the period of treatment as it is very good at preventing the emergence of drug resistance.

Ethambutol has some bacteriocidal action in the early stage of therapy but it is not a sterilizing drug. It is as effective as streptomycin in preventing the emergence of resistance and has replaced the latter in many regimens as it is given orally. Other antituberculous drugs – ethionamide, prothionamide, thiacetazone, viomycin, cycloserine and para-amino salicylic acid (PAS) – are of much lower efficacy although some, notably thiacetazone, are still used as first-line drugs in certain countries.

The 4-quinolones (gyrase inhibitors such as ciprofloxacin and ofloxacin) are active against *M. tuberculosis* and some other species *in vitro*, and there is limited evidence that they are highly effective for the treatment of tuberculosis that has failed to respond to standard therapy owing to multiple drug resistance (Tsukamura *et al.*, 1985). The 4-quinolones should not be used in combination with rifampicin as this agent antagonizes their bacteriocidal activity (Smith, 1984).

Design of chemotherapeutic regimens for pulmonary tuberculosis

Chemotherapeutic regimens should cure the patient and prevent the emergence of drug resistance. In view of the problem of mutation to resistance, it is essential to give at least two drugs to which the strain is sensitive as the chance of two mutations occurring simultaneously in a single cell is negligible. In practice, particularly in regions with a high prevalence of drug resistance, this involves giving three or more drugs.

In the early days of antituberculous therapy, it was considered necessary to give the drugs four times daily in order that inhibitory concentrations were constantly maintained. As therapy then lasted for two years, about

3000 doses were required. Subsequently it was shown that it was more effective to give the drugs daily and, in some modern regimens, they are given only thrice or twice weekly. Indeed, tuberculosis can now be cured with a mere 64 supervised doses.

For the reasons stated above, modern drug regimens are all based on the use of rifampicin and isoniazid throughout. In addition, one or two other drugs are usually added in an initial intensive phase. A regimen advocated by the British Thoracic and Tuberculosis Association (1976) consists of rifampicin and isoniazid daily for nine months, with the addition of daily ethambutol for the first two months; i.e. a total of 270 doses. The addition of a fourth drug, pyrazinamide, in the initial two-month period enables the duration to be reduced to six months (British Thoracic Association, 1984). These regimens are well tolerated and highly effective but are relatively expensive owing to the daily rifampicin. In developing countries cost is certainly important, but an even more relevant consideration is the need to give therapy under full supervision. If many doses are to be given, the cost of supervision may exceed that of the drugs. Consequently, much effort has been made to establish the most suitable drug combinations for use in short-course intermittent regimens (Fox, 1981). Various such regimens have been described. Some are intermittent throughout while others commence with an intensive phase of, for example, rifampicin, isoniazid, pyrazinamide and streptomycin daily for two months, followed by the first two drugs twice weekly for a further four months (East and Central African/British MRC, 1986). The advantage of using an early intensive phase of treatment is that it probably cures a high proportion of the patients so that there is less chance of the disease relapsing if the patient absconds before completion of the less-intensive continuation phase. The first- and second-line antituberculous drugs and their recommended doses are listed in Table 10.3.

Therapy of non-pulmonary tuberculosis

There is general agreement that the modern chemotherapeutic regimens discussed above are suitable for the treatment of all types of non-pulmonary tuberculosis, even life-threatening forms such as tuberculous meningitis. There is less agreement over the duration of therapy: many physicians continue therapy for up to 18 months. There is no real evidence that this is necessary: indeed, limited trials suggest that treatment for the standard duration is adequate. The agents used in modern regimens penetrate tissues well and the intrathecal administration of drugs in tuberculous meningitis is not necessary.

Surgery, now rarely used in pulmonary tuberculosis, is often of value, or essential, for the diagnosis and management of non-pulmonary disease. Examples include debridement or excision of bone lesions (with bone grafting if necessary), excision of non-functioning kidneys and repair of ureteric strictures, excision of lymph nodes and associated abscesses and sinuses, relief of intestinal obstruction due to abdominal disease, and relief of constrictive pericarditis.

Table 10.3 Daily doses of antituberculous drugs

Drug	Adults	Children
Rifampicin	450 mg if body wt < 50 kg 600 mg if body wt ≥ 50 kg	Up to 20 mg/kg to maximum of 600 mg
Isoniazid	200–300 mg	10 mg/kg
Pyrazinamide	1.5 g if body wt < 50 kg 2.0 g if body wt ≥ 50 kg	35 mg/kg
Ethambutol	15–25 mg/kg*	As for adult †
Streptomycin	750 mg if body wt < 50 kg 750 mg if age > 40 years 1 g if body wt ≥ 50 kg	20–30 mg/kg to maximum of 1 g
Thiacetazone	150 mg	4 mg/kg
PAS	8–12 g (divided doses)	200 mg/kg
Ethionamide	750 mg (divided doses)	15 mg/kg
Cycloserine	500 mg (divided doses)	Not recommended

*The 25 mg/kg dose should not be given for more than 2 months; additional therapy should be at 15 mg/kg
†Ethambutol should only be given to children old enough to be tested for visual acuity (see under 'Drug toxicity')

Traditionally, tuberculosis of the spine was treated by radical surgery and bone grafting. Studies in several countries showed that, in the absence of neurological complications or marked deformity, radical surgery had no advantage over more simple debridement operations or even chemotherapy alone with bed rest or a plaster jacket if necessary (Medical Research Council Working Party on Tuberculosis of the Spine, 1978). Other forms of bone and joint tuberculosis also respond well to chemotherapy and, in the absence of destructive lesions, surgery is not required.

Treatment of patients with renal or hepatic disease

It is fortunate that four drugs – rifampicin, isoniazid, pyrazinamide and ethionamide – are either metabolized or eliminated in the bile. These may therefore be used safely at the normal doses in patients with renal impairment.

Patients with impaired liver function may be treated with isoniazid and ethambutol for one year, with streptomycin for the first two or three months. The more hepatotoxic drugs rifampicin and pyrazinamide are thus avoided, although in fact there is no evidence that these are any more toxic in patients with impaired hepatic function. It is important to monitor hepatic function regularly during therapy.

The role of steroids in therapy

Steroids have been used in the therapy of tuberculosis for two reasons. First, there was an assumption that, by reducing the host's immune response, the dormant bacilli would replicate freely and thereby become susceptible to the drugs. There is little evidence to support this conjecture and it has been shown that the use of steroids does not affect the outcome of modern short-course chemotherapy (Fox, 1978). Secondly, steroids are used to suppress damaging inflammatory reactions, particularly in tuberculous meningitis. Indeed, steroids may be life-saving in the presence of cerebral oedema which sometimes develops after the commencement of drug therapy and is probably an allergic phenomenon.

The use of steroids routinely in cases of tuberculous meningitis is controversial. Although several controlled clinical trials have indicated that steroids reduce early mortality, many of the additional survivors are seriously disabled. It has even been stated that steroids merely delay death and prolong suffering (Escobar *et al.*, 1975).

Most studies were carried out before the era of modern rifampicin-based regimens, and it seems reasonable to conclude that the addition of steroids to such regimens is, as in pulmonary tuberculosis, of limited value. Steroids reduce effusions in tuberculous pericarditis but it is not known whether their use prevents subsequent scarring and constrictive pericarditis (Williams and Hetzel, 1978).

Chemoprophylaxis of tuberculosis

Strictly speaking, chemoprophylaxis is the prescription of antimicrobials to uninfected individuals who are exposed to a risk of infection as, for example, in protection against malaria. In the case of tuberculosis, the term is more usually applied to treatment of individuals who have *probably* been infected with tubercle bacilli but show no signs of active disease. Chemoprophylaxis, as opposed to therapy of overt disease, usually involves the use of isoniazid alone. The use of a single drug is justified by the assumption that the bacillary load is very small so that the chance of the emergence of a drug-resistant mutant is remote and that an effective immune reaction will destroy any persisters.

Examples of individuals who might benefit from 'chemoprophylaxis' include tuberculin-positive but radiologically clear children who have been exposed to an open case, and adults with lung shadows suggestive of inactive disease.

As infected children under the age of three years are prone to develop serious non-pulmonary tuberculosis, including meningitis, therapy should be given. Some authorities use isoniazid 'chemoprophylaxis' while others give a full course of therapy. Certainly the latter should be considered in regions where isoniazid resistance is common.

There is no doubt that reactivation of quiescent disease in adults is reduced by the use of isoniazid 'chemoprophylaxis'. On the other hand, more morbidity may result from isoniazid hepatotoxicity, particularly in individuals aged over 30 years, than from reactivation of disease. It is therefore better to observe such high-risk individuals and treat them only if activation of the disease occurs.

Supportive measures in the therapy of tuberculosis

It is well established that by far the most important aspect of the treatment of tuberculosis is adequately supervised chemotherapy. Supportive measures such as fresh air, good diet and vitamins may be of psychological benefit but they have no significant effect on the outcome of therapy. Studies in India (Dawson *et al.*, 1966) showed that patients treated in the relative luxury of a hospital fared no better than those living in squalid conditions. In addition, those engaged in heavy manual labour recovered as rapidly as those who rested in sanatoria. Accordingly, admission to hospital is unnecessary for uncomplicated cases, particularly as modern chemotherapy renders patients non-infectious very rapidly.

In some regions, patients are admitted to hospital to ensure that they receive the initial intensive phase of therapy. Supervision of the therapy of homeless alcoholics presents a particular problem as such individuals will usually neither stay in hospital nor take their drugs regularly. In the East End of London the provision of hostel accommodation for such patients has proved a success.

Monitoring and follow up of therapy

The purpose of monitoring drug therapy is three-fold: first, to ensure that the patient is taking the drugs; secondly, to determine whether the regimen is effective in the individual patient; and thirdly, to ensure that the drugs are having no harmful effects.

The need to monitor patients during modern short-course therapy by radiology and sputum culture is questionable and, in many regions, impossible. Likewise, the need for regular check-ups after the completion of therapy has been challenged. Even where such follow-up procedures have been applied (at considerable expense), almost all cases of reactivation were detected when the patient presented with a recurrence of symptoms (Albert *et al.*, 1976). Accordingly, resources should be used to ensure that every patient receives a full, supervised course of therapy rather than for any follow-up procedures (Rouillon *et al.*, 1976). Patients should then be informed that although they have almost certainly been cured they should seek medical advice promptly if symptoms recur.

Drug toxicity

Although all antituberculous drugs have some untoward side effects, drug toxicity is, in general, not a serious problem and is a small price to pay for the very real benefits of modern chemotherapy. The major side effects are hepatotoxicity, peripheral neuropathy, mental disturbances, rashes and fevers.

The principal drugs used in modern short-course regimens – isoniazid, rifampicin, pyrazinamide – are all potentially hepatotoxic. The occurrence of such toxicity appears to vary from country to country but rarely gives cause for concern (Girling, 1978). In particular the hepatotoxic properties of pyrazinamide were over-stressed in the past, probably because large doses were given (Fox, 1978). Rifampicin may cause an influenza-like syndrome but, paradoxically, this is less likely to occur if the drug is given daily rather than twice weekly.

Isoniazid may cause peripheral neuritis and mild psychiatric disturbances, both of which are minimized by prescribing pyridoxine. Although pyridoxine is not prescribed routinely, it should certainly be given to alcoholics who are receiving isoniazid. The rarely used second-line drug cycloserine also causes psychiatric symptoms, including hallucinations. Streptomycin is toxic for the eighth nerve, including that of the fetus. For this reason, streptomycin should not be given during pregnancy. There is some evidence that rifampicin is teratogenic (Furesz, 1970), so this drug should not be used during the first three months of pregnancy. For the same reason, women receiving rifampicin should avoid becoming pregnant. In this respect it is important to note that rifampicin interferes with the action of oral contraceptives. Alternative forms of birth control should therefore be used.

Ethambutol has a very important side effect; namely, ocular toxicity. Although this is rare if the drug is given for no more than two months at a daily dose of 25 mg/kg body weight, or for longer at a dose not exceeding 15 mg/kg, vigilance is indicated (particularly in litigation-conscious communities). A code of practice, suitable for use in developed countries with the requisite staff and facilities, has been recommended by the British Thoracic Society:

1. Pretreatment renal function should be investigated by assay of serum urea and/or creatinine; and ethambutol should not be given to patients with impaired renal function.
2. The recommended dose of ethambutol and duration of therapy should never be exceeded.
3. Any history of eye disease should be recorded in the notes.
4. Pretreatment visual acuity should be assessed by the Snellen test or, for those unable to read, by the Cambridge Low Contrast gratings. If the patient normally wears spectacles for distant vision they should be worn for the test. Ethambutol should not be given to patients with poor sight who may not notice further minor deterioration of vision.
5. The small risk of ocular toxicity should be explained to the patient with an admonition to discontinue the drug if vision becomes impaired.
6. A record should be made in the notes that the danger of ocular toxicity has been explained to the patient.
7. The patient's general practitioner should be informed of the instructions given to the patient.
8. Patients complaining of visual disturbance should be referred to an ophthalmologist.

9. Routine tests of visual acuity during therapy are not recommended.
10. Ethambutol should be avoided in patients who are unsuitable for objective tests of visual activity, i.e. young children and adults with language or other communication problems.

The role of drug sensitivity testing

The purpose of drug sensitivity testing, as outlined at the beginning of this chapter, is to determine whether the great majority of organisms in a culture are sensitive to levels of the drugs that are achieved clinically. Tests on pretreatment isolates will reveal whether the patient has been infected by a resistant strain, i.e. initial or primary resistance. Tests during therapy or following relapse will detect the emergence of resistant mutants, i.e. acquired or secondary resistance.

Sensitivity testing is indispensable in the assessment of new drug regimens and has epidemiological value in determining the pattern of resistance in a community. The necessity or even desirability of testing every isolate in clinical practice is contestable. Sensitivity testing is expensive and time-consuming and requires a high level of technical expertise and rigid quality-control procedures. The test is not worth performing unless high standards of accuracy can be maintained. Much harm may be done by modifying a regimen to include less effective and more toxic drugs on the basis of a false report of resistance. Many more therapeutic failures result from irregular medication due to poor supervision or an erratic supply of drugs than to primary resistance (Grosset, 1978). Even where testing is available, its usefulness has been seriously questioned. Studies in Hong Kong, when treatment consisted of isoniazid, streptomycin and para-aminosalicylic acid and where primary resistance to these drugs was fairly common, showed that the availability of the results of sensitivity testing contributed very little to the management of the patients (Hong Kong Tuberculosis Treatment Services/British MRC, 1974). It is therefore likely to contribute even less to the management of tuberculosis by short-course chemotherapy. Indeed, there is no convincing evidence that the efficacy of such modern regimens is related to *in vitro* resistance.

In some prosperous regions, where technical standards are usually high, sensitivity testing is virtually essential for medico-legal reasons. In other countries, testing is an unjustifiable luxury and priority should be given to early diagnosis and the provision of effective, supervised chemotherapy.

Leprosy

Chemotherapy of leprosy

The aims of the chemotherapy of leprosy are similar to those of tuberculosis; namely, the destruction of all bacilli and the prevention of the emergence of drug resistance. As discussed in Chapter 7, the number of bacilli in patients varies greatly and is related to the position of the patients on the immunological spectrum. Patients at or near the lepromatous pole have an enormous bacillary load and are presumed to be the most infectious. Furthermore, as they have little or no immune reactivity, the bacilli can only be destroyed by effective drug regimens. As in the case of tuberculosis, some bacilli persist for long periods of time, even in the presence of the bacteriocidal drug rifampicin (Pattyn *et al.*, 1976).

The earliest effective drug for leprosy was chaulmoogra oil, derived from the fruit of *Hydnocarpus wightiana*. This drug, the origin of which has inspired various legends, is certainly effective in tuberculoid leprosy and may act by stimulating the immune response. In 1943 Faget and his colleagues introduced promin which, although effective, was rather toxic and had to be given by intravenous injection. Subsequently the same team isolated the active principle of promin – diaminodiphenyl sulphone (DDS, dapsone) – which was cheap, very effective, virtually non-toxic and could be taken by mouth. In view of these properties, dapsone became the mainstay of antileprosy therapy, even though it is bacteriostatic rather than bacteriocidal. Side effects are uncommon with dapsone, although cases of haemolytic anaemia, hepatitis, dermatitis and psychosis occasionally occur.

Two additional drugs of great value are clofazimine (B663, Lamprene) and rifampicin. The former is weakly bacteriocidal and anti-inflammatory and suppresses erythema nodosum leprosum reactions (see page 116). It occasionally causes gastrointestinal disturbances, but it has the more serious disadvantage of causing skin discolouration which many patients find objectionable. Rifampicin is bacteriocidal and causes a rapid killing of bacilli with a concomitant rapid reduction in infectivity. Indeed, about 99.9 per cent of bacilli are killed within a week of administering a single dose of 600–1500 mg of rifampicin. Unfortunately a minority of bacilli 'persist' and small numbers of viable drug-sensitive bacilli have been isolated from lepromatous patients after five years of rifampicin therapy (Waters *et al.*, 1978). The mechanism for such persistence is unknown.

Other drugs include ethionamide and prothionamide, which are very similar, weakly bacteriocidal and show cross-resistance. Further studies are required to establish their usefulness and optimum dosage: at present they are recommended for patients who, owing to skin discolouration, refuse to take clofazimine.

Drug regimens for leprosy

Until around 1976 dapsone monotherapy was the standard treatment for all forms of leprosy; the emergence of dapsone resistance in multibacillary patients had been reported but the incidence appeared to be very low. The World Health Organization Expert Committee on Leprosy (1977) stressed the need for studies on drug combinations, such as rifampicin with clofazimine, for treatment of proven dapsone-resistant cases. At that time there were disturbing reports from some regions suggesting that the incidence of such resistance was higher than previously suspected and that it was increasing. Thus the WHO sponsored a number of studies, which confirmed the seriousness of the problem (World Health Organization Scientific Working Group on the Chemotherapy of Leprosy, 1982). Such emergent, or secondary, drug resistance is confined to multibacillary cases, as only these have large enough numbers of replicating bacilli to make it likely that a mutation will occur. On the other hand, when such bacilli are transmitted to other individuals, any form of leprosy may develop, as is the case with drug-sensitive bacilli. Such primary resistance, unfortunately, now occurs in several countries (Waters *et al.*, 1978).

In view of the increasing frequency of both primary and secondary dapsone resistance, multi-drug therapy (MDT) is now essential for the treatment of the individual and for the reduction of the incidence of the disease in the community. Different regimens have been recommended for multibacillary and paucibacillary cases.

Multibacillary cases (LL, BL and BB) require therapy for at least two years and probably longer. As there is a huge bacillary load, there is a high chance that mutation to rifampicin resistance will develop. It is therefore essential to use three drugs. A suitable regimen consists of daily self-administered dapsone with a small dose of clofazimine and a monthly supervised dose of rifampicin with a large dose of clofazimine. An alternative highly effective regimen advocated by Freerksen and Rosenfeld (1977) consists of rifampicin, prothionamide, dapsone and isoniazid. This regimen avoids the skin discolouration caused by clofazimine and is particularly useful for the treatment of patients who have both leprosy and tuberculosis.

Paucibacillary cases (indeterminate, TT and BT) are treatable by short-course regimens as any persisters are almost certainly destroyed by the patients' immune defences. As the bacillary load is small, the chance of drug resistance developing by mutation is very low. Accordingly a third drug is not essential. A standard regimen consists of monthly supervised rifampicin and daily self-administered dapsone for six months. For full details of these recommended MDT regimens, see the 1982 WHO report referred to above.

Chemoprophylaxis of leprosy

In principle, leprosy may be prevented in household contacts by the regular administration of dapsone or its injectable long-acting analogue acedapsone (provided that dapsone resistance has not developed). One intensive therapeutic and prophylactic programme did indeed reduce the incidence of the disease to zero in a circumscribed community (Russell *et al.*, 1976). Nevertheless, in view of the medical and administrative problems generated by prophylaxis programmes, as well as the increasing problems of dapsone resistance, the WHO Expert Committee does not recommend the widespread use of such prophylaxis.

Opportunist mycobacterial diseases

There are few aspects of antimicrobial chemotherapy that are more problematical than the treatment of the opportunist mycobacterial diseases. The subject is complicated by the problems of diagnosis and the difficulty in distinguishing between true disease and colonization; by the frequent occurrence of serious underlying immune defects or other predisposing conditions; by the resistance of many mycobacteria to the available antimicrobial agents; and by the lack of correlation between *in vitro* resistance and response to therapy *in vivo*. Furthermore, owing to the relative rarity of the diseases there have been very few controlled clinical trials of therapeutic regimens: many cases are treated on the basis of anecdotal experience or by trial and error.

The therapy of disseminated disease, especially that due to *M. avium–intracellulare*, has become a topic of considerable importance in view of its frequent occurrence in AIDS patients, but such disease also occurs in individuals with other predisposing conditions.

The choice of therapy depends on the type of disease, the causative organism and the presence or absence of an underlying disorder. Less emphasis is placed on the *in vitro* drug resistances. Surgical intervention, though rarely resorted to nowadays in the case of uncomplicated pulmonary tuberculosis, should be seriously considered in the case of localized pulmonary lesions due to other mycobacteria (Moran *et al.*, 1983). Localized non-pulmonary lesions, such as lymphadenitis, joint lesions, abscesses and solitary skin nodules, are likewise often treated surgically.

In the absence of treatment, localized lymphadenitis in infants almost always resolves spontaneously. In most instances surgery is undertaken for diagnostic purposes (Taha *et al.*, 1985). It is most important that the whole infected node is removed: not only is this usually curative, but simple incision often leads to sinus formation, scarring and delayed healing. Children should be examined for evidence of immune defects and should be followed up for a few years to ensure that the disease does not recur or spread.

Localized post-traumatic or post-injection lesions also usually occur in otherwise healthy individuals and tend to resolve spontaneously. Nevertheless, healing may be prolonged and complicated by the presence of discharging sinuses. Small lesions may be cured by simple excision, but more widespread lesions, including sternal infections following heart surgery, require extensive debridement and antimicrobial therapy (Kuritsky *et al.*, 1983).

Localized pulmonary disease is usually caused by *M. avium–intracellulare* or *M. kansasii* and it is with these infections that most clinical experience has been acquired. In view of *in vitro* resistance of *M. avium–intracellulare* to most of the antituberculous agents, it was considered that standard triple antituberculosis therapy would be of no value. Instead, regimens containing five or six empirically chosen drugs were advocated (Davidson, 1976). This regimen is undoubtedly successful in a high proportion of patients. Unfortunately many patients develop adverse drug reactions and the regimen has to be modified. It has now been found that the more acceptable standard triple regimen of rifampicin, isoniazid and ethambutol is as effective for uncomplicated cases, provided that therapy is continued with all three drugs for at least 18 months (Engbaek *et al.*, 1984). This regimen has also proved effective for the treatment of pulmonary disease due to other slowly growing species, including *M. kansasii* (Banks *et al.*, 1983), *M. xenopi* (Smith and Citron, 1983) and *M. malmoense* (Banks *et al.*, 1985).

In recent years there has been a growing trend to treat opportunist mycobacterial infections with drugs, or combinations thereof, that are not used for the treatment of tuberculosis. These have proved particularly useful for the treatment of infections that do not respond to the triple therapy outlined above, of disseminated disease, and of all types of infections due to the rapidly growing species *M. chelonei* and *M. fortuitum*. The use of such drugs is mostly based on anecdotal experience and few firm recommendations can be made. As the drugs are almost always used in combination, it is often difficult to assess the contribution of each agent to the result. Furthermore, the optimum duration of therapy has not been established and must therefore be based on the clinical response in individual patients.

Strains of *M. kansasii*, *M. scrofulaceum* and about half those of *M. avium–intracellulare* are sensitive to erythromycin *in vitro* and there is evidence that the drug is of use *in vivo* (Singh, 1985). It has been used successfully together with amikacin or gentamicin, for the treatment of disease due to the rapid growers (Wolinsky, 1979). Erythromycin with co-trimoxazole (the latter being trimethoprim with sulphamethoxazole) also appears to be a very effective combination for the treatment of such infections (Azadian *et al.*, 1981; Jackson *et al.*, 1981).

Co-trimoxazole has also been used in the successful treatment of several cases of disease due to *M. marinum*, *M. xenopi* and *M. avium–intracellulare* (Grange, 1984). It has been suggested that the sulphonamide component acts by enhancing the ability of the phagocytic cells to kill the bacteria (Tice and Solomon, 1979). It therefore appears to be of benefit even if the organism is resistant to it *in vitro* (Jackson *et al.*, 1981).

Doxycycline has been used with amikacin to treat infections caused by *M. chelonei* and *M. fortuitum* (Dalovisio *et al.*, 1981) and it was used with erythromycin and co-trimoxazole in the successful treatment of a dissemi-nated *M. chelonei* infection in a renal transplant recipient (Azadian *et al.*, 1981). Minocycline appears one of the best agents for the treatment of *M. marinum* infection (Arai *et al.*, 1984).

Two cephalosporins, cefoxitin and ceftizoxime, were found to be effective against some mycobacteria *in vitro* and the former was used successfully with amikacin in the treatment of *M. chelonei* sternal infections following cardiac surgery (Kuritsky *et al.*, 1983).

The antileprosy drug clofazimine is concentrated in the epithelium, bone marrow and lymphoid tissue and is effective in the treatment of disseminated *M. avium–intracellulare* infection. A combination of clofazi-mine and ansamycin (spiro-piperidyl rifamycin) has been advocated for the treatment of such disease in AIDS victims (Iseman *et al.*, 1985). Although this is probably the best of the currently available regimens for such disease, therapeutic failures are common (Pinching, 1987). It has been used, apparently with some success, in the treatment of *M. ulcerans* infection (see page 142).

Some new antimicrobial agents show activity against many mycobacterial species *in vitro*. Of these, the most promising are the 4-quinolones such as ciprofloxacin (Collins and Uttley, 1985).

In summary, localized lung disease due to slowly growing species usually responds to a prolonged course of rifampicin, isoniazid and ethambutol. Two useful regimens for disease due to rapidly growing species are erythromycin with co-trimoxazole, and cefoxitin with amikacin. Disseminated disease due to *M. avium–intracellulare* sometimes responds to clofazimine with ansamycin. Other drugs, and combinations thereof, also appear to be of value; although experience with them is limited. In view of the complexity of the subject of antimicrobial therapy of opportunist mycobacterial disease, it is strongly advised that treatment should be planned and monitored in consultation with the staff of a recognized mycobacterial reference centre.

References

Albert, R.K., Iseman, M., Sbarbaro, J.A., Stage, A. and Pierson, D. (1976) Monitoring patients with tuberculosis for failure during and after treatment. *American Review of Respiratory Disease* **114**: 1051–60.

Arai, H., Nakajima, H. and Nagai, R. (1984) *Mycobacterium marinum* infection of the skin in Japan. *Journal of Dermatology* **11**: 37–42.

Azadian, B.S., Beck, A., Curtis, J.R., Cherrington, L.E., Gower, P.E., Phillips, M., Eastwood, J.B. and Nicholls, J. (1981) Disseminated infection with *Mycobacterium chelonei* in a haemodialysis patient. *Tubercle* **62**: 281–4.

Banks, J., Hunter, A.M., Campbell, I.A., Jenkins, P.A. and Smith, A.P. (1983) Pulmonary infection with *Mycobacterium kansasii* in Wales 1970–9: review of treatment and response. *Thorax* **38**: 271–4.

Banks, J., Jenkins, P.A. and Smith, A.P. (1985) Pulmonary infection with *Mycobacterium malmoense* – a review of treatment and response. *Tubercle* **66**: 197–203.

British Thoracic and Tuberculosis Association (1976) Short course therapy in pulmonary tuberculosis. *Lancet* **ii**: 1102–4.

British Thoracic Association (1984) A controlled trial of 6 months chemotherapy in pulmonary tuberculosis. Final report: results during the 36 months after the end of chemotherapy and beyond. *British Journal of Diseases of the Chest* **78**: 330–6.

Collins, C.H. and Uttley, A.H.C. (1985) *In vitro* susceptibility of mycobacteria to ciprofloxacin. *Journal of Antimicrobial Chemotherapy* **16**: 575–80.

Dalovisio, J.R., Pankey, G.A., Wallace, R.J. and Jones, D.R. (1981) Clinical usefulness of amikacin and doxycycline in the treatment of infection due to *Mycobacterium fortuitum* and *Mycobacterium chelonei*. *Review of Infectious Diseases* **3**: 1068–74.

David, H.L. (1970) Probability distribution of drug-resistant mutants in unselected populations of *Mycobacterium tuberculosis*. *Applied Microbiology* **20**: 810–14.

Davidson, P.T. (1976) Treatment and long-term follow-up of patients with atypical mycobacterial infections. *Bulletin of the International Union Against Tuberculosis* **51**: 257–61.

Dawson, J.J.Y., Devadatta, S., Fox, W., Radhakrishna, S., Ramakrishnan, C.V., Somasundaram, P.R., Stott, H., Tripathy, S.P. and Velu, S. (1966) A 5-year study of patients with pulmonary tuberculosis in a concurrent comparison of home and sanatorium treatment for one year with isoniazid plus PAS. *Bulletin of the World Health Organization* **34**: 533–51.

East and Central African/British Medical Research Council (1986) Fifth Collaborative Study. Controlled clinical trial of 4 short-course regimens of chemotherapy (three 6-month and one 8-month) for pulmonary tuberculosis: final report. *Tubercle* **67**: 5–15.

Engbaek, H.C., Vergmann, B. and Bentzon, M.W. (1984) Lung disease caused by *Mycobacterium avium/Mycobacterium intracellulare. European Journal of Respiratory Diseases* **65**: 411–18.

Escober, J.A., Belsey, M.A., Duenas, A. and Medina, P. (1975) Mortality from tuberculous meningitis reduced by steroid therapy. *Pediatrics* **56**: 1050–5.

Fox, W. (1978) The current status of short-course chemotherapy. *Bulletin of the International Union Against Tuberculosis* **53**: 268–80.

Fox, W. (1981) Whither short-course chemotherapy? *British Journal of Diseases of the Chest* **75**: 331–57.

Freerksen, E. and Rosenfeld, M. (1977) Leprosy eradication project of Malta: first published report after 5 years running. *Chemotherapy* **23**: 356–86.

Furesz, S. (1970) Chemical and biological properties of rifampicin. *Antibiotica et Chemotherapia* **16**: 316–51.

Girling, D.J. (1978) The hepatic toxicity of antituberculosis regimens containing isoniazid, rifampicin and pyrazinamide. *Tubercle* **59**: 13–32.

Grange, J.M. (1984) Therapy of infections caused by 'atypical' mycobacteria. *Journal of Antimicrobial Chemotherapy* **13**: 308–10.

Grosset, J. (1978) Should a sensitivity test be made for each new case of tuberculosis? *Bulletin of the International Union Against Tuberculosis* **53**: 200–1.

Hong Kong Tuberculosis Treatment Services/British Medical Research Council (1974) A study in Hong Kong to evaluate the role of pretreatment suceptibility tests in the selection of regimens of chemotherapy for pulmonary tuberculosis – second report. *Tubercle* **55**: 169–92.

Iseman, M.D., Corpe, R.F., O'Brien, R.J., Rosenzweig, D.Y. and Wolinsky, E. (1985) Disease due to *Mycobacterium avium–intracellulare. Chest* **87** (Suppl. 2): 139s–149s.

Jackson, P.G., Keen, H., Noble, C.J. and Simmons, N.A. (1981) Injection abscesses due to *Mycobacterium chelonei* occurring in a diabetic patient. *Tubercle* **62**: 277–9.

Kuritsky, J.N., Bullen, M.G., Broome, C.V., Silcox, V.A., Good, R.C. and Wallace, R.J. (1983) Sternal wound infections and endocarditis due to organisms of the *Mycobacterium fortuitum* complex. *Annals of Internal Medicine* **98**: 938–9.

Medical Research Council Working Party on Tuberculosis of the Spine (1978) Seventh Report. A controlled trial of anterior spinal fusion and debridement in the surgical management of tuberculosis of the spine in patients on standard chemotherapy: a study in two centres in South Africa. *Tubercle* **59**: 79–105.

Mitchison, D.A. (1985) Hypothesis: the action of antituberculosis drugs in short course chemotherapy. *Tubercle* **66**: 219–25.

Moran, J.F., Alexander, L.G., Staub, E.W., Young, W.G. and Sealy, W.C. (1983) Long-term results of pulmonary resection for atypical mycobacterial disease. *Annals of Thoracic Surgery* **35**: 597–604.

Pattyn, S.R., Dockx, P., Rollier, M.T., Rollier, R. and Saerens, E.J. (1976) *Mycobacterium leprae* persisters after treatment with dapsone and rifampicin. *International Journal of Leprosy* **44**: 154–8.

Pinching, A.J. (1987) The acquired immune deficiency syndrome: with special reference to tuberculosis. *Tubercle* **68**: 65–9.

Rouillon, A., Perdrizet, S. and Parrot, R. (1976) Transmission of tubercle bacilli: the effects of chemotherapy. *Tubercle* **57**: 275–99.

Russell, D.A., Worth, R.M., Scott, G.C., Vincin, D.R., Jano, B., Fasal, P. and Shepard, C.C. (1976) Experience with acedapsone (DADDS) in the therapeutic trial in New Guinea and the chemoprophylactic trial in Micronesia. *International Joural of Leprosy* **44**, 170–6.

Singh, G. (1985) Treatment of septic arthritis due to *Mycobacterium kansasii*. *British Medical Journal* **290**: 857.

Smith, J.T. (1984) Awakening the slumbering potential of the 4-quinolone antibacterials. *Pharmaceutical Journal* **233**: 299–305.

Smith, M.J. and Citron, K.M. (1983) Clinical review of pulmonary disease caused by *Mycobacterium xenopi*. *Thorax* **38**: 373–7.

Taha, A.M., Davidson, P.T. and Bailey, W.C. (1985) Surgical treatment of atypical mycobacterial lymphadenitis in children. *Pediatric Infectious Diseases* **4**: 664–7.

Tice, A.D. and Solomon, R.J. (1979) Disseminated *Mycobacterium chelonei* infection: response to sulphonamides. *American Review of Respiratory Disease* **120**: 197–201.

Tsukamura, M., Nakamura, E., Yoshi, S. and Amano, H. (1985) Therapeutic effect of a new antibacterial substance ofloxacin (DL 8280) on pulmonary tuberculosis. *American Review of Respiratory Disease* **131**: 352–6.

Waters, M.F.R., Rees, R.J.W., Pearson, J.M.H., Laing, A.B.G., Helmy, H.S. and Gelber, R.H. (1978) Rifampicin for lepromatous leprosy: nine years' experience. *British Medical Journal* **1**: 133–6.

Williams, I.P. and Hetzel, M.R. (1978) Tuberculous pericarditis in South-West London: an increasing problem. *Thorax* **33**: 816–7.

Wolinsky, E. (1979) State of the art: non-tuberculous mycobacteria and associated diseases. *American Review of Respiratory Disease* **119**: 107–59.

World Health Organization Expert Committee on Leprosy (1977) *Fifth Report.* Technical Report Series No. 607. Geneva: WHO.

World Health Organization Scientific Working Group on the Chemotherapy of Leprosy (1982) *Chemotherapy of Leprosy for Control Programmes.* Technical Report Series No. 675. Geneva: WHO.

Index

Index

AIRLINE SURVIVAL KIT

Dedicated to the loving memory of my mother,
Shanti Devi, who deserved so much more

Airline Survival Kit

Breaking Out of the Zero Profit Game

NAWAL K. TANEJA

ASHGATE

Published by
Ashgate Publishing Limited
Gower House
Croft Road
Aldershot
Hants GU11 3HR
England

Ashgate Publishing Company
Suite 420
101 Cherry Street
Burlington, VT 05401-4405
USA

Ashgate website: http://www.ashgate.com

British Library Cataloguing in Publication Data
Taneja, Nawal K.
    Airline survival kit : breaking out of the zero profit game
    1.Airlines - Economic aspects 2.Airlines - Finance
    3.Airlines - Planning
    I.Title
    387.7'1

Library of Congress Control Number: 2003100752

ISBN 0 7546 3452 3

Printed and bound in Great Britain by MPG Books Ltd, Bodmin, Cornwall

Contents

List of Figures

List of Tables

Foreword

Masashi Izumi
Vice President, Corporate Planning
All Nippon Airways

Most airlines in today's world, particularly network carriers, have faced unprecedented difficulties in their industry in the wake of September 11th, 2001, as well as structural problems accumulated over the last ten years. Since the 1970s, the major carriers' response to a newly introduced liberalized framework, first in the United States and then spreading to other markets, was the development of hub and spoke systems, frequent flyer programs, and sophisticated revenue management systems. These systems should have provided the carriers with much more efficient operations and an improvement in profitability. However, the low profitability of this industry has remained unchanged as Nawal Taneja shows in his new book *Airline Survival Kit*. It is ironical that an excessive reliance on those systems and tools by network carriers resulted, in fact, in successful challenges of network carriers by low-cost, low-fare carriers.

In Japan, the tight regulatory framework has been gradually relaxed since 1986, beginning with the privatization of government owned Japan Airlines. All Nippon Airways started up scheduled international service in 1986, and almost all restrictions in the domestic market were lifted in 2000 (with the exception of slot allocations at congested airports). The Japanese economy and society are so centralized in the Tokyo metropolitan area that air traffic demand remains concentrated there. However, due to delayed airport facility improvement efforts in the Tokyo area, there are aircraft movement limits imposed by the government at both the Haneda Airport for domestic operations, and at the Narita Airport for international operations, and there are no secondary airports available in the area. Those airport constraints have largely prevented new entrants in Japan from emulating the success of the low-fare airlines as in the United States or in Europe.

It may be argued that the airport constraints in Tokyo provide protection from competition for existing carriers. However, it doesn't mean that Japanese carriers are exempt from forces that affect airlines in other areas. We have seen the continuous decline of yield, despite average growth in air traffic. Especially during the recent long recession, it is obvious that Japanese consumers have become very sensitive to price and now look for value in purchasing their goods and services.

The trend towards full liberalization of airline industry will continue. The airline business is no longer considered to be a special or unique sector. The market has started to change. The challenges to airline management have started as well. We must cope with changes in the needs of customers. How can an airline survive? It depends on the strategy that airline management takes. There can be no single solution for every airline. The solutions are as varied as the prevailing conditions.

Readers of this book, who are interested in the destiny of this exciting sector of the economy, will find the background to and a structure for the challenges that the airline industry now faces. In the *Airline Survival Kit: Breaking Out of the Zero Profit Game*, Nawal Taneja makes pragmatic analyses of strategies that have worked and those that have not worked. The reader may also find some ideas on what strategies could work in the future as well as the necessary ingredients for their successful implementation. I believe discussions made in this book will give readers clues to help them find their own better solutions.

Tokyo, Japan
20 February, 2003

Foreword

Pete McGlade
Vice President, Schedule Planning
Southwest Airlines

Since 1978, when the air transportation industry was de-regulated, air service products and services have dramatically changed and evolved. In the environment before de-regulation, consumers and producers were faced with a limited supply of seats, whose pricing was strictly controlled. Product differentiation and segmentation were in their infancy.

In 1971, Southwest Airlines innovated a new business paradigm. Its model emphasized low costs, high standards of service and extremely low prices. Consumers responded in record numbers to this newfound freedom to fly. Markets grew in size and vibrancy. More people than ever before were able to fly.

As I write this foreword in early 2003, our industry has been almost completely transformed. The "more for less," consumer revolution is at full throttle. The traditional network carriers are engaged in extra-ordinary efforts to better align their product with demand. Airports are busy with efforts to adjust to new and challenging security demands that inevitably will lead to cost pressures. Competition is flourishing across a broader spectrum of markets. It is undeniably a milestone moment in the history of air transportation.

Nawal Taneja's new book, *Airline Survival Kit*, is a well-written and cogent guide for all those interested in these changes. His book lays out the elements of this debate in such an accessible fashion that it allows the reader to reach well-considered judgments about deeply complicated matters. I found the author's insights on the roles of labor and management during this crisis particularly helpful for this reason. It is obvious that Nawal Taneja's long study of the air transportation industry provides for a steady and reasoned discussion of the powerful forces that are currently buffeting our industry. I heartily recommend it to anyone interested in the fate of one of the world's most important economic sectors.

Dallas, Texas
February 2003

Foreword

Scott Nason
Vice President, Revenue Management
American Airlines

The airline industry has, almost from its very beginning, been uniquely important, complicated, and interesting to the public at large. Many countries treat their airlines and airline service as a matter of national security and national pride. The press writes incessantly about airline safety, airline pricing, and airline service issues.

The airline business has also tended to be financially unrewarding to investors. Airlines have traditionally had sufficient—maybe even excessive—access to capital, while, for reasons that Nawal Taneja describes very well in this book, competing away most profits. As a result, there have been many bankruptcies over the years, and frequent calls to reform aspects of the business model, pricing and scheduling strategies, and labor costs. But today, the industry finds itself in a crisis unlike any it has seen before. Most airlines have a very clear challenge: adapt or vanish.

In *Airline Survival Kit: Breaking Out of the Zero Profit Game*, Nawal takes on many of the traditional industry paradigms. He uses his very broad knowledge and familiarity with the international airline business to show how and why many actions have failed, even those actions that succeeded in the 1980s and 1990s. The growth of low cost carriers in the US, Europe, and elsewhere, and the public's apparently insatiable appetite for low cost air transportation has changed the rules for airline management, airline labor, airline investors, and even airline customers.

There is not a single strategy that will work for every airline. Tempting as it might be, most large networked airlines cannot transform themselves into Southwest or Ryanair, nor should they. But all of them must transform themselves into viable, profitable companies. In some cases, like British Airways' "Size and Shape" project, they were able to implement dramatic changes consensually. In others, such as the bankruptcies of US Airways and United Airlines, it is taking at least the protection of the U.S. Bankruptcy Court, if even that works, to restructure their costs to a level that will make them viable. Most other airlines also know that they must substantially reduce their costs, and typically that includes their labor costs, in order to compete in the airline marketplace of the 21st Century.

This book doesn't have all the answers. How could it? But it does a very good job of identifying the problems, their causes and symptoms, and helping us to think about them in ways that should enable each of us to address them at our own airlines.

Fort Worth, Texas
10 February, 2003

Foreword

Carsten Spohr
Vice President, Alliances
Lufthansa German Airlines

The year 2002 has been the toughest year in history for most airlines worldwide. For the publicly traded carriers UBS Warburg estimated in Fall 2002 a cumulative net loss of 2.2 Billion USD. Some carriers like Swissair and Sabena in Europe and Midway and Vanguard in North America were no longer able to bear the losses and had to cease operations—the comeback of three of them under new brands is another story. Two of the top six American airlines, US Airways and United Airlines, filed for Chapter 11 to reorganize their businesses. Yet, there is little sign of an economic recovery. The situation remains very difficult and further demises seem probable.

The case is clear—the airline business urgently needs change. This is especially true for network carriers operating with the traditional full service model while many low cost carriers were able to make a profit despite the harsh circumstances. *Airline Survival Kit* provides useful insights into the nature of the multi-faceted challenge airlines face and the strategies airlines can choose to cope with the continuing crisis and structural impediments.

The solution to the challenge is not as obvious as just copying the low cost model over onto an "old-fashioned" flag carrier. Cost management is—at least for Lufthansa and other successful carriers—at the core of every airline operation. But because of their history, traditional carriers are facing different conditions and restraints—external and internal.

The customer need for carriers providing a comprehensive network and seamless service worldwide sets the stage for further growth by globalisation. Nowadays no carrier can satisfy this set of requirements alone. The answer is a continuous effort to further integrate your network and services within a global alliance. This alliance integration will be one key success factor of the surviving full-service carriers who will take up a major part of the playing field at the one extreme with the low-service carriers at the other extreme. To survive in the "middle" between these models will become more and more the exception.

One more thought—although it might appear that low cost carriers are doing the right thing, even for these carriers a shakeout in the industry is inevitable. The founding frenzy we are experiencing, especially in Europe, brings to mind memories of the new economy hype. The readers of this book know where that ended. The overcapacity syndrome in this market segment will result in cutthroat competition taking its toll there as well.

No matter if a full-service, global, "alliance-integrated" carrier or a low-cost carrier—the key will be to have a clear cut strategy which distinguishes the airline from its competitors. Not all opportunities or strategies presented in this book will be applicable to every situation or every company. However, I think everyone who reads *Airline Survival Kit* will get some ideas that will spark deeper thought.

Frankfurt, Germany
Winter 2002/3

Preface

In the beginning of 2002, some colleagues in the airline industry shared with me information that the industry's problems were perhaps insurmountable and for some, perhaps fatal. I share that concern. This book is as candid as I can be about the contradictions and constraints under which we operate. We are once again losing massive amounts of capital, employees, and customers. It appears that the conventional solutions are inadequate. This crisis, unless dramatically addressed, will not go away. What follows is a presentation, and some explanation and discussion of the multi-faceted challenge encountered by the airline industry—the lack of long-term profits. It also presents some opportunities and possible strategies to break out of the zero profit game.

My audience is you, my colleagues—airline executives and members of Boards, major shareholders, entrepreneurs, government policy makers, labor leadership, the investment community, aircraft manufacturers, airport operators, and the providers of other aviation products and services. It will also be of interest to those from other backgrounds who wish to gain some insight into the specific problems and opportunities of the airline industry.

My approach is to provide a dispassionate analysis and pragmatic insights into (a) the complexities of the airline business, (b) successful and unsuccessful strategies addressing the multi-faceted challenge, (c) future prospects for passenger growth and its regional distribution, and (d) alternative business models, with an emphasis on the ones that have led to enduring success for some airlines and for businesses in other industries.

First, "low-cost low-fare" is self-explanatory but for those unfamiliar with the industry, the term "conventional airline" used frequently throughout this book, requires a definition. It is meant to convey those airlines which offer a full menu of services—choices for making reservations (such as travel agents and call centers, as well as the newer websites); certain choices available before flight departure (such as advanced seat selection and special meals); numerous fare types each with their own terms and conditions; passenger agents at airports to help with check-in, baggage handling, and processing at the gate; frequent flyer programs; typically more than one class of in-flight service; and, also typically, several different types of aircraft in the fleet deployed over a wide network of routes often supplemented with interline connection options to any other carrier.

Large network, full-service carriers may resent the frequent mention of low-cost, low-fare airlines in this book. They are only one force changing the current environment. When the events of September 11th, 2001 (the second force) brought us to a standstill, we began to discover that the business model of the conventional airlines had been crumbling for some time. For example, the Internet and the power of its information had been affecting the major airlines for a number of years (the third force). Finally, some airlines began to feel severely not only the impact of this prolonged downturn in the economy, but they also began to notice that the impact of each downturn was deeper and lasted longer than the previous downturn (the fourth force). The confluence of these four forces alone was a sobering wakeup call to some conventional airlines to come out of denial. It appears that some airlines now do believe that their existence is at stake and they have begun to address some of the issues raised here. If this book has any message, it is to stress the immediate urgency to go further, deeper, and faster.

The book is divided into six chapters about the dynamics of the airline industry and how all of them must be addressed in the structural changes that are required to survive much less prosper.

Chapter 1 provides a brief description of a dozen characteristics of the industry that, taken as a group, make this industry quite unique. They cover such areas as financial, commercial, and technical structures of the industry; the competitive environment; and a number of areas of added complexity such as weather, network implications, and cyclicality. The second part of the chapter describes the multi-faceted challenge, how difficult it has been for airlines to make money in the long-term while other members of the air transportation value chain have often made money.

Chapter 2 begins with a presentation of the conspicuous successes such as technical and operational excellence; falling prices; and new business models. It then discusses areas where the jury is still out such as network design, customer reward and recognition programs, strategic alliances, and corporate diversification. The final part presents some millstones around the neck of the airline industry such as pricing strategies, overly diversified fleet, overly complex product for the price, and the suboptimal airport infrastructure.

Chapter 3 discusses six key drivers of the inevitable shakeout: changing government attitudes; consumer revolt; availability of low-fare options; flawed network structures; emerging distribution systems; and management too slow to adapt to change. The chapter then discusses some plausible outcomes such as the filing of bankruptcy by United (recently realized) and US Airways, consolidation in some countries (for example, US and UK), size reduction and structure change in others (for example, Switzerland and Greece), and carrier replacement in yet other countries (for example, Indian Airlines with Jet Airways in India).

Chapter 4 presents a description of the evolving global demographic and socio-economic patterns and the opportunities associated with these patterns. Examples include the countries of India, China, Brazil, and Mexico; and the potential travel relating to the Asian immigrant population of the US returning to their home markets. The second part talks about the potential influence of aircraft technology: (a) ultra long-haul aircraft in trans-Pacific markets; (b) the flexibility and potential market power of narrow-body aircraft in trans-Atlantic markets; (c) regional jets within the US, Europe, and Asia; (d) narrow-body aircraft in all business configuration both across the Atlantic (such as the Airbus Corporate Jetliner and the Boeing Business Jet) and within North America; and finally, (e) families of aircraft (the Airbus 320s and the Boeing 737s) that enable airlines to more effectively match capacity with demand and reduce operating costs. The last part discusses some examples of opportunities related to passenger services and operating practices.

Chapter 5 starts with a discussion about the stakeholder discipline required by management; developing a win-win relationship with other members in the value chain; and what can be learned from successful new entrants in such areas as marketing, cost control, and management science. The third part presents some ideas from businesses in other industries: flexible manufacturing—the use of the same factories for producing different cars within a short period of time (Ford); brand development (Marriott Hotels); the power of cost control combined with merchandising savvy (Wal-Mart and Target Stores); and the development of profit-and-loss models on each and every customer (Royal Bank of Canada).

The concluding chapter talks about scenarios: (a) the need for scenario planning; and (b) six examples of radical scenarios relating to: current and future aircraft; community-sponsored airline operations; two very different possibilities for alliances; international services by low-cost, low-fare airlines; airport passenger processing staff and systems; and the possibility of a conspicuous success by a single low-cost low-fare carrier. The main purpose of this section is that making the reader think about such scenarios may reduce management's risk of being unprepared for dramatic changes.

As this book goes to the publisher in early 2003, new complications and structural changes continue to surface. We see the limits to cost-effective technology as Boeing abandons, at least for the time being, the pursuit of more speed in aircraft design by canceling the Sonic Cruiser project and initiating a more economical 7E7 design. The bankruptcy of United Airlines reminds us that the second largest airline in the world is not immune to failure. Bankruptcy protection is also being actively considered by a number of other leading airlines worldwide. In Brazil, the two largest airlines (Varig and TAM) are talking about merging. In Europe, the independent low-cost carrier (Ryanair) acquired KLM's low-cost subsidiary (Buzz). The old pricing model continues to collapse. A decade ago, when the International Civil Aviation Organization last convened a world summit on liberalization, the world's civil aviation authorities were not sure whether to liberalize the aviation industry, let alone be willing to discuss openly the airline ownership and nationality issue. Now, when ICAO opens its 5th World Air Transport Conference at the end of March 2003, it seems there will be more urgency and momentum on not whether but how to liberalize. The implications raised in the book seem even more urgent in light of these and other events, as we contemplate the future of this industry so vital to the mobility of the world's travelers.

Acknowledgements

I would like to express my appreciation for all those who made this book possible, especially two contributing editors, Jim Hunt (formerly with Air Canada) and Jim Oppermann (formerly with America West) and: ABN Amro—Andrew Lobbenberg; AeroMexico—Juan Nicolas Rhoads; Airbus—Adam Brown, Paul Clark, Cronan Enright, David Jones, and Claire Labedaix; Air Canada—Rupert Duchesne, Alice Keung and Ross MacCormack; Airline Monitor—Edward Greenslet; Air New Zealand—Andrew David, Brendan Fitzgerald, Charmion McBride, Andrew Miller, Ed Sims, Paul Skellon and Norman Thompson; Alaska Airlines—Ketty Hsieh; All Nippon Airways—Masashi Izumi; AirTran—Robert Fornaro; American Airlines—John Darrah, Thomas Kiernan (recently retired), Scott Nason, Mike Pearce, and Lawrence Rosselot; Austrian Airlines—Wolfgang Prock-Schauer; ATA—John Heimlich and David Swierenga; Boeing—Douglas Ball, John Feren, Daniel Mooney, and William Swan; Bombardier Aerospace—Chul Lee; British Airways—Rod Muddle (recently retired), and Andrew Sentance; Centre of Asia Pacific Aviation—Peter Harbison and Derek Sadubin; Continental Airlines—William Brunger and Bonnie Reitz; Cranfield University—Fariba Alamdari and George Williams; DHL International—Michael Smith; Emirates—Nasib Sultan; GATX Air—Bruce Hogarth; GOL—David Barioni; IATA—Jack Adelman, Giovanni Basignani, Julian de la Camara, Peter Morris, Patrick Murphy, Koki Nagata and Richard Smithies; Iberia Airlines—Arturo Benito and Xabier de Irala; IBS—Peter Krebs, V.K. Matthews, and Stewart Wallace; ICAO—Chris Lyle, Vijay Madan and Upali Wickrama; JetBlue—Frankie Littleford; Lufthansa—Olaf Backofen and Carsten Spohr; M2P Technologies—David Palmieri; McKinsey—Lucio Pompeo; Morgan Stanley—

Kevin Murphy; NCR—Steve Dworkin, Brendan Hickman, Monica Smith, Andy Tellers, and Rick Volz; Roland Berger—Michael Beckmann; Rolls-Royce—Jim Guyette and Ken Perich; SpenserStuart—Michael Bell; Southwest Airlines—Dave Fintzen, John Jamotta, Lee Lipton, Pete McGlade and Susan Stich; Singapore Airlines—Stanley Kuppusamy; TAP, Air Portugal—Fernando Pinto; Transport Research Laboratory (UK)—Peter Mackenzie-Williams; TrueBrand—John Diefenbach; UK Civil Aviation Authority—Robert Cotterill and Trevor Smedley; Unisys R2A—Phil Roberts; Virgin Atlantic—Barry Humphreys; and Vanguard Airlines—Scott Dickson.

There are a number of other people who provided significant help in such areas as: the design and formatting of all the exhibits (Benjamin Kann, Ryan Leidal, and Matthew Whitcher at the Ohio State University); the design of the jacket for the book (Maria Ward at Colenso BBDO in Auckland, New Zealand); research regarding some material presented on the Japanese aviation market (Aya Nakamachi, a recent graduate of the Ohio State University); the production of the book (John Hindley—Consulting Editor, Claire Annals—Desk Editor, and the Ashgate in-house publication staff). Finally, I would like to thank my wife, Carolyn, for her support and patience.

Chapter 1

The Heavy Burden of Excess Complexity

The airline industry is one of the most visible sectors of a nation's infrastructure and normally one of the leading contributors to the economy of a nation. Yet, on a global and long-term basis, this sector has not only been unprofitable, it is also overwhelmingly complex.[1] For example, while it is possible to take any one of a dozen or so characteristics of the airline industry and show how it is common to some other industry, it is difficult, if not impossible, to find another industry that shares all of the characteristics at the same time. These characteristics make this industry very complex and make it difficult, as a whole, to generate any sustainable profitability, even though some individual airlines (such as Southwest in North America, Emirates in the Middle East and Singapore in the Asia-Pacific region) have been able to do so. The industry as a whole has also made money occasionally but inconsistently. Figure 1.1 shows the cumulative net profit margin of the airline industry worldwide between 1947 and 2000. During this period of more than five decades the industry achieved a net profit margin of less than 1 percent. Once the losses for the years 2001 and 2002 are added, the cumulative performance becomes even worse.

It is difficult to believe that conventional airlines can continue to function in their historical mode of operation due to the changing expectations and behavior of governments, shareholders, employees, and customers. While the business model of conventional airlines may be broken, the business model of the few successful low-cost, low-fare airlines is not. Consequently, the conventional airlines are in serious need of reform (in terms of scope and speed of change). This chapter provides an overview of the fundamentals and intricacies of the airline industry that have made it overly complex, producing a multi-faceted challenge for the industry—namely, the achievement of a reasonable net profit margin in the marketplace. It sets the stage for the chapters to follow which convey the message that, while the marketplace is expected to become even more turbulent, it is possible to break out of this zero profit game. This situation calls for stakeholders to exercise discipline and share a mindset. Moreover, the times we are living in require us to implement strategies that face critical issues at their root cause and that go farther, deeper, and wider than ever previously conceived.

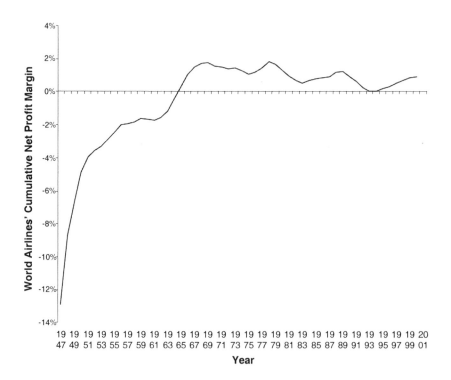

Figure 1.1 World Airlines' Cumulative Net Profit Margin (%)
Source: Constructed from data abstracted from various publications of *The Airline Monitor* and The International Civil Aviation Organization.

Industry Characteristics

Figure 1.2 shows a set of fundamental characteristics of the global scheduled airline industry. This list is by no means complete. There are other factors such as too much reliance on business traffic, over-exposure to certain geographic regions, and the highly technical nature of the business in such areas as aircraft, maintenance, information technology, and air traffic control. There are two important points about this information. First, it is the combination and simultaneous existence of these characteristics that have made this business complex and made it difficult, although not impossible, to earn a reasonable return. Second, while painful, the conventional carriers must reform by altering the influence of some of these constraints through the leadership of critical members in the value chain (management, labor, governments, airports, and so forth), through the achievement of a shared mindset, or else they risk having the changes forced on them by the heavy hand of government or the "invisible hand" of the marketplace.

Excessive Government Intervention

Airlines are not only regulated very heavily for safety's sake (aircraft and crew certification and operations) but also in areas related directly or indirectly to economics; control and ownership; mergers and acquisitions; alliances; passenger rights; protection of the environment; competition such as predatory pricing; and in the case of many international operations, capacity and frequency. Government intervention in terms of political, social and fiscal policies is understandable given the strategic importance of the airline industry. In the US, for instance, civil aviation accounted for 9 percent of the U.S. Gross Domestic Product in the year 2000.[2] Moreover, according to a recent US government report, productivity growth and the US Gross Domestic Product are directly related to an efficient and growing air transportation system. Consequently, the

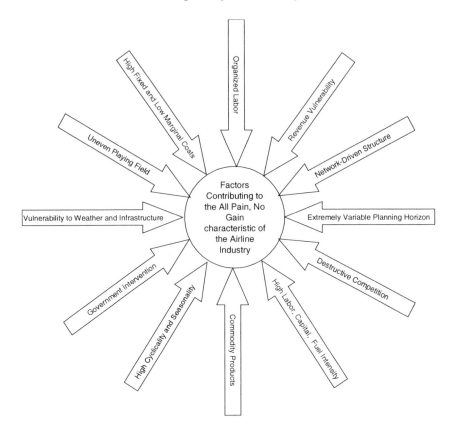

Figure 1.2 Fundamental Characteristics of the Airline Business

report goes on to say, a healthy airline industry is a national resource that should be enabled and allowed to prosper.[3] The existence of an airline is important to a country that is heavily dependent on spread-out business activity, tourism or exports of products that must travel by air. In some developing and even developed countries the need for flag carriers is tied not only to economic development but also to geographical linkages and in a few cases even to a government's political aspirations. Thus, from a given government's point of view, their airline's goal may very well be the number of destinations served rather than its profitability.

Because of the actual and perceived strategic value of an airline, most of the 187 ICAO states will not only continue to have flag carriers in one form or another but there will continue to be government involvement and intervention, not just in the areas of safety, security, and the protection of the environment but in economic regulations. The degree and area of intervention can and has varied from country to country. Examples include required service in particular markets, low fares in some markets, politically-appointed management, pressure to acquire aircraft of certain types and from certain manufacturers, and the control of competition in domestic and international markets.

Government intervention is not always bad. For example, it protects the public by enforcing safety, controls destructive competition, and prevents major strikes at times that would affect the general public good and hinder regional and inter-regional trade. It is also government intervention that has saved airlines in recent times in countries such as the United States, Switzerland, and Malaysia. It was the benevolent hand of governments that enabled some airlines to earn profits during the 20 year period between the beginning of the 1960s and the end of 1970s. However, after the events of September 11th, 2001, no government in Latin America gave support to its national carrier. Therefore, airlines in that region, presumably, experienced unfair competition from US some airlines that did receive government subsidy and did lower their fares to and from the US.

However, government intervention has not always been unidirectional or consistent. For example, some governments privatized their airlines, attempted to maximize the proceeds from their sale, and did not create a competitive post-privatization industry. Some governments sold their airlines and then bought them back (for example, Malaysia and New Zealand). In some countries, the government sold their airline, then interfered later when they did not like the private marketplace activity. Canada forced Air Canada to absorb bankrupt Canadian Airlines (through a combination of financial and regulatory pressure). In some countries restrictive regimes (such as in Scandinavia and India) have held back the development of airlines. On the other hand, government intervention stabilized the domestic aviation industry in China which now represents the third largest market in the world (after the United States and Japan). Within China, between the years 1989 and 2000, there was an increase of 400 percent in the number of routes operated and an increase of 585 percent in the number of weekly seats offered.[4]

The consequence from government inaction or insufficient action is also a reality. Consider the lack of infrastructure. Consider also the economic and personal cost of delays caused by inadequate airport runway and terminal capacity and the outmoded air traffic control system. According to the government report on the US aerospace industry cited above, commercial air transport had become (prior to the events of September 11th, 2001) unpredictable, with frustrating and expensive delays. According to the government itself, the air traffic system is based on 1960s technology and operating concepts and was approaching gridlock. Air passenger traffic in the US has increased 40 percent since 1991. Since then, one new airport (Denver International) has opened and seven major new runways have been built—representing a 5 percent increase in the number of runways at the top 50 airports.[5] Consequently, there is a critical need for the government to facilitate the modernization of the air traffic control system and the significant enhancement of airport and runway development.

Network-Driven Structure

The airline business is a network business and the network effect adds enormous complexity. Revenue is generated on an origin-destination (O&D) basis, while costs are generated on a flight segment basis. Consequently, the operating and financial information of one segment is interlinked with other segments at different levels. The performance of long-haul flights is affected by the performance of short-haul flights. Take, as an example, the short-haul segment

Mexico City-Acapulco. More than 50 percent of the traffic on this segment is connecting traffic from long-haul flights from the US and Europe. The revenue contribution from these long-haul flights to the short-haul segment is very small. A change in one area can have a tremendous impact on the costs of another area. The decision to acquire one or two high-capacity aircraft at the top end by a relatively small airline can have an enormous impact on the crew and maintenance costs. Crew members and mechanics need to be trained to move up the ladder according to the provisions contained in union contracts. Because of the network effect of the business, it is very difficult to develop a unique set of performance measures that are common to all departments and the divisions of an airline. The problem arises because of the necessity either to allocate revenue to a segment or costs to an O&D. Either way, the end result is controversial due to the lack of a unique answer. Some aspects of the problem exist even for point-to-point carriers—both conventional, such as Virgin Atlantic and low-fare, such as Ryanair.

The global interline system is another feature of the airline network business. The system provides global travelers convenient services, for example, payment in one currency, for travel involving multiple airlines and intermediaries. While convenient, and even essential for frequent global travelers on business, interlining systems have costs that are not visible to most passengers. This extra cost does not exist for point-to-point carriers who choose not to provide interline service. Consequently, large global airlines cannot duplicate the simplicity and less-networked structures of point-to-point low-fare carriers.

Organized Labor

The airline industry is highly unionized not only with respect to the percentage of the labor force that is represented but also the strength of the unions. Aircraft pilots and mechanics, by virtue of their skills and federal licenses, have negotiated not only complex wage structures but also productivity provisions in their contracts.

Productivity is impacted by the way management schedules crews; government requirements relating to work rules and rest periods; and contract provisions based on seniority. Some might say the undue influence of unionized labor, particularly pilots, on management's decision-making process is clearly evident from the implementation of scope clauses that limit the degree to which management can establish low-cost subsidiaries, as demonstrated by the experience of US Airways with respect to the use of regional jets and United with respect to the establishment of it shuttle service on the US West Coast.

Some unions are less powerful than others. Airport customer service, ramp services, and purchasing and supply are examples of unionized areas with less bargaining power because they are relatively easy to replace.

First, the power of the pilots and mechanics unions is derived, (a) from the fact that they, particularly pilots, are almost impossible to replace, and (b) to the poor negotiating strategies of airline management, exemplified by the ESOP experience at United.

Second, the level of unionized labor wages is high in the airline industry due to the results of "pattern bargaining", a procedure that has enabled a union at one airline to negotiate close to the industry-leading contract.

Third, even more disturbing is the fact that labor at some airlines collaborate with their counterparts at other airlines more than they do with their own management.

Fourth, it is interesting to note that contracts do not expire in the airline industry. At the end of the contract period, the contract continues to be in force until it has been renegotiated. In the US, this provision was included by the Congress in the Railway Labor Act to reduce the possibility of work stoppage that could be detrimental to interstate commerce.[6]

Finally, in the airline industry an intensive labor action (strike) can have an enormous impact on the economy, not only in smaller countries dependent on, for example, tourism, but also in larger countries where air transport has a major impact on the economy. In fact, even a slowdown by the labor force can have a tremendous impact on an airline, evidenced by the experience at United during the year 2000. Consequently, direct Presidential/government intervention exists in the US to forestall strikes.

Despite the strength of unionized labor in the airline industry, it is possible to have a good labor-management relationship. Consider the experience at Southwest which is heavily unionized. Southwest management has been able to keep unit costs low by achieving high labor productivity. Consider for example, that the output per employee at Southwest is 20 percent higher than at United, even though Southwest operates smaller capacity aircraft and flies shorter distances. Even more revealing is the statistic that in 2001, the average number of hours flown in a month by a Southwest pilot was 62 compared to 36 for United.[7] Conversely, poor labor-management relations raise costs but also create a poor image with the traveling public, exemplified by the experience of United during the Summer of 2000.

High Labor, Capital, and Fuel Intensity

Labor represents the largest expense of every major airline in the US. See Figure 1.3 that shows the data for US airlines for the year 2001. Labor costs have increased at the second-fastest rate in the past two decades (in spite of increase in two-crew cockpits), surpassed only by the cost relating to the acquisition of fleet. According to the Air Transportation Association of America, for US airlines labor costs increased by 209.7 percent between 1982 and the second quarter of

2002. The average cost of products and services only increased by 162.4 percent between the same periods.[8]

The capital intensive nature of the airline business can easily be seen from (a) the price of assets such as aircraft (ranging typically from about US$15 million for a small regional jet to over US$150 million for a large wide-body aircraft), (b) the number of aircraft in the fleet of the large global network airlines such as American, Lufthansa, and Japan Airlines, and (c) what routes they fly. As airlines increase the length of haul of markets served, two aircraft are required to serve long-haul routes. London-Chicago is an example of a route where an airline can achieve high aircraft utilization. The return trip requires exactly one aircraft that can rotate on itself ad infinitum or until the first disruption or maintenance input. Not only are the initial and operational costs of assets high, the replacement costs of parts are even higher. Take, for example, the replacement costs of aircraft seats. One estimate shows that the replacement cost of a seat in the economy class cabin can be US$1,500 while the cost of a first-class seat with all the paraphernalia (electronics for seat adjustment to convert it from a standard position to a flat bed and the in-seat in-flight entertainment system) can run as high as US$60,000.[9] For US airlines, the costs of aircraft fleet increased by 256.9 percent between 1982 and the second quarter of 2002.[10]

While fuel represents a significant percentage of the total costs (13.6 percent as shown in Figure 1.3), fuel costs did in fact come down from about 20-25 percent during the early 1980s. Part of the explanation is the reduction in the price of fuel and part is in the higher fuel efficiency of each generation of new aircraft. Another problem with fuel costs relates to the volatility in the price of fuel that can have a tremendous impact on profitability, particularly for airlines operating older aircraft. Some airlines have tried to lessen the impact of fuel price swings through the implementation of hedging strategies. Finally, an additional problem relating to the cost of fuel, is the price charged by certain monopolies in Latin America, where a monopoly handling the fuel may charge an extra 6 percent fee just for putting the fuel in the plane.

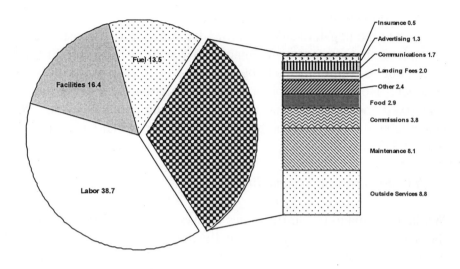

Figure 1.3 US Airline Cost Components, 2001
Source: Constructed from data contained in the publication *The Airline Monitor*.

High Fixed and Low Marginal Costs

For a typical airline, fixed costs can approach 75 percent of the total costs. In this context, fixed costs include the cost of aircraft, wages, and airport and maintenance facilities. The high fixed costs combined with low marginal costs and the perishable aspects of the product have led managements to introduce some fare structures and levels during normal times that did not, and still do not, reflect the cost of equipment, let alone fully-allocated costs. The justification typically offered is that aircraft that cost between 15 and 150 million dollar cannot sit idly on the ground. Consequently, in many cases low fares have been introduced to maintain and even increase market share. Some airlines would argue that low fares fill, supposedly, marginal-cost seats and contribute to the fixed costs. Riding on the back of the high fixed costs argument, some fares have even been introduced to simply generate cash flow.

On the negative side, it is true that massive fixed costs are an obstacle in that management cannot easily scale back in a short time period due to commitments to be a network operator and due to labor contracts. However, there is a small consolation to the problem of high fixed costs and low marginal costs. During upturns, airlines can generate significant revenue by transporting additional passengers at marginal costs.

High Cyclicality and Seasonality

While the demand for air travel has always been highly dependent on the state of the economy, historically the growth in air travel has exceeded the growth in the economy. Figure 1.4 shows that by the late sixties, the 10 year moving average ratio of the growth in passenger traffic (measured in Revenue Passenger Miles or RPM) to the growth in the economy (measured by the GDP) was more than two. This ratio has been declining and now is less than one. The other noticeable change is that in recent years, airline passenger traffic (particularly the part relating to business) has become overly sensitive to quarterly corporate profits. Businesses tend to reduce staff travel after experiencing losses. This sensitivity can be easily understood when one considers that some walk-up fares have been as much as 10 times the lowest leisure fares.

There are two other significant points relating to the relationship between airline traffic and the state of the economy. First, while the discussion above stated that the airline industry is cyclical and within the United States has become mature (as shown in Figure 1.4), it does not mean that the impact of the state of the economy is the same on all airlines. Low-cost, low-fare airlines, for example, appear to have fared fairly well during the current downturn. Second, while business travel tends to be tied closely to the state of the economy, leisure travel is tied more closely to the price, disposable income and the consumer confidence Index.[11]

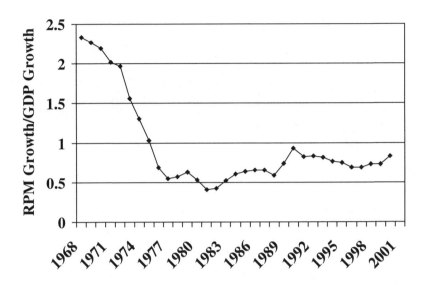

**Figure 1.4   Ratio of the Ten-Year Moving Average US RPM to
GDP**

Source: Computed from data in various issues of *The Airline Monitor*.

The other significant problem of the airline industry is the
highly seasonal nature of the demand for air travel.   First, the
demand varies substantially by month.   Consider, for example, the
disparity in the demand for air travel on the North Atlantic between
February and July. Second, there is an enormous variation in traffic
by the day of the week. Figure 1.5 shows one example of this
variation in the market between Newark, New Jersey and Houston,
Texas. Such variations in traffic lead to excessive complexity in
operations and high costs.   Should an airline buy enough aircraft to
meet the trans-Atlantic demand for November or July? Similarly,
should an airline schedule the size of the aircraft to meet the demand
for Monday morning or Wednesday afternoon? And having different
aircraft to meet the variation in demand leads to the problem of
overly diversified fleet, discussed in Chapter 2.

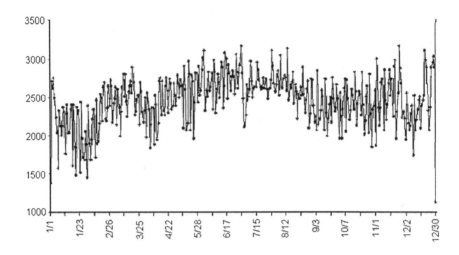

**Figure 1.5 Passenger Boardings by Day in the Newark-Houston
     Market, 2000**
Source: Continental Airlines.

Revenue Vulnerability

Revenue of the airline industry is extremely vulnerable. To begin
with, the fact that an airline's product is perishable puts pressure on
management to keep load factors high even if it means selling
capacity at marginal revenue. While, this attribute is no different
than in many other industries, it is more critical in the airline
industry. The cyclical and seasonal nature of the demand leads to
excess capacity since capacity is relatively fixed and demand is
highly variable. Airlines have attempted to minimize the revenue
vulnerability in this area by implementing sophisticated revenue
management techniques and flying families of aircraft. However,
until airlines can schedule these aircraft families and manage
capacity and their costs effectively in real time and until
manufacturers can reduce significantly the time between aircraft

orders and deliveries (through, for example, improved manufacturing processes), revenue will continue to remain vulnerable.

The other problem is that profit in the airline industry is the difference between two very large numbers, revenue and costs. However, as discussed above, a very large percentage of costs are fixed in the short term. Revenue, on the other hand is extremely vulnerable to many external factors such as entry into a market by a low-cost carrier, irrational competitive moves relating to fares and market entry, threats of terrorism and war, even the threats of strikes by labor, and an increase in taxes by the government. Take, for example, market entry. While competitors can enter a new market in any business, it is particularly easy in the airline industry in those markets that are not regulated by international bilaterals or constrained by slots. Unlike many other businesses, market entry into the airline industry in many markets is facilitated by the mobility of the aircraft and the existence of excess capacity. Consider, for example, the recent market entry by a number of US airlines into markets served by the struggling US Airways. Examples of revenue vulnerability in other areas include the threat of a war between the United States and Iraq and the increase in taxes relating to the price of an airline ticket, particularly in short-haul markets. The total taxes on a US$200 ticket can be US$40, 20 percent of the price of a ticket. Moreover, these are just the direct transactional costs; there are more "hidden" taxes.

Revenue vulnerability can also be illustrated by the high visibility and dynamic nature of the global airline industry. Given that many names of airlines are associated with the names of countries and given the visibility of the airline industry, it has been the subject of terrorism and hijackings—events that have a substantial and immediate impact on revenue. The dynamic nature of the airline industry and its impact on revenue can also be illustrated by the October 2002 bombings in Bali and the attacks in the Philippines and their impact not only on the tourism industry in the impacted areas but also on the tourism industry in the broader Southeast Asia. Finally, an issue that will be addressed in more detail in Chapter 3—consumer revolt to price service options—is certainly making revenue vulnerable for the conventional airlines.

Destructive Competition

Deregulation in some countries and liberalization in others are only partly responsible for the excessive competition in some regions. Economies of scale, density, and scope have further added to the competitiveness of the airline industry. Economies of scale and scope have led to overexpansion in some markets. Economies of density (providing lower unit costs of larger airplanes) have led to the acquisition of large aircraft in some other markets. Other attributes of the industry, such as overlapping hub-and-spoke systems that are disproportionate in size relative to the size of O&D traffic, have also led to a situation that can only be described as destructive competition. Another factor that contributes to competition reaching a destructive level is when some developing countries allow their weak airlines to operate without paying landing fees or taxes. The artificially lower costs enable these airlines to introduce artificially low fares—a common practice in some Latin American countries.

The overcapacity issue is not difficult to envision. Many airlines have always been market-share driven and the conventional wisdom has always been that market share is more or less a function of frequency share. Other factors such as frequent flyer programs have very little impact on market share (a percent or so on either side of the relationship established between market share and frequency share). For a few airlines this observation may, however, be valid on a traffic basis but not necessarily on a revenue basis. A second reason for overcapacity relates to the lack of harmonization between the times when capacity is ordered and when it is delivered relative to the economic cycles. The third factor, at some US airlines, has been the existence of scope clauses that have prevented management from replacing higher capacity jets with smaller capacity regional jets. The fourth factor for overcapacity is government intervention that in some cases has prevented airlines from exiting the marketplace. In some cases the same argument can be applied to the financing division of aircraft manufacturers and aircraft leasing companies who have helped keep struggling airlines in business rather than take back airplanes already in service. Suppose we divide the cause of overcapacity into two parts. The first is the part caused by variations in demand due to seasonality and economic cycles. However, it is the equally important second part that is under the control of management, governments, labor, and some sector of the financial community that has led to destructive competition in some markets.

Commodity Products

Whether an airline seat is a commodity product is a highly debated and emotional discussion. From a broader perspective a seat is a seat. Some would argue that there is no significant difference in an economy-class seat offered, for example, by American and United between New York and London. Both airlines fly similar aircraft, with similar schedules, with similar in-flight service, and similar loyalty programs. Consequently, this group would argue that product differentiation in the airline industry is difficult relative to products such as hard goods where, for example, one could argue that Mercedes builds better cars. This is a valid argument except that there are airlines that have succeeded in differentiating their services, exemplified by Singapore Airlines and Virgin Atlantic. Both have proved that product differentiation is not only possible but sustainable.

It could, however, be argued that some airlines who claim to have differentiated their product still match fares. In other words, while the product differentiation may not be significant enough to warrant a premium fare, it may be significant enough to obtain a higher share of the premium traffic. If this argument is valid then continuous improvements in the product have simply led to higher costs and marginal and temporary benefits in market share—not necessarily higher unit revenues or profits. Second, the inability to differentiate the product has led to not only the matching of prices in competitive markets but it has also influenced prices in noncompetitive markets. British Airways, for example, could feel the need not only to match prices in the New York-London nonstop markets with competitors such as Virgin, American, and United but it could also feel the need to compete in some indirect way with carriers like KLM offering connecting service through Amsterdam.

The second discussion relating to the commodity nature of the airline product is a misunderstanding when it is applied to different prices charged for similar seats in the economy-class cabin. It is often heard that if a "seat is a seat", then why are some passengers paying much more than others. While the amount of differential can be argued (as seen in Chapter 2), the rationale is valid. Passengers paying higher fares have fewer restrictions attached to their fares. These restrictions include the need to make reservations and pay for the ticket in advance, restrictions on the days of travel (both outbound and inbound), the need to pay for changes in reservations, and possibility of the loss of fare in case of a cancellation. The most important reason for the higher fare for some seats relates to keeping an inventory available for the last-minute reservations.

The commodity attributes of the product reduce the degree to which management can control pricing decisions. This limitation becomes much more critical when other attributes such as perishable product and network-driven business are taken into consideration simultaneously. Consequently, management must rely on exercising its power to control costs. However, as mentioned above, a large percentage of the costs have not been under management control. As other parts of the book will discuss, it is perhaps time to take back some of these controls relating to costs. The airline can then try to match the perceived value of fare and product features to the customer with the cost of offering them. Finally, the distribution system also has an influence in that travel agents and Internet search engines try to find the lowest fare for the customer.

Vulnerability to Weather and Infrastructure

The airline industry is extremely vulnerable to two factors that are outside of its control. First, a sudden deterioration in the weather can have an enormous impact on an airline, not only at an airport with bad weather but throughout an airline's network. If a major airport such as Chicago's O'Hare closes for an extended period of time due to snow, the impact by global airlines is felt worldwide. Again, because the airline business is network-driven, operations involving aircraft and crews are affected worldwide. There is a significant difference between the airline business that is affected by bad weather worldwide and other businesses that are very likely to be affected locally or regionally. Even other network businesses, such as telecommunications, can easily reroute their traffic over other hubs in the event of difficulty as long as sufficient capacity exists. The complexity of the business can easily be envisioned by the management of irregular operations of a major airline when a number of flights are delayed or cancelled due to weather, requiring relocations of crews and aircraft and re-booking passengers worldwide.

Second, within the airline industry the aviation infrastructure remains limited in capacity, is inflexible, and is strongly controlled by local interests. Airport capacity constraints at key locations have held back the expansion of airlines, exemplified by the situation at New York's Kennedy, London's Heathrow and Tokyo's Narita airports. Moreover, further liberalization of bilateral agreements has been held back in many regions by infrastructure constraints such as lack of airport slots and airspace congestion (for example, between the US and the UK). In the case of Japan, low-fare airlines were not able to penetrate the domestic market due to the lack of sufficient slots at the Narita Airport. Similarly, additional service cannot be added at Heathrow Airport due to limited slots. Another example could be when Cathay and Qantas recently increased the frequency in the Sydney-Hong Kong market to be closer to the maximum allowed by the bilateral (36 flights a week), presumably to pre-empt entry in the market by a new low-fare airline. It is also important to note that airlines often get bad publicity relating to poor operations when the situation is not under their control such as delays due to insufficient airport and air traffic control capacity.

Sometimes, the irrational behavior of an airline can easily be explained due to the constraints of the infrastructure. An airline may schedule flights with low load factors just to protect slots. Another airline may be forced to use a high capacity aircraft (and limit frequency) due to the limited availability of slots. Yet another airline may operate a very late night flight from one airport to a neighboring airport and a return in the very early morning hours due to the unavailability of overnight gates, parking, or maintenance. Unaware of such a constraint, crew and passengers are very likely to blame management for "poor" scheduling of aircraft and crews.

Uneven Playing Field

The global airline industry is made up of players at very different stages of development (for example, Cathay Pacific vs. Air China), at different levels of efficiency (for example, US Airways vs. America West), operating in very different regulatory environments (for example, within the United Kingdom vs. Japan vs. Mexico), and operating in different economic environments (for example, to and from India vs. to and from the United Kingdom), and receiving varying degrees of government subsidy. Part of the explanation of the stage of development is simply due to the evolution of the airline industry, for example, due to the progression of the liberalization movement in the past 20 years. Consider, for example, the rate of the liberalization process within the domestic market in Japan vs. the United States.

Part of the explanation relates to the role of governments. Contrast the difference between the development of the flag carrier of Singapore vs. Malaysia or the Philippines. The degree of coordination and cooperation among different members in the air transportation value chain (airline, airport, government, and so forth) in Singapore is much higher than in most countries around the world. However, while coordination may be good in one sense (if it leads to greater efficiency), it may be bad in another sense (if a government ends up subsidizing its flag carrier to the point that the subsidized carrier "dumps" low fares in a market, leading to unfair competition and market distortion). In addition, flag carriers in some countries have a tighter control of distribution outlets. Finally, part of the explanation is simply the age of the airline, for example, an older airline such as Aeroflot that is saddled with an older fleet and newer airlines such as Vietnam Airlines whose development has been phenomenal. This is due partly to the ability to start with a clean slate and partly to act quickly with, for example, the decision to replace the older fleet consisting of Tupolev aircraft with western aircraft such as the Airbus A320, ATR, Boeing 767, and Boeing 777.

The difference in the stage of development becomes significant when these airlines end up competing in the same global markets. The flag carriers of Malaysia and Singapore compete in the same markets to and from Europe, to and from the West Coast of United States, and within the Asia-Pacific region. Consequently, global airlines are competing in playing fields that are not always level. Consider an airline that is financially supported by its government competing with an airline that is self-sufficient. Consider also an airline that has spent five years developing an international route incurring significant development costs. Under the old regulatory system a new carrier could not just enter the marketplace when the "pickings are ripe". In the new liberalized environment in some regions, such as the United States and Europe, this is not only possible but it is happening. While, one could say that is what competition is all about, the problem is that ease of entry may not be accompanied by ease of exit. The inability to exit can be illustrated by a government requirement to continue service on the route or simply the desire on the part of the incumbent airline to protect a slot.

Extremely Variable Planning Horizon

In the airline industry management has always had to balance between a very long-term orientation and a very short-term orientation. At one end of the spectrum, consider decisions involving fleet and airport development. When United ordered the Boeing 777, it would be a number of years before the aircraft would be delivered and it would be in operation probably another 20 to 25 years. The same consideration applies to airlines when they ordered the Airbus 380. Consequently, management is faced with making very long-term forecasts of the economic, regulatory and operating environments. Given the high cost of these assets, a wrong decision can be very expensive. These are expensive assets and may be very difficult and time-consuming to shed. Similarly, it takes a very long time to develop not only totally new airports but also components of an airport. Consider the number of years it took to plan, build, and open the new airport in Denver. Consider the number of years it has taken to construct the second runway at Tokyo's Narita Airport and to receive the go-ahead to build Terminal 5 at London's Heathrow Airport. Adding a new runway at Toronto's Pearson Airport took seven years.

At the other end of the spectrum, management must make decisions whose impact would be felt literally within minutes. Pricing decisions are one example. Airlines change their prices not just daily but many times a day. Revenue management and pricing groups are making decisions regarding pricing and capacity virtually by the hour and around the clock. These two extremes of the planning horizon make planning extremely risky.

The Multi-Faceted Challenge

The fact that the global airline industry has not made a net profit on a cumulative basis during the past five decades is not news. What is beginning to appear as news is the strong possibility that the business model of the conventional airline industry is broken, mostly because the ability to reduce its costs is not aligned with its revenue generating capability. The misalignment was not an issue in the regulated era when airlines were able to pass increasing costs on to their customers, governments controlled competition, traffic was growing at rates well in excess of the growth in the economies, and productivity was increasing at a rapid rate due to improvements in aircraft technology (producing higher speeds, capacity, and range and reducing unit operating costs).

On the other hand, for low-cost, low-fare airlines, their model is not broken because their cost structure is aligned with their revenue structure. Figure 1.6 shows the unit operating costs of Southwest (as measured in cents per ASM) against the conventional US airlines. Not only are the costs of conventional airlines much higher even without an adjustment for the length of haul and the number of seats per departure, but the difference becomes even more pronounced when a hypothetical curve is generated for what the costs would be for Southwest at different lengths of haul. According to one recent analysis, the total operating cost disadvantage suffered by other major carriers relative to Southwest was US$18.6 billion in the year ended 30 June, 2001.[12] The curve through the Southwest point shows what the unit operating cost of Southwest would be if it was operating at different stage lengths. A revenue differential does exist, but it is not sufficient to compensate for the difference in costs. Figure 1.7 shows the difference in the operations of a given aircraft by United vs. JetBlue, substantiating once again the need to align costs. Even after adjustments are made for differences due to age of aircraft and comparable maintenance costs, United would still be paying 28 percent more than JetBlue for direct costs alone.[13]

The key challenge for conventional airlines is to adapt their business model to a rapidly and profoundly changing marketplace. The strategies required to achieve this goal will require hard decisions not only on the part of airline management but also by other members of the air transportation value chain—labor

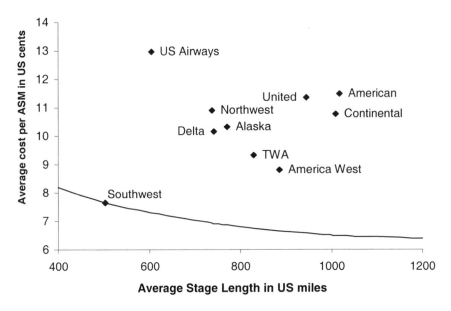

Figure 1.6 Comparison of Unit Costs for Selected US Airlines
Source: Unisys, R2A Transportation Management Consultants.

government policy makers, and providers of the infrastructure, just to name a few. Not only does each group need to exercise discipline, but there needs to be a shared mindset. It appears that while the profit margins of airlines have been moderate even in the best years, providers of products and services to the airline industry have generated substantial amounts of profit. Consider, just the case of airport operators. Tables 1.1 and 1.2 show the operating profit margin for the top 35 airports and the top 35 airlines.

There are two striking observations. First, airports worldwide have generated profits margins that are considerably higher than the levels generated by airlines. Second, the top three airlines are all low-cost, low-fare airlines. Consequently, it is clear that conventional airlines need a new business model and that they should look to the low-cost airlines for gaining insights into the components of this business model.

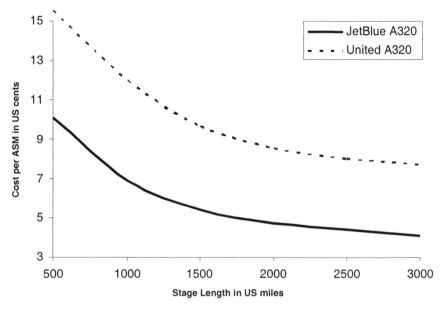

Figure 1.7 Differences in Unit Operating Costs between a Conventional and a New Low-Cost Airline
Source: Unisys, R2A Transportation Management Consultants.

Conclusion

I hope that you now see that the airline industry is complex and unique, not because of any one characteristic but because of a combination of all of the aforementioned characteristics. It is an industry that is layered like a cake, that is, one layer of complexity added on top of another. The industry shares some characteristics with other service industries such as hotels and car rentals—variable demand, perishable product, and so forth. It shares other characteristics with product industries such as auto manufacturers—high fixed costs, powerful unions, and so forth. Moreover, while the business model within the airline industry is fairly obvious, it is the existence of all these characteristics that makes the execution of the

Table 1.1 Airport Performance Indicators: Operating Profit for the Year 2000

Ranking	Airport	Operating Profit (%)
1	Auckland	56.5%
2	ACSA	49.6%
3	Washington National	48.3%
4	Melbourne	43.8%
5	Singapore	41.8%
6	London-Heathrow	40.7%
7	Calgary	38.6%
8	Perth	38.0%
9	Copenhagen	37.5%
10	London-Gatwick	36.4%
11	Honolulu	35.5%
12	Vienna	34.8%
13	BAA Group	34.2%
14	Brisbane	33.8%
15	Frankfurt	33.2%
16	Birmingham	32.6%
17	Vancouver	31.8%
18	Oslo	31.7%
19	Sydney	29.4%
20	Zurich	27.9%
21	Aer Rianta	27.8%
22	ANA	26.3%
23	Hawaiian Airports Group	26.1%
24	Stockholm	26.0%
25	Amsterdam Group	25.0%
26	AENA	24.9%
27	Geneva	23.4%
28	Osaka Kansai	22.1%
29	Los Angeles	21.6%
30	Manchester	20.5%
31	Aeroports de Paris	18.5%
32	Swedish Airports Group	18.3%
33	Aeroporti di Roma	17.5%
34	Budapest	17.2%
35	Washington Dulles	16.4%

Source: Transport Research Laboratory, UK.

Table 1.2 Airline Performance Indicators: Operating Profit for the Year 2000

Ranking	Airline	Operating Profit
1	Ryanair	22.7%
2	Southwest Airlines	18.1%
3	EasyJet	17.2%
4	Cathay Pacific	15.3%
5	Thai Airways	13.7%
6	Singapore Airlines	10.3%
7	Delta	9.8%
8	China Southern Airlines	9.5%
9	Qantas	9.4%
10	Finnair	9.2%
11	SAS	8.9%
12	Continental	7.4%
13	Air France	7.3%
14	American	7.0%
15	Lufthansa	6.6%
16	Lan Chile	5.3%
17	Northwest Airlines	5.1%
18	Austrian Airlines	4.3%
19	Varig	3.7%
20	United Airlines	3.4%
21	Air New Zealand	3.0%
22	Air Canada	2.8%
23	Japan Airlines	2.5%
24	Air Malta	2.4%
25	All Nippon	2.3%
26	KLM	1.5%
27	British Airways	0.9%
28	Iberia	0.8%
29	Croatia Airlines	0.8%
30	US Airways	-0.5%
31	Crossair	-0.6%
32	Icelandair	-1.6%
33	Alitalia	-9.6%
34	Malaysian Airlines	-14.8%
35	Virgin Express	-22.4%

Source: Transport Research Laboratory, UK.

business model very complex. However, despite the inherent complexities surrounding the airline industry, some of which we will never be able to eliminate (such as government intervention), the challenge for us all is to involve ourselves in those areas where we can make a difference. And, in order to make a difference in areas where we do have some control, it is necessary first to look back at the strategies that were successful, unsuccessful, or that had mixed results.

Notes

[1] Crandall, Robert L, "The Unique U.S. Airline Industry", in *Handbook of Airline Economics* (ed. David Bond) (New York: McGraw-Hill, 1995), pp. 3-8.

[2] *"The National Economic Impact of Civil Aviation"*, A report by DRI-WEFA, Inc., A Global Insight Company in collaboration with The Campbell-Hill Aviation Group, Inc., July 2002, p. 4.

[3] The US Government, *"Final Report of the Commission on the Future of the United States Aerospace Industry"*, Arlington, Virginia, USA, November 2002, Chap. 2, pp. 1-2.

[4] Williams, George, *and Airline Competition: Deregulation's Mixed Legacy* (Aldershot, England: Ashgate Publishing, 2002), pp. 10-11.

[5] The US Government, *"Final Report of the Commission on the Future of the United States Aerospace Industry"*, Arlington, Virginia, USA, November 2002, Chap. 2, pp. 1 and 13.

[6] Linenberg, Michael J. and Sandra Fleming, *Airline Industry: Airline Equity Investments*, A Report by Merrill Lynch, New York, 29 September 2000, p. 12.

[7] Velocci, Anthony L. Jr., "United Flying Headlong Into An Uncertain Future", *Aviation Week & Space Technology*, 16 December 2002, p. 23.

[8] Air Transportation Association of America, *Airline Cost Index*, 2nd Quarter 2002, p.3.

[9] Lunsford, Lynn J. and Daniel Michaels, "Masters of Illusion Make Jet Cabins Seem Roomier", *Wall Street Journal*, 25 November 2002, p. B5.

[10] Air Transportation Association of America, *Airline Cost Index*, 2nd Quarter 2002, p.3.

[11] Linenberg and Fleming, *op. cit.*, p.15.

[12] Unisys R2A Scorecard, *Airline Industry Cost Measurement*, Volume 1, Issue 1, October 2002, p. 15.

[13] Velocci, Anthony L. Jr., "United Flying Headlong Into An Uncertain Future", *Aviation Week & Space Technology*, 16 December 2002, p. 23.

Chapter 2

Hard-Won Lessons:
The Continuum from Success to Failure

Although the unprecedented events of September 11th, 2001 had immediate, dramatic, and widespread effects on the airline industry, the industry was already undergoing fundamental and irrevocable changes. They include, but are not limited to the proliferation of low-cost, low-fare airlines, the transparency of fares provided by the Internet, the remarkable success of regional aircraft operating in thinner markets, the consumer revolt by last-minute travelers against high fares, and a sharp downturn in the economy. The events of September 11th, 2001 simply added fuel to a smoldering fire and began the era when the conventional airlines must look for a new business model. The problems of the airline were global in nature. For example, in Mexico passenger demand began to decrease at about 6 percent almost a year prior to the events of September 11th, 2001. While it is impossible to predict the exact timing much less the ultimate outcome of these negative forces, it would be useful to review how conventional airlines might integrate hindsight with insight to develop foresight. Although this chapter divides a few important developments into three categories (successes, debatable, and failures), the divisions are to some extent arbitrary which explains the rationale for the subtitle of the chapter—the continuum from success to failure.

Conspicuous Successes

Safety

Commercial air travel is the safest mode of transportation due to the collective efforts of aircraft manufacturers, governments, airline management and operations, and the providers of the aviation infrastructure—airports and the air traffic control system. Aviation accidents account for only 2 percent of all transportation related deaths in the United States.[1] According to an analysis by the Air Transport Association, "in a typical three-month period, more people die on the [US] nation's highways than have died in all airline accidents since the advent of commercial aviation".[2] Although the airline industry has achieved a very low and stable accident rate, it continually works toward decreasing the accident rate. No industry is as rigorous or as committed.

Accessibility

The global airline industry has developed schedules, a network, and an interline system that provides incredible accessibility. A passenger can get from one point to virtually any other point on the planet in a little more than one day. A business person based in New York can attend a meeting in Los Angeles and return home the same day. Airlines offer nonstop service in long-haul markets such as New York-Hong Kong. Prior to the events of September 11th, 2001, United was contemplating a nonstop flight between Chicago and New Delhi, a flight that would take a passenger almost half-way around the world. High-density markets around the world (Boston-New York, Paris-London, Tokyo-Osaka, Taipei-Hong Kong, Kuala Lumpur-Singapore, Rio-Sao Paulo) have either shuttle or almost shuttle-like frequency. During September of 2002, there were 132 flights a day in each direction between Rio and Sao Paulo. This frequency represents a flight every seven minutes. This accessibility has made a significant contribution to the globalization process.

Reliability

The airline industry provides extremely reliable service, given the uncertainties of weather, mechanical failures, limited capacities of airports and the air traffic control systems. Most flights arrive within 15 minutes of their scheduled arrival time. The completion rate (flights operated as a percentage of flights scheduled) is in the high nineties. Such a high reliability, which we tend to take for granted, is only partly the result of advanced technology in the aircraft and the air traffic control system. It is also the result of complex operations control centers developed by airlines to run the day-to-day operations, and respect for the published schedule. Skilled and experienced personnel in these centers work around the clock to manage and integrate the numerous operational functions such as flight planning and dispatching, as well as managing the unplanned operational irregularities caused by severe weather or aircraft breakdowns.

Continuously Lower Fares

Although passenger yield has been declining since the early 1970s, the rate of decline increased in the late 1970s and early 1980s. See Figure 2.1. The higher rate of decline in passenger yield has been the result of (a) lower unit costs (due to increases in productivity), (b) the proliferation of the liberalization process, (c) the availability of other uses of discretionary dollars, for example, cruise business and other leisure activities, and (d) the development and implementation of revenue management systems that enabled airlines to introduce deeply-discounted fares to fill seats that might otherwise have flown empty. The lower fares stimulated traffic (also seen in Figure 2.1), and although capacity controlled, from a leisure passenger's point of view, these fares represented a bargain. Even today (in early 2003), the Internet shows that a passenger willing and able to meet certain conditions can travel between New York and London roundtrip for US$248, and for US$200 using connecting flights.

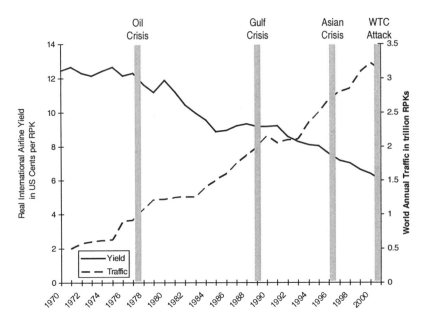

Figure 2.1 Airline Declining Yield and Traffic Growth Trends
Source: Constructed from the data contained in publications of the International Civil Aviation Organization.

Constant Innovation

Not only has the airline industry offered its products at continuously lower fares, but it also has provided continuous innovation in its products and services. Each generation of aircraft has been improved in some way, in such areas as higher speed, longer range, and more comfortable cabins that contain a broad spectrum of entertainment and service systems. On the ground, improvements include a broad range of distribution channels, fast airport check-in processes, comfortable and well-equipped airport lounges, and automated information systems that keep the passenger informed of the status of the flight. For premium-class passengers, some airlines provide ground transportation and facilities at airports to freshen up upon arrival.

Lower-Cost Feeder Airlines offering Jet Service

Although feeder airlines have been around for many years, their success skyrocketed when they introduced jet aircraft into their fleets. Passengers liked the comfort and speed of these aircraft while airlines benefited from their lower overall costs and extended range (relative to turboprops). These feeder airlines with jet service enabled the network carriers to enhance their hub-and-spoke systems by extending the length of haul for feeder flights, to replace larger jets with smaller jets in thinner markets, and to by-pass the hubs in some markets that had sufficient O&D traffic to warrant nonstop service.

The network conventional carriers went further, developing their own lower cost subsidiaries. The first attempts did not go far enough. Continental Lite, MetroJet of US Airways and United Shuttle on the US West Coast did not succeed for a variety of reasons, such as customer confusion with respect to the product and the fact that their lower costs were not low enough in order to compete with Southwest and still realize a profit. Later, the reorganized lower-cost subsidiaries were able to realize profits, but only when the smaller regional jets were flown. The elusive equation is perhaps best put by affirming the following. Although a regional jet may per se be more expensive to operate on a unit seat-mile cost basis than a larger aircraft, it can produce a profitable combination of load factor and yield where a larger aircraft could not. The success of these systems should not be under-estimated. Even the low-cost, low-fare airlines are beginning to establish their own even lower-cost affiliates. Consider the recent announcement by AirTran to work with Air Wisconsin.

Economic Impact on Countries and Regions

As mentioned in the first section, scheduled air transportation provides a significant impact on the economy of the region it serves, not to mention the people and the businesses. The extent of the benefit depends on the mix of traffic (business vs. tourist), the type of scheduled service (domestic vs. international, long-haul vs. short-haul, point-to-point vs. hub-and-spoke), the relative isolation of the region, and the availability of other modes of transportation. The impact of the operations of an aircraft alone creates employment and economic support activities. For example, a single nonstop daily flight with a 200-seat wide-body aircraft between a mid-size city in the USA and a major city in Western Europe (say, London) is expected to contribute a couple of hundred million dollars per year to the economy of the region. Similarly, a 400-seat wide-body aircraft providing a daily nonstop flight to a major city in the Asia-Pacific region (say, Tokyo) can easily double and possibly triple the value of the economic benefit. The economic benefits are derived from three major areas, (a) those relating to the impact of visitors, (b) those relating to the new jobs at the airport, and (c) those relating to new trade. It is the economic value added to a region by scheduled service that has recently prompted smaller cities to pursue both conventional as well as the new low-cost, low-fare airlines. An airline like Southwest has literally dozens of cities in the US domestic market on the waiting list. In Europe, Ryanair is reported to have received economic subsidies from some of the airports it serves.

Low-Fare Airlines

A real success in the past decade has been the success of a few low-cost, low-fare airlines modeled after Southwest in the USA. Examples include JetBlue and Westjet in North America, GOL in Brazil, Ryanair and easyJet in Europe, AirAsia in Malaysia, and Virgin Blue in Australia. In general these carriers began providing point-to-point, low-frill service in short-haul markets. However, some of the original characteristics of service are changing. Southwest now provides service coast-to-coast within the United States and includes nonstop service in some transcontinental markets. Moreover, an increasing number of its flights now connect at its major cities. For example, during the year 2001, 18.3 percent of Southwest's passengers made connections.[3] Similarly, Ryanair now serves medium-haul markets such as London-Rome and London-Stockholm while easyJet serves similar markets such as Liverpool-Malaga and London-Athens.

The common denominator among all these carriers is the deployment of streamlined operating processes and systems to keep productivity high and costs low. Examples include a single type of aircraft in the fleet, the use of varying types of secondary airports, and direct distribution. Additional information is available in Chapters 3 through 6. The success of these low-cost, low-fare carriers is amply demonstrated by their phenomenal growth in the market place and their financial performance reflected in the increase in the price of their stock. In the past five years, the value of Southwest stock increased significantly relative to the DJIA while the price of United's stock performed worse than the DJIA. See Figure 2.2. Similarly, in the UK, the price of Ryanair's stock increased considerably relative to the FTSE 300 while the price of British Airways' stock decreased relative to the FTSE 300. See Figure 2.3.

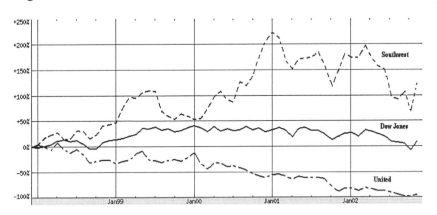

Figure 2.2 Stock Price Performances of United and Southwest Relative to the Dow Jones Index
Source: WWW: Yahoo Financials, Date: Nov 29, 2002.

The success of the surviving low-cost, low-fare airlines (keeping in mind that many did not survive) is having an enormous impact on the airline industry worldwide. These airlines have not just introduced a new price-service option in the marketplace, but their unparalleled success is perhaps the single most important contributor to the hypothesis that the business model of conventional airlines is no longer valid. Once you begin to question the merit of the overall business model of the conventional airline, then you are forced to examine and reassess all the components which make up that model.

There are two important points that need to be made about the comments in this section. First, the preceding list of examples of success stories in by no means complete. Other success stories include reduction of non-labor costs, substitution of technology, and reduction in distribution costs. The last example is truly impressive in that major airlines in the US have been able to reduce their distribution costs by 50 percent in the past five years. Second, the success of the few surviving low-cost, low-fare airlines must not conceal the fact that a very large percentage of the low-cost, low-fare new entrants failed. According to one analysis, 80-85 percent of mostly low-cost entrants in the United States since deregulation did

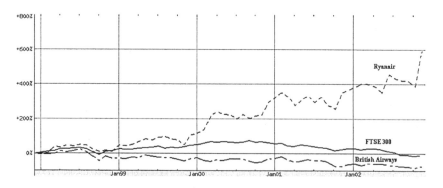

Figure 2.3 Stock Price Performances of British Airways and Ryanair Relative to the UK FTSE 300
Source: WWW: Yahoo Financials, Date: Nov 29, 2002.

not survive. Similarly, 60 percent of these types of airlines did not survive in Europe.[4] Finally, failures have not been limited to just the US and Europe. Canada 3000 (and Royal) failed in Canada and Impulse Airlines failed in Australia.

The Jury is Still Out

Hub-and-Spoke Systems

Developed in the early 1950s in the United Sates, the hub-and-spoke concept became the foundation of most US airline networks in the 1980s after the domestic market was deregulated. The concept enabled an airline to increase frequency in thin markets by connecting the thin markets (the spokes) to a central airport (the hub). The fundamental tradeoff was an increase in the frequency with a stop at the hub to make a connection. The original concept of the hub-and-spoke system is still valid for the network of some airlines, for example, those who want to serve a large number of thin markets. Theoretically, such systems are still an efficient way of serving thin markets for both airlines and passengers. For airlines they produced a lower-cost operation. For passengers they provided higher frequency.

It is not that the hub-and-spoke system is a wrong strategy. Too many hubs are the issue. Some airlines have gone too far by insisting on never flying over a hub. The result is excessively large complexes, with inappropriate spacing between them. An airline wishing to serve a large and a broad set of markets needs hubs. How else can an airline the size of American or United serve 150-200 cities many of which are very small? As powerful as Southwest is, it still only serves about 80 airports whereas United serves closer to double that number. On the other hand, one must ask a question: Is the objective to serve as many cities as possible (requiring the establishment of hubs to accommodate service from small cities) or is the objective to make money?

However, just because it is a cost-effective way to serve the market for one airline or at one location, it does not mean that such systems are cost-effective for all airlines or for one airline at multiple locations. A hub must meet certain requirements such as a reasonable balance between O&D traffic and connecting traffic, minimal overlap between competing hubs (even worse if the competing hubs belong to the same airline), a critical mass (particularly important if there is more than one network carrier at the hub), and a reasonable geographic location and catchment area. Moreover, a hub city must not have extreme seasonality of price sensitive passengers or inconsistent patterns of time sensitive passengers. Finally, even if a hub makes good sense for a particular airline at a particular city, there is still an important question to be answered. Is the hub city worth "owning"?

Here are some examples of airlines for which the hub-and-spoke system does not or did not prove to be cost-effective. Take, for example, a hub-and-spoke that represents mostly leisure travel such as the hub at Las Vegas (an overwhelmingly leisure destination) for America West. It would be difficult to use the high frequency justification at such a hub to attract business travelers and charge them high fares. The lack of critical mass was illustrated by American's hubs at Raleigh Durham, North Carolina and Nashville, Tennessee, and America West's hub at Columbus, Ohio. Overlapping hubs is illustrated by American's hub at San Jose, California (competing with San Francisco), Sabena's hub at Brussels (squeezed between KLM's hub at Amsterdam and Air France's hubs at the two airports in Paris), Swissair's hub at Zurich (sandwiched between Lufthansa's two hubs at Frankfurt and Munich), and Malaysia's flag carrier hub at Kuala Lumpur competing with Singapore Airlines at Singapore. However, the strategic location of an international gateway coupled with the strength of its hinterland can make it a viable hub despite the low level of O&D traffic. Examples include Amsterdam, Dubai, Singapore, and Hong Kong. The classic example of overlapping hubs being difficult to work with for the same airline is that of Heathrow and Gatwick for British Airways as a result of the lack of expansion capability at Heathrow.

The decision to never fly over a hub has also proved to be less cost-effective. There is nothing wrong with a strategy to fly a passenger from a medium size airport nonstop to a major destination (say, from San Antonio, Texas to New York). A small portion of this market (say, out of San Antonio) may, in fact, be to major cities (such as New York and San Francisco) that might justify nonstop flights. And, the remaining larger portion of this market can be adequately served through a hub. While it is true that to add another passenger hardly impacts the costs of any one scheduled flight, it significantly raises costs when an airline chooses to compete for market share. The typical tactic to increase market share is to schedule greater frequencies of service. Not surprising, but still often overlooked is the excessive seating capacity that results. The only way to sell the excess capacity is to discount price. While the revenue from the discounted pricing might cover marginal costs, it cannot ever cover the larger true costs when more aircraft, crews and resources are being utilized.

Again, there is nothing wrong with the original concept of the hub-and-spoke system. It is just that some airlines went too far in such areas as the size of the hub (relying too heavily on the connecting traffic) and the number of fleet types. While it is perfectly reasonable for United to feed its Chicago hub with traffic from nearby markets such as Grand Rapids destined for the West Coast, it does not make sense to feed the traffic from Boston through Chicago destined for the West Coast. If, however, the object is to feed the low-fare traffic out of Boston through Chicago, you then compete with the low-fare, connecting service offered by Southwest from airports near Boston such as Manchester, New Hampshire to the north and Providence, Rhode Island to the south. These airports compete with Boston, being more convenient for passengers in the northwest and southern suburbs. Thus, while the concept of a hub-and-spoke system is just as valid today as when it was first developed in the 1950s, it is the lemming-like fruitless replication and invalidated practices that have raised a question about its costs and benefits. There are too many and some are too big for their locations.

Frequent Flyer Programs

American introduced its frequent flyer program in 1981. Everyone copied them to retain the customers they already had. The success of the frequent flyer programs as a form of loyalty was so powerful that a number of other businesses such as hotels copied it. Even Southwest buckled in 1987 having first tried a different approach by offering rewards to secretaries for booking seats for their bosses instead of rewarding the actual traveler. The power of these programs is overwhelming. Nearly everyone has one.

Given the enormous success of these programs, what is the issue? First, since most of the conventional carriers offer programs that are very similar and most frequent travelers are members of multiple programs, does one carrier ever have a significant advantage over its competitors? Second, the industry as a whole, at least for the US carriers, has built a very large contingent liability in terms of the obligation to provide free travel for the unused accumulated miles. Third, the conventional carriers have made the program so complicated that it has led to significant frustration for the customers they wish to reward. The worst cases involved carriers raising expectations that could not be delivered—too few seats in the redemption category. Fourth, until very recently, the reward system was inequitable in the sense that it provided the same reward to a passenger who paid a full economy-class fare as one who purchased a deeply-discounted seat. This deficiency enabled many frequent flyers to earn enormous credit by buying discounted travel and the more miles they accumulated the more rewards they received. Fifth, there is a significant cost involved. Some carriers served certain markets with high frequency and capacity not because they were cost effective, but more because they were "required" to ensure that the program remained attractive. One example would be flights between the US mainland and Hawaii. There is also the significant cost involved in the administration and maintenance of these programs.

In recent years, conventional airlines have begun to fine-tune this loyalty system. For example, some airlines now provide mileage credit based on the price of the ticket. Some airlines are not even offering credit for travel purchased on deeply-discounted fares. In the case of low-cost, low-fare airlines, they either do not participate in such programs (Ryanair and easyJet) or they have kept it very simple. Southwest, for example, has not only kept its frequent flyer program quite simple but has in fact made it more valuable by placing no limits on available seats to the users in the program. If a seat is available, a frequent flyer can always obtain it.

In the final analysis, whether a frequent flyer program is an advantage or a disadvantage depends on the carrier. For some, it has not been cost effective. For some, it has become overly complex when the carrier allows its members to use their miles to purchase other things such as tickets for sport events or lower interest rates for items financed through credit cards. The complexity arises from the need to calculate the "exchange rate". For some carriers, either because their networks warrant it or because they have managed it more effectively, the frequent flyer program has been very cost-effective. Air Canada has just spun off its frequent flyer program into a separate company, and in addition to air travel, plans to offer redemptions in such items as High-Definition Television. Through the judicious sale of its seats to other businesses, one carrier generates more money from its loyalty program than from the transportation of its cargo. Thus, like hubs, it is not the concept, but the lemming-like behavior that is the difficulty. It also raises the issue whether this type of a loyalty program can be considered an asset or a liability. In the final analysis, even if an airline decides to end its frequent flyer program, the task will not be easy. There are analogies with override commissions that got out of control in the 1980s. Some carriers, such as Pan Am, kept raising the stakes. We know the end result, at least for that airline.

Strategic Alliances

The major purpose of alliances, whether it is the very old type known as interline agreements or the more contemporary ones such as those involving code-shared flights, is to enable airlines to expand their network—either from the viewpoint of cost-effectiveness, or simply to get around regulatory constraints. The alliance between KLM and Northwest is probably the first one of the contemporary type. As with many alliances, it was created to: (a) get around the bilateral restrictions relating to expansion in international markets; (b) get around the restrictions relating to mergers and acquisitions involving international airlines; (c) enhance the value of their individual networks by increasing the number of destinations and frequencies; (d) improve the quality of connections; and (e) capitalize on the market power of Northwest in the United States and KLM in the Netherlands. Alliances that include antitrust immunity (ability to coordinate capacity and prices) have proven to be exceptionally valuable. Most people would agree that strategic alliances have produced benefits in the area of revenue enhancement. However, the question remains whether an alliance has stimulated new business or simply taken market share away from another alliance.

The major debate, however, involves the cost side. First, alliances so far have produced relatively little in the area of major savings in costs. Second, there is the question of not just the extent of savings in costs but the costs of joining and exiting alliances. There are substantial costs involved in converting information technology platforms, integrating systems, aligning business processes and procedures, and spending time and energy in coordinating meetings. The costs of joining an alliance can be substantial for small carriers, particularly if they have to change their information technology platforms and systems. Some would argue that big costs savings can only come from markets where the partners operate as if merged. Unfortunately, serious and extensive joint operations have not taken place either due to lack of commitment on the part of management or lack of support of unionized labor. Management in each of the partners tend to think "we do it right". Often neither partner is willing to look for a new, better way. Ironically, the old Swissair-Sabena Airline Management Partnership came the closest to receiving a number of the benefits of a merger without really having one. In an alliance, the partners may not go as far, particularly if they are not sure of their status in the partnership. Consider airlines that have changed partnerships—Air Canada, Air France, Alitalia, Austrian Airlines, British Airways, and Singapore Airlines, just to name a few.

In an actual merger, the situation would be different in that a well-managed acquiring airline would drop routes, eliminate certain types of fleet, eliminate some senior management positions, and close some maintenance bases, crew domiciles, reservation centers, and computer operations. There have been failures to act on the opportunities that a merger presents. Examples include Continental and Eastern, TWA and Ozark, US Airways and Piedmont, and Republic and Northwest. Even worse, there are cases when the stronger acquiring airline has been fatally weakened by acquiring a weak airline, exemplified by the situation when Pacific Western Airlines in Canada acquired CP Air. Conversely, there are cases where a weak airline somehow acquired a strong carrier to the ultimate demise of both (Pacific Southwest Airlines/Piedmont and US airways).

There are a couple of other issues besides the costs of joining an alliance and the supposed benefit in cost reductions. First, some airlines began to depend too much on alliances and neglected to focus on their own core strategy (network, fleet, and frequency). Second, many strategic alliances are unstable at this time. Carriers in one alliance are openly talking with both aligned and non-aligned carriers. Partners are often afraid to share competitive advantage and management information in case the relationship does not last. Third, it has been difficult to integrate systems, particularly information technology. Everyone agrees that it is a good idea and then everyone offers their own system as the core to which others can link. The only way it would work is if a holding company owned the different carriers and forced integration and joint operations. Finally, there is the branding issue. Some carriers have found it difficult to be part of a super brand and at the same time maintain their own brands. This problem becomes more serious for carriers who want to maintain flexibility in the event they want to switch partners.

Diversification

At various times, various airlines have owned (and some still do) hotel chains, car rental companies, computer reservation systems, information technology businesses, catering companies, other airlines, and so forth. Diversification in the aviation industry, as in other industries and businesses, has had mixed success. It seems to have worked for airlines such as Lufthansa and failed with others such as United and partially for the old Swissair. For Lufthansa, diversification seems to have worked because the company has managed it correctly.

One key success factor seems to have been for the airline diversifying to make sure that it has a competitive advantage, core competencies, and a leading market position in the activity being diversified. Lufthansa certainly has achieved such a position in its catering, maintenance, ground handling, and information technology businesses. Although the old Swissair managed to achieve leading positions in some of its businesses such as catering, information technology and duty-free sales, it did not achieve the same level of success for its diversification into other airlines, that is, having a leading or close to a leading share.

The second success factor seems to be to diversify into businesses that are less sensitive to economic cycles (or possibly are even counter cyclical). It also helps if the businesses into which the company is diversifying are less capital intensive. Since hotel businesses seem to have similar cyclicality and capital requirements, it would not make sense to add hotel chains to an airline's portfolio. Some of these reasons could explain the failure of United's initiative into diversifications—old ones such as Allegis and more recent ones such as the Avolar project that involved a subsidiary operating business jets in various roles. Some airlines continue to diversify into businesses that do not meet the criteria for success. The SAS group, for example, includes different hotel chains. Consequently, the jury is still out on the value of diversification. After all, if the objective of diversification for an airline is to hedge its risks for the shareholders, they can achieve the same objective more effectively by diversifying their own portfolios.

Millstones

The first section discussed some conspicuous success stories in the airline industry while the second section presented a few areas where the jury is still out as to their position on the continuum from success to failure. This section takes up some topics that fall on the side of failure.

Pricing Strategies

One could argue on either side as to whether pricing strategies have been successful or unsuccessful. From the viewpoint of stimulating traffic, building load factor, maintaining or even increasing market share, being everything to everyone, and coping with economic downturns, airline pricing strategies have been quite successful. However, from the viewpoint of growing revenue at rates faster than the growth in the economy, generating enough revenue to cover costs, providing an adequate return for shareholders, and building customer trust, pricing policies have been unsuccessful.

In the previous chapter, Figure 1.1 showed the historical information on the cumulative net profit margin achieved by the global airline industry. Figure 2.1 showed the declining trend in the global airline passenger yield. One could draw the conclusion that one reason for the lack of profitability is the uneconomic fares implemented by the industry. Let us examine this hypothesis further. The top line in Figure 2.4 shows the increase in productivity achieved by airlines as a result of numerous factors such as higher productivity of each new generation of aircraft and an improvement in the utilization of aircraft (both flight hours per year as well as load factor). The second line shows the changes in the real input price (cost of labor, maintenance, fuel, and so on). The bottom line shows the trend in airline passenger yields.

If we consider the line showing the ratio of revenue to expenses as relatively flat and as the benchmark, then it appears that even after paying for the increase in the price of goods and services used by airlines to produce their service (marked Real input price on the chart), there was a small amount of productivity benefit left for the shareholders. However, the line showing the decline in real yield is far below the revenue-to-expense ratio line, indicating that the industry not only gave away left over gains in productivity (after paying for the real input prices) in uneconomic fares, but it went far beyond that and priced some seats at artificially low fares. These fares did not recover anywhere near the fully-allocated costs but they did produce growth—growth that was profitless. The standard justifications given related to (a) the need to offer competitive fares, (b) the need to protect market share, (c) the existence of a wide variation in the opinions on fully-allocated costs vs. marginal costs, long-term vs. short-term costs, and (d) the need to utilize excess capacity.

The uneconomic nature of pricing policies can be observed in two other ways. Just about the time the US airline industry was almost deregulated with respect to its domestic operations, the average passenger fare represented the fare that was paid by most travelers—in other words, the distribution of fare reflected a bell-shaped curve. See the dotted line in Figure 2.5 that represents the year 1981. Sixteen years later, in 1997, in a completely deregulated environment, the average fare became meaningless in that few passengers paid the average fare.[5] The fare no longer represented a

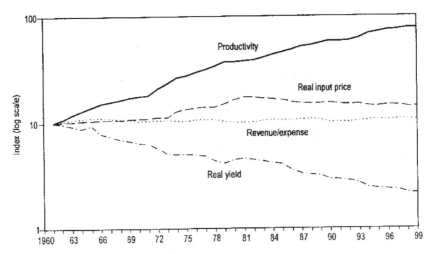

Figure 2.4 Trends in Airline Industry Performance
Source: International Civil Aviation Organization.

bell-shaped curve. Most passengers pay a fare much lower than the average fare and a few pay a fare that is much higher than the average fare. In another study of one-way, economy-class fares between Los Angeles and New York during December 1998, the actual fares paid ranged between US$50 and US$1,200. While the average fare was US$330, 67 percent of passengers paid less than the average fare. It is interesting to note that the average fare paid by "business passengers" (passengers traveling on unrestricted or less restricted tickets) was US$918. The median fare (that is more representative) was US$188.[6]

In the deregulated environment, the fare curve tends to be represented by a "two-hump camel". There is a larger hump representing the larger number of passengers paying a fare much lower than the average fare and a smaller hump representing passengers paying a much higher fare than the average fare. Translating this information into the total number of passengers utilizing the low fare (see Figure 2.7), 54 percent of all fares purchased by US domestic travelers from network airlines are more

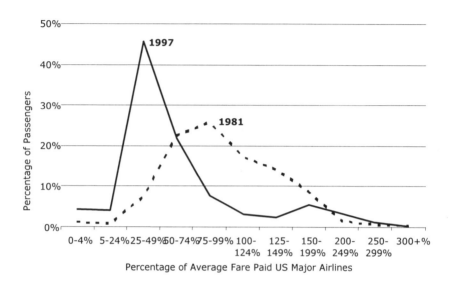

Figure 2.5 Distribution of Passenger Fare Paid around the Average Fare

Source: Unisys R2A Transportation Management Consultants.

than 50 percent below the industry's average fare. When this information is combined with low fares purchased at low-cost airlines, 78 percent of all tickets are purchased at low fares. Translating the current passenger fare curve for the network airlines into the new passenger revenue curve shows two peaks, revenue earned at or below 50 percent of the average fare, and revenue earned at fares more than 50 percent above the average fare. The 54 percent of the network airline passengers purchasing low fares produce 26 percent of the total revenues. The 13 percent of network passengers purchasing high fares produce 38 percent of the total revenues.[7] It is this later category of high fare passengers that are not necessarily disappearing from the network airlines; rather, they are paying lower (uneconomic) fares. See Figure 2.7.

Figure 2.6   Total Percent of all Domestic Passengers Utilizing Low Fares, Year 2000

Source: Unisys R2A Transportation Management Consultants.

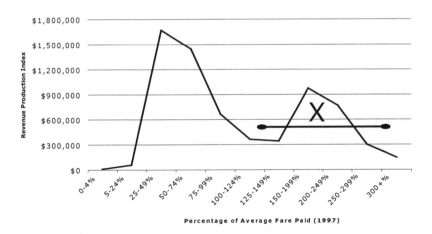

Figure 2.7   New Revenue Curve Showing the Disappearing High-Yield Customer

Source: Unisys R2A Transportation Management Consultants.

The final evidence that pricing policies have got out of control is simply to glance at the information about the domestic operations of US airlines. In 1980, the full fare yield was 15 cents while the discounted yield was 8.56 cents. More than half of the passenger traffic (57.5 percent—measured in RPM terms) made use of discount fares and the average discount was 42.9 percent of the full fare. In 2002, the full fare yield was 39.39 cents while the discount yield was 11.13 cents. For the year 2002, 96.1 percent of the passenger traffic (measured in RPM) made use of discount fares and the average discount was 71.7 percent of the full fare.[8] Airlines have continuously increased the number of passengers receiving discount fares as well as the amount of the discount so that now people do not pay higher fares even if they are justified on the basis of costs. To change this consumer behavior, it is necessary to go back to a properly fenced market segmentation strategy that matches the economics of customer demand.

Overly Diversified Fleet

The second millstone around the necks of major conventional airlines is an overly diversified fleet, with respect to both the number and age of aircraft types in the fleet. Table 2.1 shows the fleet of American Airlines as of 31 December, 2000 (a date selected to avoid issues relating to the events of September 11th, 2001). American had more than a dozen kinds of aircraft, not even counting the aircraft in the fleet of its regional feeder carriers. While such a diversified fleet had the advantage of matching more closely aircraft capacity and range to meet the needs of various markets, it also produced enormous costs relating to the fleet itself.

Table 2.2 shows similar information for British Airways. Although it is true that some types of aircraft have been eliminated from the fleet and some types are operated only by subsidiaries, the point is that airlines still operate too many types resulting in costs that are higher than benefits. The decision to reduce fleet types will be cost effective in the long term even though in the short term fleet changes will bring about complexity during transition that can take many years to complete. During this period, airlines operate both previous and future types and, in some cases, purely temporary aircraft, exemplified by the operation of Boeing 737-300s by British Airways. Some airlines have also kept too many types of aircraft in their fleet due to the "short-term" difficulties relating to their disposal or due to the poor condition of their balance sheets. In the long term, fleet rationalization will be cost effective, particularly if it is accompanied by true fleet commonality where pilots, for example, can transition easily across the full spectrum of aircraft, from the smaller narrow-body to the largest wide-body.

The successful low-cost, low-fare airlines have remained extremely focused not only in the type of service they provide but also with respect to the use of a single type of aircraft with different versions. It is important to note the difference in the mindset of the two types of carriers with respect to fleet. A conventional airline is likely to say: "We want to offer services from city X to city Y, what kind of an airplane do we have that will do the job, and if we do not have it, let us get a few." A low-cost, low-fare airline is likely to say: "Where can we fly with our type of airplane and make money?" It is a very different mindset that keeps the low-cost, low-fare airlines focused on costs and margins.

In some cases, strategies of a number of carriers were built around the egos of the senior executives at the helm. Egos determined such basics as the number of aircraft, the size of aircraft, and the number of destinations served. These decisions had serious consequences. For example, the scheduling department ended up serving thin markets, or dense markets with low frequency and poor utilization. Low frequency, in turn, led to a poor product for the business passengers. In some cases, egos led management to lose touch with reality (customers and employees), causing them to expand too much and too fast. Some rewarded themselves by buying

Table 2.1 Diversity of American Airlines' Fleet, 31 December 2000

Airbus A300-600R	35
Boeing 727-200	60
Boeing 737-800	51
Boeing 757-200	102
Boeing 767-200	8
Boeing 767-200 ER	22
Boeing 767-300 ER	49
Boeing 777-200 ER	27
Fokker 100	75
McDonnell Douglas MD-11	7
McDonnell Douglas MD-80	276
McDonnell Douglas MD-90	5
Total	717
AMR Eagle Aircraft	
ATR 42	31
Embraer 135	33
Embraer 145	50
Super ATR	43
Saab 340	79
Saab 340B Plus	25
Total	261

Source: AMR Website.

Table 2.2 Diversity of British Airways' Fleet, 31 March 2001

Concorde	7
Boeing 747-200	13
Boeing 747-400	56
Boeing 777	40
Boeing 767-300	21
Boeing 757-200	45
Airbus A319	21
Airbus A320	10
Boeing 737-300	8
Boeing 737-400	34
Boeing 737-500	10
De Havilland Canada DHC-8	15
Embraer RJ145	7
Total	287
Deutsche BA, CityFlyer Express and Go	
Boeing 737-300	30
Avro RJ100	15
ATR 72	6
Total	51

Source: British Airways Website.

larger airplanes. Some management actually expanded their fleet to lower unit costs. US Airways, for example, tried to reduce its unit costs by increasing its share of long-haul flights, not only within the US domestic markets but also across the Atlantic.

Some carriers "bought" fleet complexity through mergers. If the acquired carrier had a significant number of aircraft with a new fleet type not previously flown by the acquiring carrier, it was often difficult and expensive to shed them and then have to replace them. The maintenance complexity of an overly diversified fleet adds cost, as more spares, test equipment, and technical specialists are required. Continental is one example of an airline that dramatically reduced its fleet types and sub-fleet types over the past 10-15 years.

Overly Complex Product for the Price

Although the airline business itself is complex (as illustrated in the first chapter), there is a question as to whether conventional airlines have made their product overly complex by, for example, trying to be everything to everyone, developing confusing pricing systems, and adding features that they think are desirable, but for which passengers are not willing to pay. The acquisition of an overly diversified fleet and implementation of overly complex pricing systems discussed above are just two examples of developing an overly complex product. Sophisticated and costly in-flight entertainment systems are an example.

There are two fundamental questions. Initially, many of these systems were "sold" to airlines with the justification, "don't look at them as cost centers; rather, look at them as profit centers". How many airlines have been able to make money directly from such systems? Even in the case of JetBlue there was a discussion at the beginning of charging US$5 for access to the live TV at every seat. Eventually, the idea was abandoned and the service is provided free. Second was the justification that these systems will increase market share. Again, while these systems are desirable, and do give advantage to the innovator, is the increase in market share, first, sufficient and, second, sustainable. In any case, even if the justification is valid for some airlines (based on the composition of its traffic, by age, ethnicity, and so forth), it is not necessarily warranted for all airlines. If the conventional wisdom is that the three most important criteria for selecting an airline are schedule, price, and frequent flyer program (with the order of importance depending on the purpose of the trip), then an in-flight entertainment system must be considered as a factor that can influence a passenger decision about the choice of an airline, not be the driver of the decision.

Figure 2.8 shows the typical product thrusts for a major conventional airline. Safety, reliability, and a competitive price are not only assumed by the passenger but they represent just the minimum capabilities needed to get into the business. All other features, while desirable to varying degrees, make the product complex and add costs. There are three fundamental problems. First, many passengers are no longer willing to pay for these features. Second, they have got so used to receiving them free, that they are not willing to give them up. Third, since all airlines end up offering these features, product differentiation is lost.

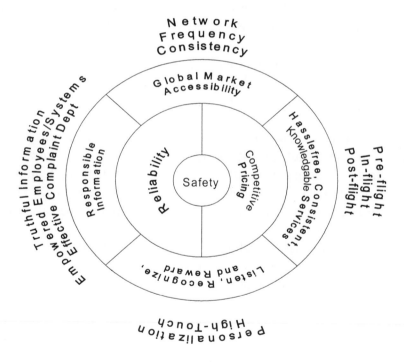

Figure 2.8 Major Product Thrust Areas for a Major Conventional Airline

Airlines worldwide have long attempted to differentiate their product. A few have succeeded (Virgin and Singapore are often quoted as good examples), but product differentiation has been overestimated. For example, does it really make a difference to offer excellent lounge facilities and cater to individual tastes, such as having a golf magazine placed in the seat pocket for the avid golfer? The answers seem to be yes and no. If all else goes well, meaning the core components of the product do perform (on time performance of aircraft and reasonable time for the delivery of baggage), then yes, the tertiary aspects of product can make a difference. However, if the plane is hours late and bags are lost then having a golf magazine does not matter.

Second, at airports where a carrier has a dominant position, secondary aspects of the product may be less valuable when the passenger has little choice. Passengers living in Pittsburgh have little choice other than US Airways. However, when an airline is offering service from an airport where passengers have a choice these differences could matter. The real issue remains. How much more is a passenger willing to pay for a secondary feature? Consider flat beds in business class. How many passengers living in Pittsburgh would choose to make a connection to fly British Airways rather than a nonstop out of Pittsburgh, even though seat comfort is important? The longer the flight the more important the feature can be. A passenger in Pittsburgh flying to Johannesburg might choose to fly to London on British Airways and then connect because of the bed/seat feature.

The point of the above discussion is not to eliminate all product features and go to the lowest common denominator, but rather to select features that are cost effective. Consider the case of Silkair, a subsidiary of Singapore Airlines. It is not a low-cost, no-frills subsidiary. It is a lower-cost and lower-frills (two class service) subsidiary. It has one type of aircraft, the Airbus 320. Management and operational staff are paid less and the staffing process is different. Pilots are paid less than those on the larger Singapore aircraft but they are paid competitive salaries to other pilots flying similar aircraft in the region. Costs are kept low by having crews fly out and back (no stays outside Singapore) and they fly point-to-point. The airline flies to secondary points such as Phuket in Thailand, Cochin and Trinvandrum in India and similar secondary points in China, Indonesia and Malaysia. The airline offers a fairly basic product that is cost-effective. Their passengers are willing to pay for the limited features and the airline is able to recover the costs of the features in the product. A similar explanation holds for the services offered by AirTran—basic features (two-class service, reserved seats, and conventional airports) with small fees for each feature.

Inefficient Airports

Even as the airlines have implemented new customer service and productivity-improving devices and methods into the airport, overall the quality and speed of customer service has continued to deteriorate. A recent trip through a Canadian gateway onto a transborder flight to the U.S. required the passenger to line up thirteen times. The stops were a combination of automated self-service check-in, baggage drop off, Airport Improvement Fee payment and ticket collection, customs, immigration, security, gate check-in, duty-free goods pick-up, jetway, and aircraft boarding. It is ironical that despite the inefficiency of many airports worldwide, they continue to post high margins. While airports in Latin America did not show up in Table 1.1, even they are reported to be achieving margins in the 30 percent area while several airlines in the region are losing money.

Airport arrival times in North America can be 1 hour for short-haul domestic up to 3 hours for international flights. No matter how well equipped the businessman is with laptops and cell-phones, you can never be as productive in an airport environment as at work or at home. These wait times are having a tremendous impact on the nation's productivity. They have the unintended impact of also reducing the attractiveness of hub-and-spoke systems where frequent flights are undermined by long terminal wait times—an increasing part of the total trip time. Fewer nonstop flights become competitive with more frequent hub-and-spoke operations when you must also subtract an extra hour or two from your work day spent waiting around in airports. Hopefully, this problem will become less critical once the 100 percent bag-check requirement has been implemented.

The response of some airports of trying to capture a passenger's attention or pocketbook with such diversions as shopping concourses cannot mask the underlying blatant inefficiency of passing through the modern airport. In addition, the larger airports also tend to be inefficient from the viewpoint of airlines in that it takes an extremely long time to launch a construction project. The rebuilding of Toronto's Pearson Airport complex is a ten-year project requiring a minimum of operational disruption and much temporary gate shuffling while new terminal buildings are constructed and old ones torn down.

Conclusions

In the past airlines achieved significant efficiency gains most of which they gave away to other members in the value chain. Now it is becoming increasingly difficult to gain anything. It is difficult to significantly reduce costs or generate additional revenue through conventional methods. The achievements discussed in this chapter (successful, unsuccessful, and with mixed results) occurred during an environment of relative stability (some exceptions being the two major oil crises and the Gulf War). However, the current environment is far more threatening than anything experienced in the past. The struggle is with the severity (length and depth) of the current economic downturn, the technological developments (the Internet), the consumer revolution (wanting more for less), and the powerful, low-cost, low-fare old and new entrants with different business models. In essence, the margin for error has become even thinner and, as the next chapter will discuss, survival is not guaranteed.

Notes

[1] National Transportation Safety Board (1999), Aviation Safety Statistics (online information), quoted in Vasigh, Bijan, "Airline Safety: An Application of Empirical Methods to Determine Fatality Risk", *Handbook of Airline Economics*, 2nd edition (New York: McGraw-Hill, 2002), p. 505.

[2] Air Transport Association of America, *Airline Handbook*, Washington, D.C., 2000, Chapter 6, p. 1.

[3] Harris, Brian D., "2002 Hub Factbook", A SolomonSmithBarney Report on Airlines, 15 April, 2002, p. 12.

[4] Harbison, Peter, "Low Cost Airlines", A Report by the Center for Asia Pacific Aviation, Sydney, Australia, September 2002, p. 24.

[5] Information based on a communications with Phil Roberts of Unisys R2A Transportation Management Consultants, November 2002.

[6] Linenberg, Michael J. and Sandra Fleming, *Airline Industry: Airline Equity Investments*, A Report by Merrill Lynch, New York, 29 September 2000, p. 21.

[7] Personal communication with Phil Roberts of Unisys R2A Transportation Management Consultants, November 2002.

[8] *The Airline Monitor* January / February, 2003, p. 27.

Chapter 3

The Shakeout Continues

With the Chapter 11 bankruptcy filing of United on 09 December 2002, we are seeing the beginning of another major restructuring in the industry. With its huge fleet (555 in service, 92 stored, and 49 on order); major hubs at Chicago, San Francisco, Denver, and Washington-Dulles; and immensely valuable worldwide route structure, the question becomes is it more valuable dead or alive? Will United go the way of Pan Am, Eastern, and Braniff or reemerge successfully? Either way, the US airline industry in particular and the global airline industry in general are in the midst of structural change and challenge unseen since the beginning of the airline deregulation in the US. The deregulation movement resulted in a shakeout within the US airline industry with some ramifications within the global industry. The confluence of a number of vital forces discussed in this chapter is expected to bring about a major shakeout in the US industry with major ramifications for the global airlines. These forces will separate the weak carriers from the strong. As stated in the first chapter, because of the actual and perceived strategic value of an airline, most of the 187 ICAO states will continue to have airlines. However, as discussed in the second part of this chapter, the number, size, and structure of surviving airlines will change, with the amount of change varying by region.

Drivers Reshaping the Marketplace

Figure 3.1 shows a list of half a dozen forces affecting the landscape of the global airline industry. Although all of these forces are inter-related to some extent, the increasing availability of low-fare airlines and the overwhelming acceptance of their services is by far the most influential force.

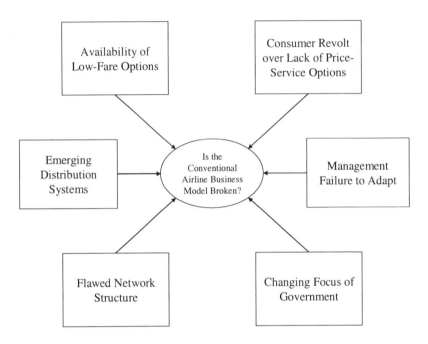

Figure 3.1 Major Drivers Reshaping the Marketplace

Management Failure to Adapt

The fundamental, and in some cases potentially fatal, problems facing the airline industry are not new. These include overcapacity, flawed network, fleet, and fare structures, high costs (not just labor but also distribution and infrastructure), very high debt ratios, and poor branding strategies. Second, management agreed to reward various members in the air transportation chain (labor with high wages, travel agents with high commissions, airports with high facility charges, and a large segment of the traveling public in the form of artificially low fares)—all at the expense of its shareholders.

Figure 3.2 shows one recent analysis of the extent of overcapacity in the US domestic markets. It may be helpful to keep in mind that the industry is not suffering from overcapacity if load factors are in the mid-70s. Anything higher in scheduled service would lead to spill. The problem arises when high load factors are achieved through the sale of artificially low fares. In that sense there is overcapacity. With this in mind, let us see the examination of one analyst who assumes that demand and supply were in equilibrium in the 1995-1996 time period when all airlines except TWA were generating profits. The major airlines increased capacity at an average rate of 3 percent per year between 1996 and 2001. To fill the capacity airlines lowered their fares at the low end. To compensate for the reduction in yield they increased the fares at the high end. The strategy worked while the economy was strong. The whole situation changed when the IT, dot-com and telecommunications bubbles burst.

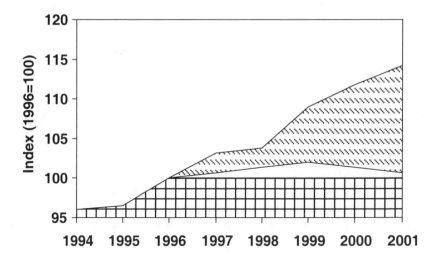

Figure 3.2 Estimate of Overcapacity in the US Domestic Market
Source: *Aviation Strategy*, July/August, 2002, p.10.

Depending on the base year selected and whether a reasonable average growth in capacity should have been zero or one percent per year, the overcapacity (prior to the events of September 11th, 2001) ranges from 10 percent in the best case to 20 percent in the worst case.[1] Even by the end of 2002 and even with aircraft retirements, the overcapacity is still at least 10 percent. Unfortunately the overcapacity situation will not go away even after the bankruptcy declared by United. First, the bankruptcy process has not taken the capacity out of the marketplace. United will continue to operate under Chapter 11 protection. Their operations, in fact, could be considered to be effectively subsidized by the airline's creditors, providing an unfair advantage against competitors, making the overcapacity situation worse not better. Second, if one looks at overcapacity being caused by aircraft and not airlines, the failure of an airline will be followed by the sale of its aircraft, possibly even at low prices under distressed circumstances. Under this scenario, the airline acquiring the aircraft could enjoy a cost advantage over surviving airlines that do not need more aircraft, but the strategy would maintain the overcapacity situation.

The industry now faces the consequences of the slow reaction or inaction of management to fundamental changes in the environment. Here are some general and some specific examples. Management at major US airlines was far too slow to react and respond to the expansion of low-cost, low-fare airlines. United, instead of proactively trying to reduce its' costs to get them closer to those of the low-cost carriers, agreed to a significant increase in the wages for pilots (partly due to the ESOP). Presumably, the higher wages would be followed at other airlines and therefore United would not be at a cost competitive disadvantage in the long run. Airline management has been abnormally slow at reducing the high debt to equity ratio. The cost of raising capital is now not only prohibitive, but it has also (a) limited the ability of some carriers to acquire desirable assets (such as aircraft) at the appropriate times, and (b) forced some airlines to use off-balance sheet financing that can be more expensive. Operating leases may appear to be financially attractive, but some airlines have fallen into the trap of operating leases for almost the whole fleet rather than for the portion needed to maintain flexibility.

Air India stood by and watched the international traffic to and from India be diverted to the US West Coast by Singapore Airlines and to Europe by Emirates. Figure 3.3 shows that Singapore Airlines and its subsidiary (Silkair) have been collecting traffic from seven major cities. Emirates has been siphoning traffic from four major cities in India to Dubai to connect with flights to other parts of the Middle East as well as major destinations in Europe. Management of the flag carrier of Malaysia has been too slow to proactively challenge competition from the flag carriers of Singapore, Hong Kong, Australia, and the UK.

In the US, when major carriers began to shift their focus to major reductions in costs, some lost sight of the value of brands and branding strategies. For some carriers, the product became even more commoditized and customers became loyal to frequent flyer programs rather than to airlines. For a few, the "promise" was not significantly differentiated. For some others, the customer experience did not even reach the level promised, let alone exceed it. And for a few others, the brand promise was not embedded into the corporate culture and as such did not have the "buy in" of employees. There is another aspect of brands that should not be overlooked. It can take years to build an airline's brand but it can be destroyed in literally minutes. Consider the case of Swissair, one of the finest brand names in the airline industry. The news of Swissair's aircraft being detained at a London airport by creditors (fuel suppliers) virtually destroyed the powerful image of a great airline. Canadian Pacific, the forerunner of Canadian Airlines International, was one of the first carriers into the Pacific in the 1940s. Its strong franchise there was eroded with a series of gaffes until its demise and eventual absorption by Air Canada.

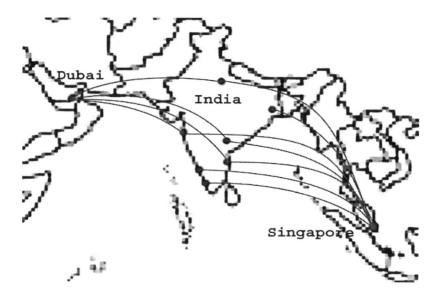

**Figure 3.3   Cities Served in India by Singapore Airlines and its
                Subsidiary Silkair and by Emirates Airlines**

Source: Websites of Singapore, Silkair and Emirates Airlines.

Availability of Low-Fare Options

The first two chapters indicated a number of constraints under which
an airline operates. Some of these constraints are inherent to the
industry while others exist as a result of management action or in
some cases inaction.   Take the case of the explosion of low-fare
airlines. The largest and one of the oldest, Southwest, got started by
the strong vision and conviction of its founders (based in part on the
experience of Pacific Southwest Airlines based on the US West
Coast) and changes in government regulations. However, its
expansion was also helped by the fact that almost all the
conventional airlines did not change their value proposition,
presumably, because they did not take Southwest seriously. The
public liked the idea of a different value proposition—that is, a
different price-service option. The airline not only grew rapidly but

its high productivity systems and processes enabled it to offer lower fares and make money. Tragically, some conventional carriers continued to ignore it thinking that it was just a tiny short-haul operator serving a few regional markets and the needs of leisure travelers and some "backpackers". Some others viewed it as an airline "selling its product to a different customer". In recent times, the foolish confidence of conventional carriers that low-fare airlines are not a serious threat can be seen by fact that the major Brazilian carriers did not even match GOL's fares at the beginning.

Southwest has not only proved to be a consistently profitable airline in the world, but it has also grown to an enormous size. See Figure 3.4 that shows the current routes flown by Southwest. The airline provides significant service across the United States and has recently started serving transcontinental markets on a nonstop basis.

Southwest's value proposition is also different from conventional airlines in two other aspects. First, it focuses on point-to-point service and avoids the development of hub-and-spoke systems. During the summer of 2002, for example, 76 percent of their passengers flew on nonstop flights, 10 percent on direct flights

Figure 3.4 Southwest Airlines Route Structure, December 2002
Source: Southwest Airlines.

(flights with very brief stops but no change of plane) and 14 percent made connections.[2] The other major difference is that Southwest provides high frequency in the markets it serves. This strategy appealed to business passengers who not only prefer nonstop flights but also high frequency. Consequently, Southwest's basic two-prong strategy (low fares for leisure travelers and high frequency with nonstop flights for business travelers) has proven to be one of the most successful strategies in the airline industry. In the case of GOL, in less than two years, flying 20 aircraft, concentrating its flights in the South and Southeast part of Brazil, it connected those regions with the capital Brasilia. This area surrounding Sao Paulo, Rio de Janeiro, Belo Horizonte and Brasilia produces not only about 75 percent of the enplanements in Brazil (representing the wealthy part of the country), it is also the part of the country where the conventional airlines did not offer any meaningful price-service option.

The Southwest strategy has been so successful that it has been copied by new entrants all over the world, although with some differences. Examples include AirTran and JetBlue in the US (even though AirTran is in some ways more like a conventional airline), Westjet in Canada, Ryanair and easyJet in Europe, Jet Airways in India, AirAsia in Malaysia, Virgin Blue in Australia, and GOL in Brazil. In Canada, Westjet has suffered recently from declining yields and the effect of seat sales and very high airport security charges, but has still managed to remain profitable. It enjoys a stock market cap three times that of its much larger full-service rival Air Canada. This is just the beginning—a number of other initiatives with some variants are on the way. The German conglomerate (Tui), a travel group with one of Europe's largest charter airline fleets, is planning to enter the marketplace with a low-cost scheduled carrier, Hapag-Lloyd.

There are four other points worth noting. First, the public's desire for low-fare service is not restricted to just short-haul domestic markets. The public desires an alternative to low-fare charter service in long-haul markets. Consider the passengers who want low-fare scheduled service between London and New York and were willing to take cheaper connecting flights via Amsterdam or even Frankfurt. Similarly, those who wanted cheaper scheduled service between London and Seattle were willing to take connecting flights via Copenhagen. Consequently, the public's appetite for low-fare service as seen by the short- to medium-haul operations within North America and Europe is just the tip of the iceberg. Second, low-fare airlines are no longer restricting their service to secondary airports and avoiding competitive incursions with conventional carriers. Figure 3.5 shows that during December of 2002, there were 50 destinations served by low-fare airlines out of Chicago, home of major hubs of two of the largest airlines in the world.[3] easyJet in Europe and AirTran in the US are serving conventional airports. Westjet in Canada has avoided the very expensive and congested Toronto Airport in favor of nearby Hamilton, but uses conventional airports in the Western part of the country.

The third point is that every low-fare airline is not a no-frill airline. Take AirTran as a case in point. For a small additional fee for each feature (in the neighborhood of US$20), the airline offers such desirable features as an upgrade to business class, assigned seats, and service to and from conventional airports. Similarly, while Jet Airways is not a low-fare airline in the same way as Southwest is in the US or as Ryanair is in Europe, it charges comparable fares to those charged by the conventional airlines, but offers a much higher level of service in the form of reliability and in-flight service as well as service on the ground. It is a conventional airline modeled on Singapore Airlines but operating in a domestic market. Even some of the staff are former Singapore Airlines staff. It is a success because it is everything that Indian Airlines is not. It identified an un-served element of the market and delivered the right product that has enabled it to expand its network across the country (see Figure 3.6) and take away almost one half of the share in domestic markets in less than ten years.

The fourth point is that while the carriers discussed above (and some others) have been highly successful, a number of other carriers have not succeeded in the marketplace. These include, for example, UltraAir, AirOne, and MGM Grand in the US, Debonair and City Bird in Europe, Canada 3000 and Royal in Canada, and Vanguard and Midway in the US. And the failures have not been due

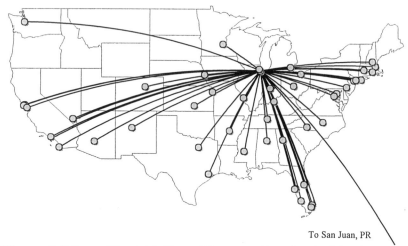

To San Juan, PR

Figure 3.5 Low-Cost Airlines, Nonstop Service in Markets Served by Hub Airlines at Chicago
Source: Southwest Airlines.

to the lack of demand for different price-service options but more due to a broad spectrum of factors such as mismanagement, lack of capital, and lack of sufficient infrastructure. Similarly, the dismal experience of low-cost subsidiaries of conventional carriers was based more on lack of focus, mismanagement of operations and brand as well as poor labor-management agreements rather than the lack of public interest in low-fare service.

The overwhelming acceptance of alternative price-options in general and the low-fare service in particular, coupled with the business model that has provided a high margin of return to the operators has won high praise from the financial community. The market capitalization of Southwest in the US is just phenomenal. See Figure 3.7. During the month of November 2002, Southwest had a market capitalization of more than US$12 billion. This is more than twice the market capitalization of eight major US airlines combined!

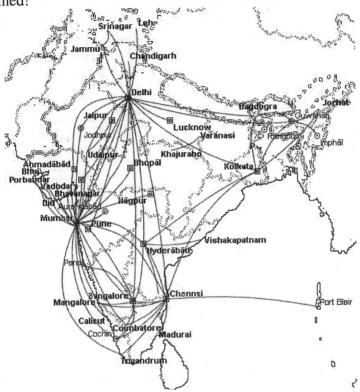

Figure 3.6 Route Map of Jet Airways, Summer 2002
Source: Website of Jet Airways.

Emerging Distribution Systems

Prior to the Internet era, distribution costs within the airline industry ranged between 15 and 25 percent, depending on the airline. For airlines such as Southwest that were less dependent on travel agents the number was on the lower side. For airlines such as Alitalia that were more dependent on travel agents and whose international operations represented a larger proportion of the total operations the number was on the higher side. There are four major components of the distribution cost: commissions paid to agents (6-15 percent of the fare); the cost of processing the ticket (2-5 percent of the fare); fees for the use of computer reservation systems (2-5 percent of the fare); and credit card fees (typically 2-3 percent of the fare).

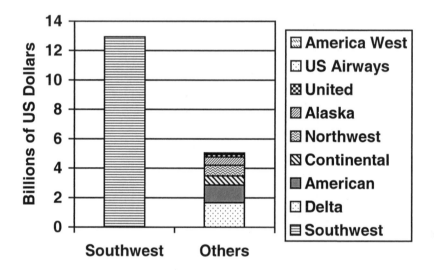

Figure 3.7 Market Capitalization of US Carriers, November 2002
Source: *Aviation Daily*, December 3, 2002, p. 7.

Although airlines have recognized the need to reduce distribution costs for a long time, they neither had the technology nor the will to take on the group that accounted for the largest component of distribution costs. Although some small airlines had been successful with simpler forms of tickets (Pacific Southwest Airlines that operated within the State of California) and even ticketless travel (Morris Air that was eventually acquired by Southwest), real opportunities to reduce the costs of distribution did not emerge until the emergence of the Internet, airline websites, electronic ticketing, online agencies (such as Expedia, Travelocity, and Orbitz), and online auction outlets such as Priceline.com. Significant strides have been made by the conventional airline industry in North America to reduce its distribution costs. It costs, for example, less than US$5 if a passenger traveling in a US market buys a ticket through an airline's own website. Some of the new low-cost, low-fare airlines only sell their tickets online or even only through their own websites. This channel reduces significantly the cost of distribution.

Figure 3.8 shows the difference in the unit operating costs of three major conventional airlines in Europe and selected low-cost, low-fare airlines. To keep the comparison somewhat realistic the data for the three major conventional carriers relates to their short-haul operations within Europe. The single component that explains the largest difference in costs between the three major airlines and Ryanair is the cost of distribution, 3.8 vs. 0.7 cents per mile. Ryanair depends completely on the availability of the Internet, a channel that is not only inexpensive, but it is available 24 hours a day in increasing number of households worldwide. Moreover, it provides access to some information that was very difficult to obtain in the past, namely, the lowest available fare. Consequently, while on the positive side the Internet has made it possible for airlines to reduce their distribution costs, on the negative side it has also made competition among airlines more intensive. It produces transparency, and in some ways, places "too much valuable information" into the hands of consumers and third parties. This channel of distribution coupled with the information provided by airlines has almost perfected the electronic marketplace encouraging consumers to take full advantage of it.

Slowly emerging is the process of electronic consolidation, whereby intermediaries bundle travel requests using the Internet, and then present this potential business to the airlines and negotiate the best deal. This process of aggregating demand could develop leverage over time, particularly if frequent flyers—airlines' best customers, start to link in to such options in significant numbers.

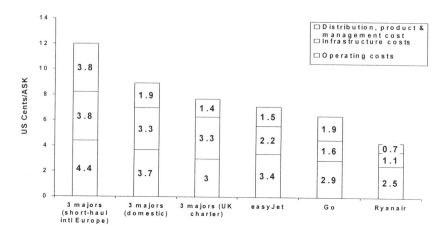

Figure 3.8 Unit Operating Costs of Selected Conventional and Low-Fare Airlines based in Europe

Source: *Aviation Strategy*, June 2002, p.17; AEA, CAA, annual reports, McKinsey analysis.

Conventional airlines are trying to reduce their costs in every area possible. However, distribution costs may well be the single area where conventional airlines can achieve the greatest savings using technology. Although most airlines have made significant progress in the sale of electronic tickets, relatively little progress has been made in increasing the amount of travel sold through the Internet. For example, some of the largest US airlines derive less than 10 percent of their revenue through Internet sales. Moreover, of the reservations that are made online for conventional airlines, less than 10 percent are made through the airlines' own websites. Online sales provide a vehicle for conventional airlines to not only reduce their distribution costs by 50 percent or more, they are also a way to generate additional revenue through advertising on their websites, building brand loyalty through their website by getting closer to the customer and, possibly, extending the frequent flyer program to earn and burn miles by developing relationships with other businesses.

Consumer Revolt over Lack of Price-Service Options

In recent years there has been a consumer revolt (by both business and leisure travelers) to price-service options. Let us take the case of business travelers first. Passengers on business have been complaining for years that the difference between walk-up fares and deeply discounted advance purchase fares is too much. Despite these complaints they kept paying the higher fares partly because of the booming economy of the mid-1990s, partly because of the height of the information technology and telecommunication revolutions (particularly the dot-com years and the rush to solve the 2K problem).

This trend, however, began to reverse toward the late 1990s. First, full Y fares in some markets reached a level twice and some times three times that of the deeply-discounted fares. Second, in the extended downturn in the economy and the burst of the dot-com and telecommunication bubbles, business passengers were no longer willing to pay the high differentials between walk-up fares and discount fares. Third, the Internet provided information to business travelers of the availability of lower fares on airlines such as Southwest, JetBlue, and AirTran—airlines that offered reasonable alternatives in many US domestic markets. Southwest and JetBlue, for example, were offering nonstop service in medium- and long-haul markets. AirTran was offering service to and from conventional airports, assigned seats, and a two-class service. Finally, as stated in the previous chapter, while there is nothing wrong with the hub-and-spoke concept per se, most of the full-service carriers went too far with it. It is one thing to charge a very high walk-up fare to a last minute business traveler in hub city where the passenger at least gets a nonstop flight. It is another thing to ask for a fare of an order of magnitude higher than the deep discount fare and still force the passenger to go through a hub with the extra trip time involved, and the possibility of missed connections and lost baggage. Lower-fare airlines, on the other hand, offered lower prices and provided nonstop service. Just glance at the route network of Southwest shown in Figure 3.4 and the availability of low-fare service in markets to and from Chicago shown in Figure 3.5.

Consequently, not only was there a decline in the amount of business travel but some passengers who were paying the extremely high full fares were no longer willing to buy such expensive tickets. Figure 3.9 shows the decline in the net yield, month by month since the beginning of 2000—an illustration of how the marketplace is changing. For example, the percentage of passengers purchasing premium-fare tickets on United—often, referred to as business travelers—decreased from 41 percent in 1999 to 22 percent in 2001.[4] It is a very real possibility that the business passengers who have shifted to the lower-cost airlines may not return even when the economy does turn around, especially if these lower-cost airlines continue to expand their networks at a rapid rate and the financial community continues to highly rate their market capitalization. Surveys show that only 45 percent of the frequent business travelers perceive that they have received good value compared to 80 percent for the low-fare airlines.[5]

Take now the case of leisure travelers who want not only lower fares but also simpler fares. They want the assurance that they are getting the lowest fare, and a full explanation of the terms and conditions. Let us first consider fare simplicity. How can the airlines possibly c laim that their fares a re simple to understand when their own reservation agents cannot explain them? Passengers expect that an airline will take care of them in case of unforeseen circumstances. How can airlines claim to be service oriented when the airline service experience consists of uncaring airport personnel (as a result of working under pressure and inconsistent management policies), inconsistent application of assistance or amenities, and inaccurate or partial information about delays or cancellations? These stories are too numerous to even begin to list. To add insult to injury, passengers believe that only when airlines are losing money do they pay attention to their customers. One need only recall the frustrations everyone had with airlines in the 1990s when they were making money.

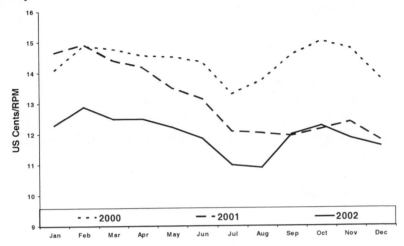

Figure 3.9 Yield Trends in US Domestic Operations
Source: Constructed from the data contained in *The Airline Monitor*, January / February 2003, p.27.

In the final analysis, a good part of the blame for the consumer revolt can be related to some conventional carriers' development and implementation of non-viable branding strategies. They exemplified the total disconnect between expectations and performance, the failure to establish the brand's or a particular product's unique value, falling far short of trying to be all things to all segments. It is this disconnection between expectations and performance that is turning passengers to low-fare airlines. Many passengers traveling in the economy class of conventional carriers feel that there is no service difference within the US domestic markets. In some markets, both have begun to install high-density seating, eliminate meals on short-haul and in some cases in medium-haul markets, and have created controlled chaos in the check-in, security and boarding processes at airports. They might as well take the flight on the low-fare airline where their reduced expectations are met, compared with the full-service carrier where they are not. Customers now distinguish between carriers along the expectations/performance dimension.

Flawed Network Structure

A number of airlines worldwide have been on unsustainable paths for a number of years. United and US Airways were at the point of bankruptcy well before the events of September 11th, 2001. The flag carriers of Belgium, Italy, Greece, India, Indonesia, Malaysia, and the Philippines have been suffering financial losses for long periods of time. Philippines Airlines declared bankruptcy in 1998; Sabena, Swissair, TRANSBRASIL, and Ansett followed in 2001. Although there are many reasons for the many financial difficulties experienced by conventional airlines, a flawed network structure (coupled with a flawed fleet structure discussed in the previous chapter) is another important contributor to the problems experienced by some of the weak conventional airlines.

Take the case of US Airways. During the 1980s, the airline's costs were not out of line relative to its network and fleet. The old Douglas DC-9s had higher unit costs relative to say the Boeing 757s, but the smaller airplanes were more appropriate for the markets they served. Then US Airways developed a massive hub at Pittsburgh which had a relatively small O&D base. It competed with other hubs in the area between Pittsburgh and Chicago. The carrier overbuilt its Charlotte hub (inherited from the acquisition of Piedmont Airlines) that had a small O&D base and competed with a giant neighboring hub at Atlanta. Figure 3.10 shows the location of major hubs in the US with the size of the circle representing the size of passenger traffic carried by the hub carriers with the black portion of the circle representing the part of the total traffic that represents O&D traffic while the white portion represents the portion of the total traffic that is making connections at the airport. Just to be sure that the overall size of the circle is not misunderstood, take the case of San Francisco and Minneapolis. The fact that the circle representing San Francisco is smaller than one for Minneapolis does not mean that San

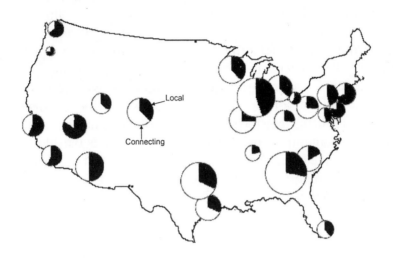

Figure 3.10 Local and Connecting Passenger Traffic Base at US Hubs

Source: Based on the data extracted from, "2002 Hub Factbook", *A Report by SalomonSmithBarney*, New York, April 2002.

Francisco handles less traffic than Minneapolis. The size simply shows the traffic handled by the hub carrier, United in the case of San Francisco (not counting all the traffic carried by the other airlines at San Francisco) and Northwest in the case of Minneapolis (again, not counting the traffic carried by the other airlines). Thus, not only do hubs such as Pittsburgh and Charlotte have comparatively small O&D bases for the massive hub-and-spoke systems built by US Airways, they have overlapping hubs, and they have very large percentage of traffic that is connecting, traffic that could just as easily connect over many alternative hubs. Under such circumstances, airlines usually end up providing discounts to encourage passengers to connect through their hubs, discounts that cannot be substantiated for carriers such as US Airways with its very high cost structure. For many years, the carrier was able to overcome this dilemma because it charged higher fares to local passengers, a strategy that could not be supported when low-fare airlines such as Southwest entered the markets directly or indirectly through neighboring airports (Baltimore to divert traffic from Washington National). The flawed network structure became worse when the carrier added longer-haul markets to reduce its average costs. The key point is that longer- haul routes lowered the average costs through arithmetic, not through any significant changes in the cost structure. Another example of the flawed network is the addition of a large percentage of markets in Florida—markets that are highly price sensitive and highly seasonal, and therefore not cost effective for high-cost carriers.

The problem of overlapping hubs is not restricted to the North American region. Consider the European and the Asia-Pacific regions. Figure 3.11 shows the location and size of the traffic base at major hubs in Europe. Concentric circles in London and Paris represent the two airports in each city (London Gatwick and London Heathrow and Paris Orly and Paris Charles de Gaulle). From a purely connecting point of view, is there is a need for Brussels to have a major connecting hub-and-spoke system when there are two very large and efficient connecting hubs at Amsterdam and Paris? According to one analysis, 31 percent of the total traffic at Brussels in 2001 was connecting, traffic that could just as easily have made connections at other nearby airports.[6]

The same logic applies to the hub-and-spoke system at Zurich where 43 percent of the total traffic in 2001 was connecting traffic.[7] Again, there are many convenient hubs in the neighborhood, Frankfurt and Munich, just to name two. And just as it was the flawed network structure that contributed to the problem at US Airways, it also played a role in the financial difficulties experienced by the old Swissair whose O&D traffic base was too small for the size of the hub-and-spoke system developed at Zurich. The airline ended up flying 36 intercontinental jets to support its hub at Zurich. After declaring bankruptcy in 2001, the airline reduced the size of its hub slightly. However, with 26 intercontinental aircraft, the size of

Figure 3.11 Overlapping Hubs in Europe
Source: Constructed with data extracted from various issues of the *World Report*, Airports Council International, Geneva, Switzerland.

the hub is still too large for the size of the O&D base much less the existence of multiple hubs in the neighborhood. Figure 3.12 shows the intercontinental routes flown by the new Swiss International Air

Lines. This intercontinental network is virtually identical to the one operated by the old Swissair, overextended again given the overlapping hub situation just discussed.

Figure 3.12 Intercontinental Routes of Swiss International Air Lines, Summer 2002

Source: Constructed from the schedule data available through the airline's website.

Finally, the contribution of the flawed network structure is equally valid in the Asia-Pacific region. Figure 3.13 shows the location and size of major hubs in that region. A perfect example of the overlapping hub situation exists between Singapore and Kuala Lumpur, two major hub-and-spoke systems separated by less than 200 miles. Not only do these two hubs overlap in terms of geography but they are highly dependent on connecting traffic relative to the size of the hubs and the size of the fleet, particularly in the case of Singapore Airlines. However, in the case of Malaysia, one could even question the network of its flag carrier—a network that included such destinations as Beirut, Buenos Aires, Cairo, and Vienna. In the case of the hub at Jakarta, while its geographic location may be appropriate for its national traffic, it is inappropriate for intra-regional connections.

Every airport cannot be a major hub. As mentioned in the previous chapter, for an airport to be a successful hub it needs to have a good geographic location with respect to connecting traffic (such as Chicago for US transcontinental service, Charles de Gaulle for trans-Atlantic service, Dubai for the Middle East, and Tokyo for

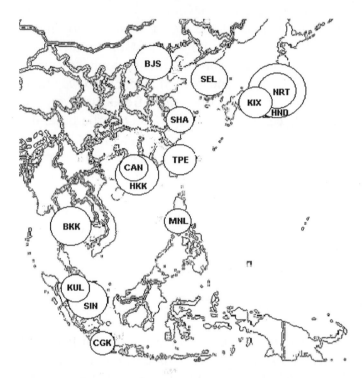

Figure 3.13 Overlapping Hubs in the Asia-Pacific Region
Source: Constructed with data extracted from various issues of the *World Report*, Airports Council International, Geneva, Switzerland.

trans-Pacific service). It needs to have sufficient infrastructure capacity (now and to expand in the future—such as the case of Charles de Gaulle), and the availability of intermodal connections (as in the case of Frankfurt and Charles de Gaulle). Finally, it needs to have a strong local O&D traffic base (such as London, Paris, and

Frankfurt in Europe; Chicago, Dallas, and Miami in the US; and Hong Kong, Taipei, and Tokyo in the Asia-Pacific region). If this last condition is not met, it must then compete as a reliever airport against the airport nearby which has the large local market O&D. Syracuse, New York, for example, has tried with marginal success to operate as a reliever airport to New York City.

In all three major regions of the world, there are too many major airlines competing for the non-local O&D traffic via connections through their hubs. The two major tools to attract connecting traffic are lower fares and high frequency. Lower fares work if the unit operating costs are low. High frequency works if there is an alignment between the size of the O&D traffic base, the capacity of the aircraft used, and the stage lengths of flights. In some markets this misalignment has led to overcapacity that in turn has led to even lower fares. Consequently, airlines with weak hubs (such US Airways at Pittsburgh and the new Swiss International Air Lines at Zurich) will find it difficult to survive, unless they restructure. The situation in the Asia-Pacific region is slightly different in terms of distances involved, large bodies of water in between neighboring countries, and the lack of a viable alternate form of transportation.

Changing Focus of Government

The first chapter described governments' interest in their airlines and various types of intervention to protect their airlines. In fact, the global airline industry made more profits during regulated periods. Consider, for example, the 20 year period from the beginning of the 1960s to the end of the 1970s. There were, of course, economic downturns relating to such events as the first oil crisis. Governments not only protected their airlines from excessive competition but also provided financial support when necessary—whether the airlines were government owned or private. Governments even revived airlines that declared bankruptcy—exemplified by the situation in Switzerland.

Governments' policies are, however, changing (although slowly and reluctantly) with respect to protection from new entry as well as financial support. Let us first take the case of protection from competition. The UK government authorized British Midland to enter the trans-Atlantic market. Brazil allowed GOL, the low-fare airline, to enter the country's most lucrative market. Malaysia allowed a low-cost, low-fare airline (AirAsia) to compete with Malaysian Airline System (a government owned and supported airline that subsequently dropped prices to compete with the new entrant). India allowed Jet Airways to compete with Indian Airlines in domestic markets. Indonesia awarded licenses to almost a dozen new entrants to compete against its flag carrier—Garuda Indonesia. Japan allowed two low-cost, low-fare airlines (Skymark and Air Do) to enter domestic markets as well as to allowing Japan Airlines to acquire Japan Air System. Since the market is not conducive for competition to increase within and to and from Japan, the government is promoting the expansion of existing airports (an additional runway at Haneda and the extension of the newly built second runway at Narita) as well as the development of new airports (one in the Osaka region and one in the Tokyo region) to develop competition. In Canada, however, additional route approvals not backed by government financial support were not sufficient to prevent Canadian Airlines' collapse.

As illustrated by the above examples, the attitude of the governments is changing (although at a snail's pace), at least in democratic countries, from the protection of the airline's interest to promotion of consumer considerations. This change is likely to accelerate not only the certification of low-cost, low-fare airlines but also the authorization of second-tier flag carriers, along the lines of EVA in Taiwan, ASIANA in Korea, All Nippon in Japan, and Dragonair in Hong Kong. Moreover, even where a government has authorized a new second-tier airline to serve domestic markets only (for example, Jet Airways in India), eventually the new domestic carriers will be allowed in international markets. This will, most certainly, bring about competition for the flag carrier, for example,

Jet Airways vs. Air India. The negotiators of the Canada-US "open skies transborder" agreement of the late 1980s considered and rejected the idea of free trade in commercial air transport both within and between the two countries. The Canadian government has asked the US and now Mexico to consider once again an "open skies North America" regime. This would allow all three countries' carriers to fly anywhere they wished subject to the technical and commercial considerations of available airways, runways, gates, and slots.

The preceding list of factors is by no means comprehensive. Here are examples of two other factors relating to the full-service carriers. First is "unrealistic over-expectation" of some stakeholders. While most people talk about unionized labor's high expectations, let us not forget about management's expectations, and travelers' expectations. Passengers still complain that fares in general are still too high despite the fact that they have been coming down dramatically. The second factor is "access to capital markets", which will undoubtedly dramatically change behaviors in the future.

Some Plausible Outcomes

The previous section discussed six drivers shaping the marketplace for the global airline industry. While change has been a constant in the airline industry, the above forces not only accelerate the pace of change but also the depth of change. There is competition on all fronts—not just on the low-cost, low-fare front. Consider two extreme examples. First, just think about the enormous size and rate of development of Emirates' hub at Dubai. The Emirates region has a relatively tiny O&D market. Yet, Emirates is becoming the next Singapore Airlines. Just a quick glance at Figure 3.14 shows the degree of competition that many global airlines will face in the near future when Emirates begins service on the dotted lines shown in Figure 3.14 and uses high-capacity aircraft to do so. This airline has already proven that it can make money by flying connecting passengers.

Consider another example at a different end of the spectrum. The hassle factor (caused by increased security procedures) coupled with a disproportionate increase in the taxes relating to the price of a ticket is decreasing the amount of travel by air in short-haul US markets. Figure 3.15 shows, for example, a decline in traffic between the first quarter 2002 and first quarter 2001 of 26 percent and a decline in yield of 8 percent, leading to a decline in revenue of 33 percent in the markets under 250 miles. If this trend continues, it could have a significant impact on the long-haul flights of US carriers since they depend heavily on the feeder flights from short-haul markets. Some of these feeder flights would be weakened if some of the local O&D passengers decide to travel by other modes of transportation. Some hub-and-spoke carriers will suffer because of the network effect. Just as a carrier like American can add a spoke that can be instantaneously successful due to the availability of connecting traffic, an elimination of a spoke weakens other spokes.

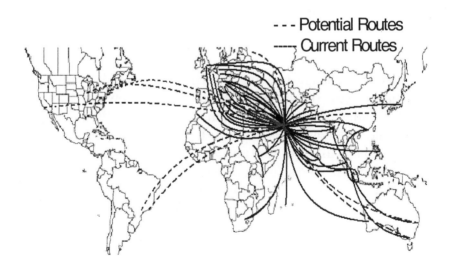

**Figure 3.14 Selected Current and Potential Intercontinental
Routes of Emirates**

Source: Emirates' Website and Announced Routes in the Aviation Press.

Some Failures over a Long Period

Although it is possible that a few airlines may cease operations (along the lines of Eastern, Braniff, Pan Am, TWA, Canadian, Ansett, and TRANSBRASIL), the process could take a long time. Large airlines will take a long time to go out of business for at least two reasons. An airline like United has franchises at Chicago and its Pacific routes are worth billions. Despite any large carriers' financial conditions, assets like these still provide a cash flow. Unfortunately, when a carrier bleeds for a long time, it can create a significant problem for an entire segment of the industry in its region. Just consider the experience of and the path taken by Pan Am and TWA. Lending institutions have a history of keeping alive, large weak carriers. Consider the financial assistance provided by lessors of aircraft and manufacturers to make every effort for a bankrupt or near bankrupt airline to continue rather than leave the marketplace. This situation is not the same for small, weak carriers, such as Vanguard.

Regrouping

One area of a change is a regrouping of the carriers—some with large global networks at one end and low-fare, low-cost airlines at the other end. There would be some variations within each group such as the difference between easyJet and Ryanair in Europe and between Southwest and AirTran within the US. Similarly, global network carriers may take on different postures depending on the geographic location, size and make-up of their home-based traffic, the capacity of the available infrastructure, and the degree of corporate diversification. While both are global network carriers, the scope and structure of British Airways is very different from Lufthansa. Some of the smaller network carriers, while they may not become low-cost, low-fare airlines, may become lower-fare carriers for passengers who are price sensitive but prefer to have higher quality service. For example, an airline such as Continental has floated the idea of thinking like the retailer Target, a low-cost retailer whose products have proven to be distinctive enough to command modest premiums in a price-sensitive world.

Hub and Fleet Rationalization

The third change is likely to be the rationalization of hubs. Smaller hubs would most likely be closed—for example, Charlotte and Pittsburgh in the US, Brussels and Zurich in Europe. Other hubs could be rationalized. London Heathrow could easily become a smaller connecting hub with more focus on nonstop flights in premium long-haul markets. Some of the connecting traffic at Chicago's O'Hare could move to St. Louis leaving the limited capacity at Chicago to handle more high-yield point-to-point traffic.

At the same time, fleets are going to be radically simplified. This process began soon after the events of September 11th, 2001 but must accelerate further. No longer will the question necessarily be what aircraft can I buy to serve a particular market or route, but rather what markets and routes can I profitably serve with a

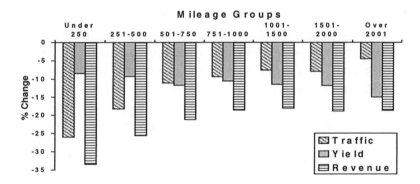

Figure 3.15 Reduction in O&D Passenger Traffic by Distance within the US Domestic Markets, First Quarter 2002 over First Quarter 2001

Source: *Aviation Daily*, August 2002.

simplified fleet? Already American has dropped the DC-10, the Boeing 727 and sharply reduced the MD-80s in its fleet. Air Canada dropped the DC-9, reduced the number of old 737s, and parked some 767-200s. There will be a general speeding up of retirements of such types as the D C9/MD80, the 737-100/200; and the DC-10/MD-11. Some large but relatively newer aircraft such as the 767-300 will be parked or see reduced flying.

Consolidation and Changes in Size and Scope

The fourth area of change will be in the area of consolidation. Although the process has been slow in the past, its pace is likely to pick up as some governments decide to allow weak carriers to consolidate rather than provide financial support or face angry local politicians and workers. A variation within the consolidation process may be the breakdown of an airline with the parts consolidated into different airlines, e xemplified by the process of P an Am who sold different divisions of its network to different airlines.

Also, some of the smaller carriers in Europe c ould become even smaller in size and scope. Carriers such as Aer Lingus, TAP, the new Swiss, Austrian, and Olympic could focus on intra-European routes and leave the intercontinental flying to strategic alliance partners. Such a strategy would enable these airlines to maintain their own brand and maintain presence in intercontinental markets in a cost-effective manner.

Wider Adoption of the Low-Fare, Low-Cost Strategy

Another area of change is likely to be the low degree of success of the low-cost, low-fare fare subsidiaries of conventional carriers in competing with independent low-cost, low-fare airlines. Independent low-fare airlines typically have low costs due to one type of aircraft, higher utilization (shorter turnaround times and in some cases longer operating day), one-class service, high-density seating, selling through direct channels (eliminating agent fees and in some cases

GDS fees), no frills (catering, in-flight entertainment), no lounges, no interlining, in some cases no seat assignment, less restrictive labor contracts, and lower infrastructure costs (low landing fees and ground handling costs at secondary airports). Besides these operational characteristics, low-cost, low-fare airlines also have a culture and a focus to their activities seldom found in conventional carriers.

An interesting experiment is underway at Air Canada, which is struggling with all these issues. It is currently dividing itself up into five sub-brands: (a) The mainline carrier, still branded Air Canada; (b) Jazz—a regional carrier flying short-haul routes, mostly with propeller aircraft but some small jets; (c) Tango—a discount carrier which still uses Air Canada staff at mainline wages, flies longer routes, and sometimes duplicates mainline routes but with no connections; (d) ZipAir—a true discount carrier with a separate operation including contracts and wage clauses, with flights under two hours, and a fleet constrained to a maximum of 20 Boeing 737-200s under a start-up agreement with Air Canada; and (e) Jetz—a charter operation flying Boeing 737s and Airbus 319s with only 48 seats and catering to business people and sports teams.

Radical General Cost Reduction

All areas must be on the table. We already have reductions of meal service and some airlines are charging for in-flight entertainment. Such features as power plugs for personal computers and multi-choice video may be scaled back or eliminated, as they are proving to be very expensive to maintain.

Related stakeholders must be asked to reduce their share—airports; security companies; and suppliers of various goods and services as well as the above attempts to lower labor costs for both hourly and management staff. Effort will go into boosting sales via the Internet and preferably via the airline's own website. American is pushing ahead with making a sophisticated booking engine which can respond to different consumers' needs—both price and time sensitive.

Conclusions

A broad spectrum of tidal wave forces will accelerate the current shakeout in the airline industry and some airlines will not survive— both conventional and new entrants, at least not in their current form. Unlike in the past, an airline may not be bailed out by its government. Some conventional airlines may not be able to raise capital in light of their poor or non-existent returns. They may not be able to rely on off-balance sheet financing. They may no longer be able to get financing with an empty balance sheet and no real assets. How are they going to raise equity? The confluence of forces discussed above demands two changes in behavior not only to survive but also to capitalize on some opportunities discussed in the next chapter.

The first behavior change will be to become much more quickly adaptive and much more disciplined. Focus will have to shift to those areas with maximum leverage (network, fleet, schedules, and brand management) with less attention paid to the peripherals (in-flight entertainment, frequent flyer programs, and revenue enhancement).

The second change is to collaborate with members of the value chain, for example, astute use of information technology, (a) to manage more effectively capacity and demand, and (b) manage more effectively the costs of logistics, such as crews, maintenance, locations where aircraft end up at night, and aircraft maintenance green time (maintenance equalization). For example, a few years ago, AeroMexico developed an effective supply chain management procedure by establishing a process involving key areas that dealt with the development of the final product: crew scheduling, maintenance, planning, union relations, and fleet planning. Using the system, the airline has been quite successful in planning its medium and longer-term needs. Failure to collaborate will surely result in a switch to confrontation as the stakes rise and good outcomes become less likely.

Notes

[1] Horan, Hubert, "Is the traditional Big Hub model still viable?", *Aviation Strategy*, July-August 2002, p. 10.

[2] Information provided by John Jamotta, Director of Schedule Planning, November 2002.

[3] Information provided by Phil Roberts, VP, Unisys R2A Transportation Management Consultants, December 2002 and updated by Southwest.

[4] Velocci, Anthony L. Jr., "United Flying Headlong Into An Uncertain Future", *Aviation Week & Space Technology*, 16 December 2002, p. 22.

[5] Levere, Jane, "Room at the front", *Airline Business*, October 2002, p. 41.

[6] SalomonSmithBarney, "2002 Hub Factbook", New York, April 2002, p. 96.

[7] Ibid.

Chapter 4

Healthy Outcomes: Seizing Opportunities

The previous chapter highlighted the shakeout that is underway in the global airline industry. While management cannot do anything directly to deal quickly with the forces responsible for the shakeout, they must also not overlook the longer-term opportunities on the horizon. This chapter presents some of these opportunities that are divided into four areas: (a) globalization and evolving demographic patterns; (b) mature markets; (c) evolving technologies; and (d) reshaping marketing and operating practices. This discussion will help conventional airlines to develop new business models based on (a) the changing needs of leisure travelers and price sensitive business travelers, and (b) the overwhelming acceptance of the low-fare airlines and airlines offering higher service at the fares charged by conventional airlines. The low-fare airlines in turn will also benefit by fine-tuning their business model, partly to differentiate among themselves and partly to position themselves to compete with the responses about to be unleashed by the conventional carriers to compete with them.

Globalization and Evolving Demographic Patterns

Rapidly changing demographic and socio-economic patterns present significant implications for the demand for air transportation services. This section provides just a glimpse of the opportunities available to airlines worldwide. Let us start with two developing regions—Asia and Latin America. There are a number of ways that one could examine the potential impact of demographic and socio-economic changes on air travel. One is the increasing degree of urbanization. Figure 4.1 shows that as the percentage of the population living in urban areas increases, there is a corresponding and exponential increase in air travel. It is, however, important to keep in mind that it is not the increase in urbanization per se that causes an increase in air travel; rather, it is the attributes associated with urbanization that may influence air travel. These attributes include education, affluence, and international trade. The populations of Singapore and Hong Kong, being 100 percent urbanized, already have a high propensity to travel. The increasing level of urbanization in other countries such as Malaysia and Thailand should produce similar results with respect to an increase in air travel.

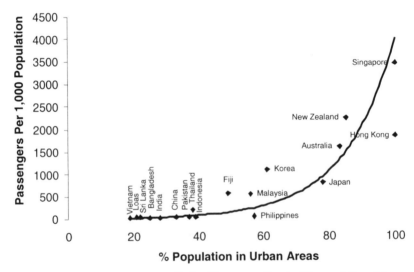

Figure 4.1 Urbanization and the Propensity to Travel by Air
Source: World Book Indicators, 2001 and Centre for Asia Pacific Aviation, Sydney, Australia, September 2002.

An increase in the household income can be a second measure of the increase in travel activity. In Asia, countries that are clearly heavy generators of air travel are also the countries with high incomes. Table 4.1 shows the median household income in US 2001 dollars for selected countries in Asia. After the four traditional "Asian Tigers", the median income drops off very rapidly. For example, the sixth country in Table 4.1 is Malaysia with a median household income of US$5,898. The low level of median income may not necessarily be an indicator of a low potential for air travel.

Table 4.1 Median Household Income in Selected Countries in Asia (US$2001)

Japan	$ 56,136
Singapore	$ 30,486
Hong Kong	$ 28,309
Taiwan	$ 26,141
South Korea	$ 24,186
Malaysia	$ 5,898
Philippines	$ 2,423
Urban China	$ 2,369
Thailand	$ 2,305
Urban India	$ 1,345
Indonesia	$ 585

Source: Dr. Clint R. Laurent, "A report on Major Demographic and Socio-Economic Trends of Asia", Asian Demographics Ltd. (website), May 2002, p. 37.

Consider, for example, Figure 4.2 that depicts a typical distribution of median household incomes in large developing countries such as India and China. The information contained in Table 4.1 may lead to the conclusion that only the populations of the four traditional tigers (Singapore, Taiwan, Hong Kong, and South Korea) and Japan can afford to travel by air. Other countries are too poor even when data is compensated for Purchasing Power Parity

(PPP). Moreover, some would argue that even when the median figures have increased significantly each decade, the numbers are still too low for the average person to purchase an airline ticket. However, there is an alternative interpretation if one moves away from the typical peaks that reflect very low incomes to the tail ends of the distribution. A small percentage applied to a very large population still produces a substantial number of potential airline passengers. So, even if one grants that only 1 percent of the households in urban India and China have the financial resources to buy a low-fare ticket, the absolute number of households could still be reasonable for the air travel industry. If one takes these segments and then projects them for the next 10-15-20 years, the result is a significant number. It may also be important to identify the locations of the prosperous segments of the total population. Take the case of China. One analysis shows that the top 10 percent of urban households in China earned over Rmb 33,000 during the year 2000 and that the average income for these top households was Rmb 43,000.[1] Moreover, it is likely that a small segment of this group could afford a low-fare ticket.

Figure 4.3 shows the location of the top 10 percent of the population. In the case of Urban China, about 45 percent of these households are located in or around the Guangdong region, about 30 percent in and adjacent to the Shanghai region, and about 15 percent in and around the Beijing region. Consequently, about 90 percent of the well-to-do Chinese households are within these three areas, pointing out clearly the potential location for aviation activity.

Besides the increasing degree of urbanization and household incomes, there are many other economic indicators of the potential growth in the aviation activity. Examples include the sudden increase in the number of cars sold and the growth in the sectors of the economy related to technology (software development, information management). These people are doing very well. They will travel. Consequently, China, and for similar reasons India, will likely be the leading aviation markets. There are a few cautions, however. First,

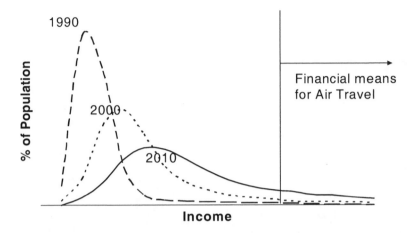

Figure 4.2 Urban Household Income Per Annum for a Typical Developing Country with a Large Population

the product they want or need could be very different. Within domestic markets, for instance, the product could resemble the services offered by different types of low-fare airlines within the United States—ranging from no frills to low-frills. Second, there are two serious obstacles to the growth of aviation activity in China. The aviation infrastructure (airports and the ATC system) is inadequate and the three major Chinese carriers are relatively small to serve the needs of the potential growth in the aviation activity.

Let us take an example in a different part of the world—Brazil. Even though air transportation is an old industry (having started in the 1920s), it is nowhere near close to maturity. If one assumes a coefficient of price elasticity of about 2 (meaning for every percentage point drop in price, a 2 percent increase in demand will result), there is plenty of room for organic growth. As with the example of China, one needs to examine the location of the markets relating to the Estado de Sao Paulo region are probably the most overpriced and most underserved in Brazil. Consequently, this region is ripe for low-cost, low-fare airlines such as GOL.

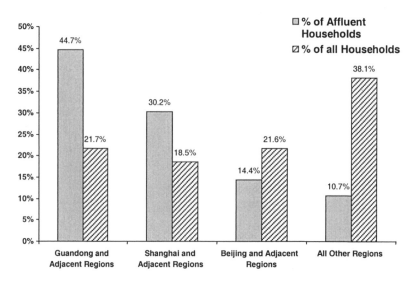

Figure 4.3 Affluent Regions of China

Source: Constructed from the data obtained from: Dr. Clint R. Laurent, "A report on Major Demographic and Socio-Economic Trends of Asia", Asian Demographics Ltd. (website), May 2002, p. 37.

Moving to another part of the world, consider the opportunities available in Mexico. In this country there are basically two modes of public transportation, buses and airplanes. Unlike some other parts of the world, there is no efficient rail system. The lack of an alternative viable mode of transportation, coupled with the large size of the country (almost two million square kilometers) as well its geography, provide significant opportunities for air travel. The demand for low-fare air travel was demonstrated a few years ago by the new entrant—Taesa. Unfortunately, poor management (for example, artificially low fares such as US$30 for medium-haul trips), lack of financing, and a fatal accident led to the demise of the carrier. Not counting the 25 percent growth achieved by Taesa with artificially low fares, the domestic market has only achieved a compound average annual growth rate of less than 2 percent per year for the past 15 years. Consequently, one could assume that the right price-service options could stimulate some segments of the domestic market. The Mexican international market has grown at a compound annual growth rate of about 12 percent per year. The economy is relatively strong (as in the case of Chile) compared to a number of other countries in Latin America such as Argentina and Brazil. The international potential is significant given that Mexico is the second largest international market for the US, after the NAFTA partner Canada. The international passenger market between the US and Canada is relatively mature compared to that between the US and Mexico. Consequently, according to the forecasts of the International Air Transport Association, the international passenger market between Mexico and the US will surpass that between Canada and the US by the year 2004.

Although attention worldwide is focused on the low-cost, low fare services, there are plenty of opportunities in other segments of the marketplace. Take the case of Dubai. It is becoming a very jet-set destination for up-scale clients. One just has to look at the type of hotels that have been built and that are being planned. The number of visitors is expected to increase from 3 million last year to 15 million in ten years. The upscale nature of expected visitors is reflected in the international quality of staff being recruited and hired. These luxurious facilities soon to be offered to upscale tourists are clear evidence of Dubai's determination to diversify the economy. The country only has a population of 900,000 while only about 12 percent are actual citizens. The remaining 88 percent are on working visas with the largest segment coming from the Indian subcontinent.[2] This explains the reason why the government's motivation and resources committed to develop tourism are also reflected in the plans of the flag carrier Emirates that has announced its plans to offer services worldwide. See Figure 3.14 in the previous chapter.

Mature Markets

Let us move to opportunities in developed and mature regions. Although Japan's market is mature, operating costs are high, and slots are limited at major airports, there are nevertheless abundant opportunities. Prior to liberalization within the domestic market, the marketplace was characterized by two main factors. First, the train was the only competitor to the three major airlines (All Nippon, Japan Air Lines, and Japan Air System). Train fares were in some cases higher than airfares. Competition among the three carriers was limited to a few routes. To the degree that competition did exist, it was based on brand image and distribution rather than prices.

Liberalization provided an opportunity for new entry and two airlines—Air Do and Skymark—did start service at lower fares. Both plunged into high-density markets, for example, Haneda-Sapporo in the case of Air Do and Haneda-Fukuoka in the case of Skymark. Although the domestic incumbents introduced limited discount fares, they did not become as prevalent as they have in other countries such as the United States and the United Kingdom. It is interesting to note that while the incumbents matched the new entrants, they did not address their main competitor, the train. The lower fares introduced by the new entrants did stimulate the markets but they were not able to offer high frequency service due to the limited availability of slots. The degree to which they could offer lower fares was limited due to the high costs of operations at the large airports. Leaving aside the high cost of operations at the major Japanese airports, the new entrants also could not get convenient locations for their check-in facilities. It is possible that within the Japanese culture there is a stronger tie between the brand name and quality and as such a greater willingness to pay a higher price for a brand name.

Despite the initial poor experience of the two new entrants, there still remain a number of potential opportunities for stimulating a mature market. Following are some examples of developments that might facilitate the stimulation of the Japanese domestic and international aviation market:

(a) a third airport in Tokyo (especially, if it is built in the Tokyo Bay area)
(b) a fourth runway at the Haneda Airport, combined with the relaxation of government rules to allow additional international flights into the airport
(c) the extension of the recently-built second runway at the Narita Airport
(d) a new airport at Kobe
(e) the availability of higher capacity aircraft, such as the Airbus 380
(g) a significant increase in the use of regional jets such as those operated by the affiliates of All Nippon Airways and Japan Airlines to such cities as Sendai
(h) the subsidization of thin markets outside Haneda by local authorities.

The domestic Japanese market has some unique characteristics. First, as of late 1997, less than 6 percent of the total passenger transportation was via air. Almost 60 percent was via auto and almost 30 percent was via rail. Second, Figure 4.4 shows the location of the 14 major cities in Japan. Most of the traffic is concentrated in a few trunk routes, which may explain why the two new entrants decided on major trunk routes instead of secondary routes. Opening up secondary airports to serve markets in the 14 cities listed in Figure 4.4 may help to relieve the problems of limited slots, and the high cost of aircraft maintenance caused by the excessive use of wide-body aircraft in short-haul markets.

In the United States the largest opportunity is to continue to capitalize on, and to expand low-fare services. Currently, about one-fourth of the seats offered in domestic markets are offered by the low-fare airlines. The product offered by the successful carriers in

Figure 4.4 Top 14 Major Cities of Japan
Source: Japan Statistics, Bureau and Statistics Center.

this group provides a healthy profit margin and its acceptance is phenomenal by the general public. Moreover, the business model of a few is so strong that the low-fare airlines are able to compete quite effectively in the backyards of the major, and, until recently, powerful full-service airlines. Given this customer acceptance and their success, some of the low-fare carriers will inevitably grow dramatically in the next decade.

Consider the data shown in Table 4.2. The first part of this table shows the information (revenue and passengers) on the domestic markets served by American and United (individually and combined—the total column) out of their powerful hubs at Chicago during the second quarter of 2002. The second part of the table shows that five low-fare airlines have been able to penetrate 47 markets from the two Chicago airports. Figure 3.5 in the previous chapter showed 50 such routes. Three more were added since the second quarter which explains why Table 4.2 contains the information on 47 routes. Low-fare airlines have been so successful operating out of Chicago that, compared with the American and United departures, they now generate 72 percent of the revenue and 75 percent of the passengers. It is interesting to note that the average passenger fare of American and United is only 8 percent higher than the average fare of the five low-fare airlines (US$162 vs. US$150). The problem, therefore, arises when American and United have to match fares but do not have the cost structure of the low-fares airlines. Midway Airport has potential for up to 50 gates. At present, Southwest occupies 15 gates. What if a Southwest type of service was provided—10 flights per gate—at all 50 gates? Under this scenario, that would mean 500 flights per day out of Midway Airport alone. This scenario assumes that the Midway Airport provides the necessary infrastructure that goes with 50 gates.

Another reason for the high forecast of the potential capacity of low-cost, low-fare airlines is the fact that 50 percent of the total O&D market within United States is in markets under 900 miles. See Figure 4.5 that provides the cumulative distribution of the O&D traffic by distance for the calendar year 2000. This traffic, and much more, can easily be transported additionally by the low-fare airlines. Moreover, the low-fare airlines can also stimulate many of these US domestic markets by introducing low fares, high frequency, or a combination of the two.

Table 4.2 Overlapping Revenue of Conventional Airlines and Low-Cost Airlines in Chicago Markets (Second Quarter 2002)

ORD to All US Domestic Markets

	American	**United**	**Total**
Revenues:	$339,677,230	$386,136,360	$725,813,590
Passengers:	2,211,860	2,433,210	4,645,070

Chicago to 47 Nonstop Markets
Served by Low Cost Carriers*

	In American's Markets	**In United's Markets**	**Total**
Revenues:	$239,613,540	$282,841,270	$522,454,810
Passengers:	1,627,860	1,865,900	3,493,760
Revenues:	**71%**	**73%**	**72%**
Passengers:	**74%**	**77%**	**75%**

*Low Cost Carriers –
AirTran, American West, ATA, Frontier, Southwest

Source: Data constructed by Unisys R2A Transportation Management Consultants, based on USDOT O&D Passenger Survey, Second Quarter 2002, Official Airline Guide, December 2002.

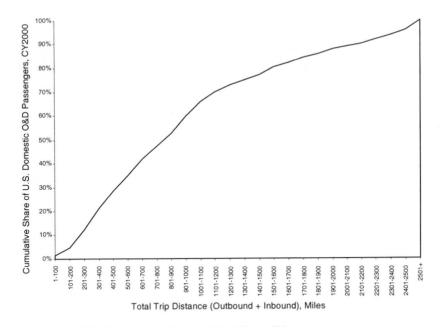

Figure 4.5 US Domestic Travel by Trip Distance
Source: Constructed from the data available from the US Department of
Transportation.

Technology

Aircraft

Evolving aircraft technology provides a very broad spectrum of
opportunities for airlines. At the low end are regional jets (and
advanced turboprop aircraft). At the top end are high-capacity as
well as ultra long-range aircraft. There are the aircraft families that
enable airlines to cost-effectively harmonize the relationship
between capacity and demand and reduce the maintenance and crew
costs, particularly if true commonality enables pilots to fly any
aircraft with minimal training. Aircraft family utilization will also
facilitate airlines more closely matching capacity with demand in

mild recessions and enable an airline to remain profitable or minimize losses.

The financial performance of regional carriers in the US (independents as well as those who are partially or wholly owned by the majors) has been far superior to the performance of the major carriers. Higher profit margins of regional carriers can be partially explained by three major factors. First, regional carriers typically need fewer passengers to break-even because their higher unit seat mile costs are offset by their higher unit revenues (a traffic-mix containing a higher percentage of business passengers paying the higher fare). Second, regional airlines operate in markets where there is very little competition from other airlines. This situation could change as low-fare airlines start operating in nearby airports. Third, regional jets have proven to be very cost-effective ways to serve thin short- and medium-haul markets connected to a hub.

Figure 4.6 shows the four major roles of regional jets since their introduction in 1993. Their most important role (44 percent) has been to supplement the services provided by major airlines. The second major role (33 percent) has been to introduce new nonstop service in thin markets. The third role (15 percent) has been to replace large aircraft in thin markets. The fourth role (8 percent) has been to replace smaller turboprop aircraft with regional jets in thin markets. Regional jets have proven to be valuable for major airlines to test questionable markets. In some markets, the success of regional jets led major carriers to upgrade service with the use of larger aircraft. In other markets, it led airlines to downgrade from larger jets. In the US, most of the regional aircraft are used in hub-and-spoke systems with less than 100 passengers per day each way. Passengers prefer these jets compared to turboprops due to their (a) higher comfort level, (b) higher cruise speed, (c) ability to fly above the weather, (d) better safety record, and (e) ability to move quickly through congested ATC airspace.

The use of regional jets represents a very cost-effective way to expand the relationship between majors and regionals. There are,

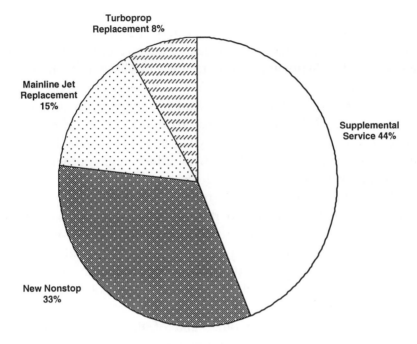

Figure 4.6 Main Uses of Regional Jet Aircraft in the US
Source: Bombardier Aerospace.

however, three key issues to resolve in order to capitalize on the available opportunities: (a) the relaxation of the scope clause (mostly a US phenomenon); (b) the integration of pilots of regional jets with pilots of mainline jets; and (c) the method by which independent regionals get paid by the major carriers—revenue proration or fixed-fee per departure. The resolution of these issues—particularly the scope clause—can have tremendous opportunities for an airline such as US Airways after labor was willing to agree to the scope clause and management was willing to expand its regional operations in the eastern part of the United States.

The United States is not the only country where the use of regional jets can present opportunities. Other countries have markets where regional jets can add value. Consider, for example, short-haul domestic markets within South Korea shown in Figure 4.7. There are only two markets where regional jets are not appropriate, not because of the distances but because of the size of the market. These are Seoul to Cheju and Seoul to Pusan. These two markets are currently served with high frequency and with wide-body aircraft. All other markets are served by standard narrow-body aircraft, such as the Boeing 737 and the Fokker 100. Most of these markets are either too thin or the passenger yield is low, or both. It would be better to serve these markets with either regional jets or even advanced turboprops. In some of these markets, in fact, turboprops might be more cost-effective because of their lower costs and because there is virtually no difference in block times since the distances are small. In some cases the difference in block times between a regional jet and an advanced turboprop would be as little as five minutes. Moreover, the regional jets may play an even more important role in the Seoul-Pusan market after 2004 when there is high-speed rail service.

Let us turn to the other end of the spectrum of aircraft technology to illustrate the opportunities of ultra long-range aircraft (the Airbus 340-500 and the Boeing 777-LR), the trans-Pacific market in general, and to comment on some carriers on both sides of the Pacific.

Prior to the economic crisis in Asia in 1997, Thailand was in some ways taking on the characteristics of an "Asian Tiger", with respect to the growth in its economy. With a population of 60 million and a developed tourism infrastructure, Thailand was one of the most visited countries in South East Asia. One reason for the high volume of visitors was that Bangkok was used as a base to travel to neighboring countries such as Vietnam. Another reason was the brand image of Thai Airways that competed well with the three other major airlines in the region—Cathay, Malaysia, and Singapore. While Bangkok is still the busiest airport in South East Asia (see Table 4.3), both Thai and Malaysia have shrunk. Cathay and Singapore have expanded.

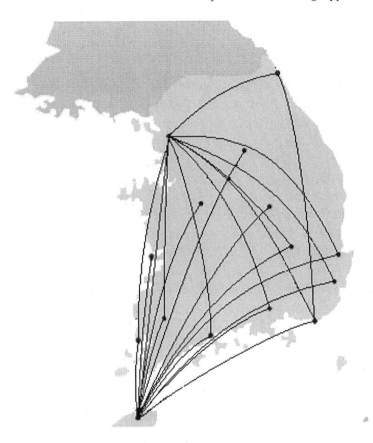

Figure 4.7 Route Map of Domestic Markets within South Korea served by Korean Air and Asiana

There are two explanations. First, while Thai and Malaysia did not develop their core products (network, fleet, and schedules), Cathay and Singapore aggressively did so. Thai serves only one gateway in the US—Los Angeles once daily, stopping alternately in Osaka or Tokyo. Singapore on the other hand serves six gateways (Los Angeles, San Francisco, Las Vegas, Chicago, New York-Kennedy, and New York-Newark) and is about to start nonstop service between Singapore and the US. It is interesting that while Bangkok-Los Angeles and Singapore-Los Angeles markets are about the same size, Thai has decided not to offer nonstop service from Bangkok to the US.

Table 4.3 Top Twelve Airports in the Asia-Pacific Region Ranked by Number of Passengers* in 2001

Tokyo (HND)	58,692,688
Hong Kong (HKG)	32,553,000
Bangkok (BKK)	30,623,764
Singapore (SIN)	28,093,759
Tokyo (NRT)	25,379,370
Sydney (SYD)	24,303,024
Beijing (PEK)	24,176,495
Seoul (SEL)	22,062,248
Osaka (KIX)	19,341,525
Taipei (TPE)	18,460,827
Melbourne (MEL)	17,019,571
Kuala Lumpur (KUL)	14,707,125

* Preliminary data showing total passengers enplaned and deplaned with passengers in transit counted once.
Source: Airports Council International, "World Report", May/June 2002, pp.10-11.

Thai, perhaps, has a chance to regain its lost lead by using new technology aircraft, especially if it is coupled with the new airport at Suvarnabhumi, with a planned opening date of 2005. It will take a lot of coordinated effort on the part of the airline, the airport, and the government to position Thai to be an effective competitor to Singapore. It does have two advantages: Thailand has a population of 60 million vs. 4 million for Singapore and the Bangkok Airport is almost 1000 miles closer to both North America and Europe than Singapore.

Malaysian Airline Systems has a similar problem, competing with major airlines based in the region such as Singapore, Cathay, and Qantas, as well as those outside the region such as British Airways. Malaysia is politically stable and is a relatively rich country and was doing well before the Asian Crisis. Malaysia's flag carrier, as in the case of Thai, has been losing ground to Singapore. Again, aircraft technology can help. One way out would be to improve the core product—nonstop flights and higher frequency. As in the case of Thailand, Malaysia has some advantages over Singapore. Malaysia has a domestic market and a potentially strong tourist market. Again, while emerging aircraft technology can help the airline to gain back its lost territory, the key is for the airline, the airport, and the government is to make it all work together, as exemplified by the cooperation in Singapore and Dubai.

The use of new aircraft technology in trans-Pacific markets is not without risk. For example, markets are either thin or have low yield, the competition is cutthroat, or Asian carriers have near monopolies of distribution channels. This situation is different from the trans-Atlantic, where the majority of airlines on both sides tried to fragment the markets. In the case of trans-Pacific markets, there are carriers on both sides with considerably less interest. Thai and Malaysia are just two examples. For the two major US trans-Pacific carriers (Northwest and United) there is less interest partly because of their strong hubs in Tokyo and partly because they are more risk averse. The situation has become even more critical for United.

Emerging aircraft technology presents new opportunities for some airlines to strengthen their market position and for others to capitalize on the changing demographic and socio-economic patterns discussed above. The opportunities for Thai and Malaysia to strengthen their market positions have already been discussed. For US carriers, it is an opportunity to further develop their strong hubs and serve, for example, San Francisco-Bangkok and Los Angeles-Bangkok. The viability of ultra long-haul has already been demonstrated by the opening up of such routes as New York-Hong Kong, New York-Taipei, and Los Angeles-Guangzhou. The concept can now be extended to routes to and from South East Asia.

The other point mentioned above was to take advantage of changing demographic patterns. While the preceding discussion was related to demographic changes within developing countries, it is also valuable to examine the demographic changes taking place within developed regions—partly the result of immigration-related movements. There has been an enormous increase in the Asian population of the US, particularly immigrants from India, China, the Philippines, and Vietnam. These segments are concentrated in Massachusetts, New York, New Jersey, Pennsylvania, Maryland, Illinois, Florida, Texas, and California. Consider the case of India. It is the third largest country providing emigrants to the United States. The median family income of this ethnic segment is higher than for all US households and 40 percent of the Indian immigrants have taken up US citizenship.[3] This group earns sufficient income not only to meet its own air travel needs but to also send money to relatives in India to visit the USA. Similar demographic patterns and behaviors exist for other countries such as the United Kingdom.

Here is another example where emerging aircraft technology can help an airline to strengthen its market position and capitalize on changing demographic patterns. Air India could easily fly nonstop from Delhi and Mumbai to secondary cities in the UK such as Birmingham, Manchester, and Leeds. As was pointed out in the previous chapter, Emirates is flying from four major cities in India (Chennai, Delhi, Hyderabad, and Mumbai) to Dubai to connect these passengers to London Heathrow and Gatwick. As for the service to the US, one needs only to look at the connecting traffic between India and the US via France, the UK, and Germany. The UK and the US are key markets for Air India. It flew 3.3 million passengers last year with about 25 percent of the revenue coming just from the UK and the US.[4]

The use of narrow-body aircraft in thinner trans-Atlantic markets represents another example of the use of aircraft technology to align costs with fares that are coming down. The Boeing 757 is being used by Continental between New York-Newark and Ireland and England and in a somewhat different case by Iberia between

Madrid and Malabo. A carefully selected narrow-body aircraft in thinner markets can produce higher margins with lower traffic much in the same way regional jets do so domestically.

The final example of the use of evolving aircraft technology to capitalize on emerging opportunities relates to the use of Airbus and Boeing business jets. In June 2002, Lufthansa started a nonstop flight with a 737-700 between Düsseldorf and Newark in a 48-seat all business-class configuration. The service is being provided through a wet-lease agreement with PrivatAir, a Swiss company that has experience in the VIP and Executive Charter business. The schedule is timed to allow connections from Hamburg, Stuttgart, and Berlin. Previously, Lufthansa had provided nonstop service on the route with an Airbus 340. The nonstop service was discontinued after the events of September 11th, 2001. However, given that 40 of the top 100 German companies are based around Düsseldorf and more than 400 German companies are represented in the New York area there remained enough demand to tailor and renew the service with a 48-seat all-business-class configured aircraft.[5] The service should be a success given that a passenger gets a 55-inch seat pitch (compared to 48-inch pitch in regular Lufthansa Business class), and an unrestricted roundtrip fare of US3,700,[6] considerably less than the published business-class fare and closer to the net business-class fare of about $3,500. In Fall 2002, Air Canada announced a new all-business-class division called Jetz to fly business people and sports teams around North America and internationally, including Toronto-London-Heathrow, using the newest aircraft in their fleet—the Airbus A319 (configured at less than 50 seats). This strategy has the potential to not only service this particular kind of market, but also to capture and control it ahead of any competitor.

Passenger Services

While airlines have already achieved substantial benefits from cost reductions in distribution, there is room to achieve more. Figure 4.8 shows the trend in distribution costs that have been reduced from about 20 percent of the total revenue to about 10 percent. The biggest and the most widely known change occurred in the reduction of fees paid to travel agents. The "other charges" category shown in Figure 4.8 refers to the support services provided to a customer. For example, a customer makes a reservation through one of the standard channels and then rings up the call center to ask for a seat assignment or with some other question. There are costs involved not only relating to the agent answering the call but also the associated overhead in supporting such activities.

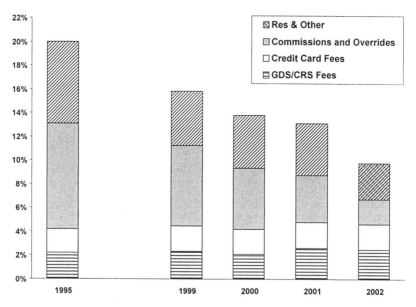

Figure 4.8 Total Distribution Costs (North America) as a Percentage of Revenue for a Typical US Major Airline

One area which should see an increase in its percentage of revenue generated is online reservations. See Figure 4.9. It has increased from about one to about 15 percent in the past five years. In this figure examples of online agents are Expedia, Orbitz, and Travelocity and an example of an auction agent is Priceline.com.

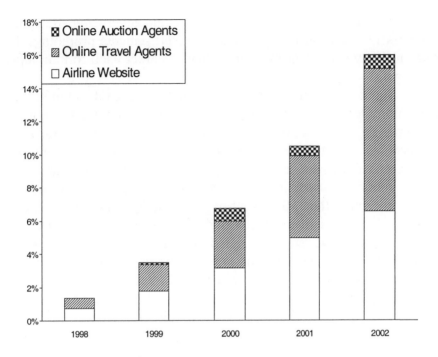

Figure 4.9 Online Revenue as a Percentage of Total Revenue for a Typical US Major Airline

To develop more customer-focused and user-friendly booking engines will further increase the percentage of reservations made online. Until recently online booking engines typically searched by day and by flight, producing a low probability of selecting the best combination of low fare, desirable date, and flight. Some airlines have now developed online booking engines that show first the best fares available and then allow passengers to make their own scheduling decisions. Eventually, it will be possible to make more decisions during booking to include such things as price, date, time of travel, change of plans, restrictions, and routings.

Online reservations between a passenger and an airline not only reduce costs but also provide the added benefit of direct contact between an airline and the passenger. This connection affects marketing strategies, promotion strategies, and advertising strategies. There is little reason to have generic advertisements promoting the total airline when the airline can virtually market itself to individual customers.

Finally, airlines are now using significant amounts of technology to reduce processing costs on the ground and to improve customer service. Take the case of self-service check-in kiosks. They are now as easy to use as a banking machine, they are front and center, a nd the smart and fast way to check in. F or one major US airline, about 95 percent of the domestic passengers now use e-tickets and more than 50 percent of them check in through e-service machines. As passengers increasingly turn to these devices to save time, airlines have steadily added baggage drop-off counters to support them. Baggage identification technology will also help to reduce costs by reducing the number of lost bags. Voice recognition technology can help to automate any rebooking and confirmation process, freeing up reservation and airport agents for more complicated cases. Passport "smart card" control systems are also becoming a potentially time-saving application. It is a fast track through passport control just like through an underground train station when a passenger puts his or her ticket through the machine.

Reshaping Marketing and Operating Practices

The business model of the full-service (so-called "conventional") airline definitely seems broken. The president of a major North-American-based carrier recently said anyone who doubts this is "in denial". The events of September 11th, 2001 (merely the catalyst or the straw that broke the camel's back), the dot.com meltdown, the stock market swoon, and the success of the low-cost, low-fare model have changed air travel in North America for ever. As a major contributor to worldwide travel, markets everywhere will feel these effects. When you add in the changes in communications, including the Internet, broadband, and High Definition Television for conference purposes, business travel is changed for ever. It has not disappeared. It will take a different form and value.

Airlines must face up to the fact that certain choices matter more than others, and customers are only willing to pay for certain things. Of interest to business people will continue to be such things as flight frequency, flexibility to change plans, on-time performance, and short trip times. Of interest to the leisure traveler will be the ease of acquisition of services and a clean and comfortable aircraft interior. Both groups will share an interest in seat size and pitch, speed and ease of airport processing, and to some degree price sensitivity. Many other product features such as some food and beverage service, in-flight entertainment, and exotic seating options will have to be closely examined as to their viability in the marketplace.

Chapter 2 listed a number of millstones around the neck of conventional airlines. An examination of those areas provides reasonable clues to the potential areas for improvement. One glaring example is the misalignment between prices and product features. One can clearly judge from the sentiments expressed in the press and the experience of high-paying passengers moving to the low-fare airlines that there is an imperative for airlines to develop alternative price-service options. The process has already started. To begin with a passenger has a choice of traveling on an independent low-fare airline or a conventional airline. Then, within the category of low-fare airlines, there are no-frill airlines such as Southwest and some-frill airlines such as JetBlue and AirTran. Then, within conventional airlines, there are full-service airlines with capacity-controlled low fares such as American and there are airlines with multiple brands such as Air Canada, Qantas, and Air New Zealand. While, the price-product options have begun to appear, the process must proceed to a much deeper level and at a much faster rate. Figure 4.10 shows one strategy for either one airline to develop separate brands or for one airline to separate its brand from others. From a consumer's point of view, it provides a clearer choice.

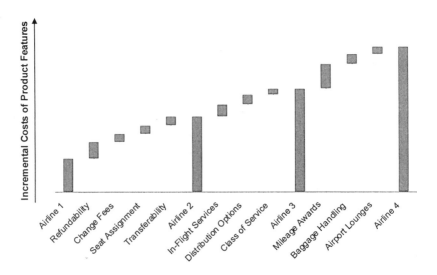

Figure 4.10 Incremental Costs of Airline Product Features

The second area relates to the positive experience of other industries in mass customizing their products—for example, having customers "build" their own computers. Is there an opportunity for such a strategy in the airline industry? For example, a fare depends not just on advance purchase times (14 days, 21 days, and so forth) but also by day, type of seat (window, aisle, middle), variation in the cost of changing reservations based on the type of change, and distribution channel used. Some marketers call this "a reversal of some marketing practices" (pricing, product design, advertising, and so forth).[7] Instead of a fare "secretly" containing these features, could not passengers buy or not buy these features?

The third area relates to communication strategies. Consumers must learn that they cannot expect something for nothing. They must be willing to accept only the service for which they pay and pay only for the service which they receive. There is a real need for "re-education of consumers". The problem facing conventional airlines is that consumers complain when they purchase a lower price ticket (the same level as charged by low-fare airlines) but do not receive the services that they have received in the past.

The challenge for airlines will be to root out and expose, disclose, and price all of their product elements. Mass customization presents a good example. The goal is to use mass production (or in this case, services) techniques to lower the cost of production, while at the same time offering the customer additional choice. It is one thing to do this by building a customized high-feature bicycle on a CAD-CAM machine in a factory, and another to develop differentiated services at check-in or in the in-flight environment. The most highly leveraged value-added products will be those relatively cheap to produce and which do not add undue complexity to service delivery. Many of these features will lie in the electronic realm in such areas as reservations, and automated airport passenger and baggage handling. The most difficult areas will lie in in-flight service and passenger service on the ground, where already overburdened staff will have difficulty absorbing any additional complexity and then consistently delivering the product correctly, professionally, and on time.

In this value continuum, an important tool will be ever-more powerful booking search engines, which will extend the value chain by allowing choices of product elements when making a reservation, appropriately priced in real time. Not only will the system recognize the individual and his value to the airline, but also provide such choices as alternate routings at even lower prices which would reduce excess inventory.

In terms of operating practices, there is clearly a need to reduce costs significantly. Figure 4.11 shows the difference in costs of operating a standard Boeing 737-500 by three different airlines. The per block hour cost varies from US$1,562 for Southwest to US$3,010 for United. The cost difference can be explained by many factors such as operating hub-and-spoke systems vs. point-to-point, operating from lower-cost airports, and accounting treatment of engineering and maintenance overheads. The last factor can play a major role, especially for airlines with large intercontinental fleets. Some airlines simply apportion these costs on an equal basis across the fleet. The arbitrary allocation process can completely change the profitability analysis. However, the key differences reflect not so much the higher costs of labor and facilities but the higher productivity of resources—labor, aircraft, gates, and so on. Again, while conventional airlines have started such processes as the "de-peaking" of their hubs, they need to go much deeper and to move at a much faster rate.

The impact of improved operating practices can be viewed in this planning framework: (a) tools, structures, and management information stream; (b) long-range—more than 18 months away; (c) medium term—6 to 18 months away; (d) short term—less than 6 months away; and (e) day of flight.

Much more sophisticated computerized planning tools are available for use through all phases of the planning cycle. They can begin with fleet and airport development planning, progress through long-range scheduling and market development, product and revenue planning, all the way to detailed financial analysis of profit and loss for the entire airline network (not just individual legs) down to the fully-allocated cost level. Few airlines view planning as an integrated activity, encouraging a competitive stance, rather than a cooperative one, between such areas as fleet planning, scheduling, marketing research, forecasting, product planning, and financial planning. While these areas need not all report to the same executive, increasingly they need to see the merit of cooperating. This cooperation must extend to agreeing on the data that will be used to plan, measure, and then move the airline forward. Everyone needs to start using the same language in planning, whether it is dollars per block hour, cents per available seat mile, or any other measure.

More than a year and a half out is the time for rethinking fleet mix, retirement strategies, optimum aircraft size, future airport choices and ground side development, and improved productivity clauses in contracts. In this time frame, companies can also develop "super brands" and negotiate interline cooperative agreements that go much beyond the current movement of passengers. These could include employees from one airline staffing the flight, in-flight and

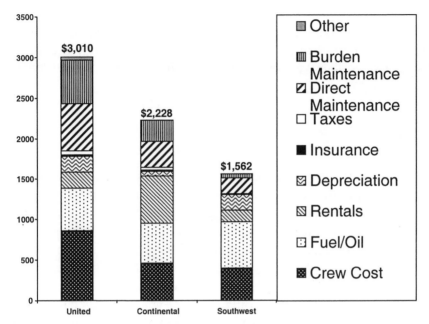

**Figure 4.11 Aircraft Operating Cost in Dollars per Block Hour
for Boeing 737-500, Second Quarter 2002**
Source: *Aviation Daily*, 10 December, 2002.

ground operations of another to improve productivity. There is no
longer any excuse for 12- and 18-hour turnarounds for aircraft that
could be operating new extensions in that time and producing
additional revenue from down time. In this time frame also there can
be negotiations between airlines and various air traffic control
regimes for improved airport and en-route traffic flows and upgraded
equipment. The short- and mid-term is the time for innovative
scheduling of both aircraft and crews, improving work processes to
improve turn times on the ramp, and negotiating with government
aviation authorities for longer check times for aircraft as they
become more reliable, particularly in the areas of power-plants and
avionics systems.

Even the day of operation itself will provide opportunities,
particularly as aircraft manufacturers provide cockpits which allow
crews to move not only within families of narrow-body fleet and

families of wide-body fleet but also between the narrow-body and the wide-body families. Real-time equipment scheduling to meet demand by time-of-day is now feasible and will become a reality as fleet sizes and the availability of compatible equipment, particularly at hubs, permit.

Conclusions

The problem faced by incumbent carriers is not that they are not aware of opportunities but rather that they cannot start with a clean sheet. They are saddled with a given fleet, network, and labor contracts. They must change to capitalize on the emerging opportunities discussed in this chapter. These changes cannot occur overnight and the costs are high. Nevertheless, change must begin, painful as it might be. Even the carriers that are implementing changes appear so far to be doing too little, and in some cases doing it too late. Take the changes recently announced by major carriers relating to the reduction in the types of fleet, the complexity of operating structure, the decision to close a few cities, and the reduction in the size of the workforce. Although all correct, it is not enough to make the difference in competing effectively in the new environment.

In order to capitalize on any of these opportunities, the toughest part is the last piece—how to make it happen, that in turn means two things, (a) moving from glossy charts to actions, and (b) exercising discipline, learning from the experience of (not copying) successful carriers (new entrants or conventional) as well as other businesses in other industries. These topics are addressed in the next chapter.

Notes

1   Laurent, Clint R., "A report on Major Demographic and Socio-Economic Trends of Asia", Asian Demographics Ltd. (website), May 2002, p. 34.

2   Keenan, Steve, "Dubai bound for premier status", *The Times*, United Kingdom, 14 September 2002, p. 8.

3   Handbook for Asian Indians, 1997-1998 Heritage edition.

4   Fernandes, Edna, "Air India to upgrade fleet as part of expansion", *Financial Times*, United Kingdom, 19 September 2002, p. 17.

5   Spaeth, Andreas, "Lufthansa and BBJ: Business fare with flair", *Airways*, October 2002, p. 27.

6   Ibid., p. 29.

7   Kotler, Philip, Jain, Dipak C., and Suvit Maesince, *Marketing Moves: A New Approach to Profits, Growth and Renewal* (Boston, MA: Harvard Business School Press, 2002), p. 43.

Chapter 5

Survival Strategies and Execution in an Uncertain World

As discussed in the previous chapter, despite the current obstacles, there are enormous long-term market opportunities in the air transportation industry. These opportunities exist not just in the emerging markets in which less than 10 percent of the adult populations travel by air, but also in the developed markets as evidenced by the stimulations achieved by carriers such as Westjet in Canada, JetBlue in the United States, Ryanair in Europe, AirAsia in Malaysia, and Virgin Blue in Australia. These opportunities are not limited to domestic operations or to new entrants.

Change within the airline industry is occurring so rapidly that before the industry can capitalize of any of these opportunities, some major airlines must first survive. Survival, however, is becoming a challenge in the emerging marketplace in which the pendulum is swinging. To varying degree in different regions, we are moving from (a) regulation to deregulation and back to re-regulation, (b) airlines controlled by governments to privatization and back to governments, (c) stable investment sources to stock investment groups who capitalize on the volatility and cyclicality of the airline share price, driving the industry into a continued high state of instability, and (d) airlines subjected to "survival of the fittest" environments and then moving back to being protected. This swing of the pendulum is expected to spread. In the US, there now are open discussions in the aviation community about letting the "Air Behemoths" (those with inappropriate networks, fleet, and products; poor management; and unproductive labor contracts) go bankrupt and be replaced by "Air Nimbles" (those in which management, labor, and products reflect the realties of the marketplace). In

Europe, there is discussion about forcing flag carriers to restructure their network. The question is: Are different players in the air transportation value chain willing to change so that incumbents can survive and thrive? This chapter discusses survival strategies in three areas: (a) the need for different stakeholders, individually and collectively, to exercise discipline and to have a shared mindset, (b) the need for management of incumbent airlines to learn from (but not copy) successful new entrants, and (c) the need for the management of incumbent airlines to profit from multi-industry insights.

Stakeholder Discipline and a Shared Mindset

Management

The first requirement for management is to recognize the existence of the multi-faceted challenge—how to make money in a very complex industry that is constrained by at least the characteristics described in the first chapter. It is not easy to manage contradictions—the need to transport multiple types of traffic with different service requirements and different abilities to pay for such services. The second requirement is to admit that key parts of the existing business model of conventional airlines are broken. The third requirement is to recognize that major surgery is needed and it is needed now.

Protecting Shareholder Interest The first area of change needed is to pay more attention to the shareholder. Between the period 1990 and 2001, none of the major carriers in North America, except Southwest, made a sufficient return on investment to recover their average cost of capital. Southwest achieved a return of 1 percent above its average cost of capital. US Airways missed it by more than 5 percent for the entire period.[1] The fact that the capital markets now want a reasonable return is evident by the lack of available investment for United's Avolar project as compared to the availability of funds for JetBlue. Unless shareholders begin to receive reasonable returns, lack of capital could be a real barrier to entry in future.

There are many reasons for the poor financial performance of conventional airlines. Perhaps the most glaring is the lack of discipline. Since the beginning of the 1990s, domestic airline yields have been decreasing in the US markets, yet management increased capacity 3 percent per year between 1996 and 2001.[2] To achieve reasonable load factors, airlines decreased fares. To decrease distribution costs, they encouraged passengers to use the Internet but offered fares that were even lower, further depressing the yield. The business model broke when they increased their full fares to compensate for the deeply discounted fares.

Adaptability The second area of change needed is to develop strategies to finally adapt to the high volatility and uncertainty associated with the airline business. Management must focus on managing uncertainty—an increasingly serious problem associated with downturns in the economy, the rapid and deep proliferation of low-cost, low-fare airlines, and the transparency of airlines pricing ushered in by the Internet. Management at some airlines has not provided the leadership necessary to manage uncertainty. In addressing uncertainty at some airlines, executive management has not been consistent, objective, and demanding, not just of labor but also of the whole management team. At others, management teams were not given the chance or the time to address uncertainty. When they had the time to meet these challenges, management needed to set challenging objectives and not accept arguments as to why such goals were not achieved. One example is the reduction in turnaround times. Management teams that did not achieve results were not replaced in a timely manner. It is ironical that some management could only see uncertainty to be a problem when in fact they could have viewed it to provide clues to opportunities.

Flexible Strategies The third area of change needed is to develop dual strategies, those that focus on long-term sustainability, and those that focus on short-term survival and crisis management. The development of dual strategies requires, in turn, reasonable return on capital raised, retained, or invested and thinking across boundaries to examine new products, pricing, distribution, and operational business models. Thinking across boundaries requires breaking down counterproductive and self-defeating organizational silos. One required key lever is information in the sense that breaking down the walls of organizational structures requires the availability of timely, accurate, and consistent information across historical organizational boundaries—marketing, scheduling, finance, and operations. Another key lever is the development of human resources (empowerment, competencies, and reward systems).

Airline management has had to deal with the variability issue that has been getting worse with time. Figure 5.1 shows the impact on profitability of the past three downturns in the economy. In each case, the impact was deeper and lasted longer. Based on this analysis, the picture is quite grim for the impact of the current downturn. Airlines have used a number of different strategies for dealing with the volatility in the airline business:

(a) Some management tried to find ways of influencing demand by such means as changing the travel patterns of corporations.
(b) Some airlines developed sub-brands to provide flexibility to alter the capacity offered to suit the changing needs of the marketplace.
(c) Some airlines tried to use the concept of aircraft families.
(d) Some airlines implemented risk management instruments and procedures such as those developed by financial institutions (for hedging against movements in currencies and fuel prices, and evaluating aircraft financing alternatives).
(e) Some airlines tried to diversify their routes. British Airways, for example, has had a greater percentage of its routes in

business markets and in markets that are more competitive. So, during downturns, it is likely to suffer more. Southwest is an example of a highly diversified airline. It suffers less, or not at all during downturns.

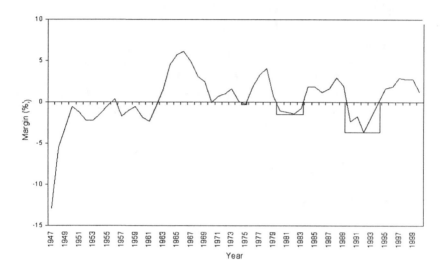

Figure 5.1 World Airlines' Net Profit Margin (%)
Source: Constructed from data abstracted from various publications of *The Airline Monitor* and The International Civil Aviation Organization.

Differentiation   A fourth area of change needed for airlines is to differentiate themselves from each other. This is one area where progress has been minimal. As shown in Figure 5.2, there are a set of full-service airlines at one end and another set of limited-service airlines at the other. The limited-service airlines have differentiated themselves not only from the full-service airlines but also among themselves. The full-service airlines (with the exception of a few such as Virgin Atlantic, Emirates, and Alaska) have had less success in their differentiation strategies. In fact, in most cases airlines have portrayed a lemming-like behavior. It is not a matter of one or the other. There are viable opportunities not only within the end circles but also in between them.

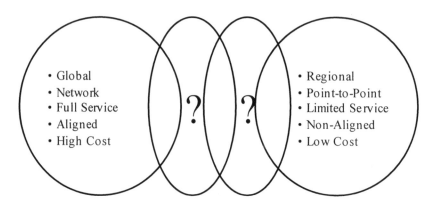

Figure 5.2 Carrier Segmentation

Alaska Airlines is one carrier that has attempted to differentiate itself in terms of products and costs. One differentiation feature is that the carrier offers fares that are competitive with the fares of low-fare airlines but offers higher class services such as seat assignment and meals. The airline has 20 percent of its customers booking through its own website, a number considerably higher than other major airlines. Alaska has also tried to focus on operational improvements. Alaska was the first to sell tickets online and the first to allow passengers to check-in online.[3] Management has also worked on branding itself because it is a relatively small airline. Moreover, because of the remote communities served by the carrier, it also carries a proportionately higher amount of cargo, using the Boeing 737-200 combi aircraft. The carrier also has a reasonably-sized hub at Seattle. See Figure 5.3.

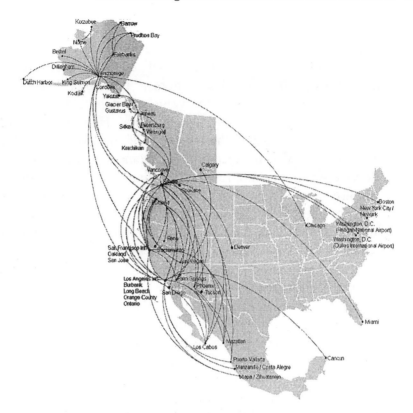

Figure 5.3 Alaska Airlines' Route Structure, Summer 2002
Source: Based on the information contained in Alaska Airlines' website.

Optimization of the business model   The fifth area of change is for carriers to not only right-size themselves but also to make themselves more productive by high utilization of assets.   Many airlines became too big: US Airways, the old Swissair, and the old Philippines Airlines, just to name a few. The reasons for airlines' overexpansion range from the ego of management at one extreme to the availability of government support at the other extreme, with the worship of market share and excessive reliance on connecting traffic in the middle. Following are some examples of how a few successful airlines have right-sized themselves:

(a) Transparent and contemporary concepts of financial accounting were used to compute product and customer profitability. For example, what are the true costs of operating hub-and-spoke systems, providing meals, or carrying cargo? What is the value of intangible assets such as routes, hubs, slots, and brand equity? How do you value your people? What is the true cost of low wages and high turnover? How can executive management espouse customer service when it cannot rely on frontline employees? Frontline employees are almost in a constant lose-lose environment. Despite their best efforts, they lack the resources, the time, and the necessary compensation to provide the expected services. Turning back to other assets, approaches that need further examination include, for example, revenue per square foot of cabin space.

(b) Careful analysis is needed on the degree of diversification. There are major airlines that focus only on the core product (exemplified by the operations of American Airlines and British Airways). Then, there are airlines that are involved in a conglomerate of businesses providing a broad spectrum of aviation services (exemplified by the operations of Lufthansa and the old Swissair Group). Although diversification does have the advantage of spreading around the revenue base, unless extreme care is taken, diversification can also result in disadvantages such as a loss of focus. The process has worked for Lufthansa because of one necessary prerequisite for successful diversification—the development of independent business units (for example, Lufthansa Cargo).

(c) Meaningful measurements of airline efficiency and benchmarks are needed as they relate to financial performance, operational performance, and marketing performance. Would it be realistic for a carrier like United to look at its costs of 11 cents vs. Southwest's 7 cents? The two airlines are very different. Trying to reduce its costs at 11 cents to be closer to 7 cents does not make sense if in fact the comparable costs (adjusted for length of haul, network, fleet, and so forth) may be much higher than 11 cents. It may be more reasonable to examine the difference between actual and break-even load factors since the latter take into account the differences in unit costs and unit revenues. Figure 5.4 shows this analysis for three airlines.

(d) Mergers and acquisitions must be managed very critically. Pacific Western Airlines (based in Canada) was a small but a very healthy airline. It acquired a number of other regional airlines and one large financially weak airline. The process led to its demise. On the other hand, US Airways (a weak airline) acquired some financially sound airlines (such as Pacific Southwest based in California and Piedmont based in North Carolina), destroyed both in the process, and is now in bankruptcy.

(e) Just as in the case of mergers and acquisitions, a much more realistic analysis of alliances is needed, keeping in mind, for example, the realities of human psychology, of organizational structures, and inflexible aviation infrastructure. Here are a couple of key questions. Is the alliance generating new revenue or simply moving revenue around? What is the extent to which alliances have achieved dramatic savings in cost?

(f) There is a need for realistic alignment of business strategy with available resources (technology, skilled employees, brand name, high costs, and so forth). Consider, for example, British Airways that has high costs but a stranglehold on the highly coveted landing slots at London's Heathrow Airport, a good brand name, and the know-how to implement technology to both save costs and, at the same time, understand and manage the behavior of its' premium customers. Here is an example of an airline that is right-sizing itself based on a plan that rationalizes fleet, products, and customers to capitalize on strengths and compensate for weaknesses listed above.

(g) The key to right-sizing may well be to go back to basic planning; strategies, network, fleet, schedules, and products. The core product decisions (network, frequency, and schedules) affect not only the revenue side of the equation but also the costs side through such aspects as aircraft and crew utilization. For example, does the limitation on slots at Heathrow for British Airways call for some frequencies to be operated with high-capacity aircraft such as the Airbus 380? Flights from Singapore, Hong Kong, and Tokyo might justify the aircraft on grounds of demand, congestion, and limited operational windows. In order to keep the utilization high, upon arrival in London, the aircraft could be scheduled for a flight to the US West Coast assuming that the size of the marketplace justified it.

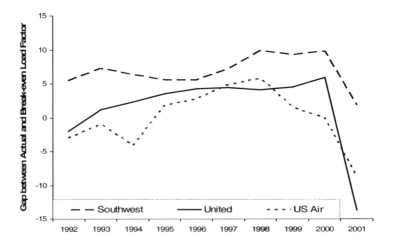

Figure 5.4 Actual and Break-Even Load Factors
Source: Constructed from the data contained in various issues of *The Airline Monitor*.

Branding Strategies The sixth area of change relates to developing branding strategies that are better aligned with business strategies. Virgin Atlantic and Emirates have achieved success in building their brands as innovators. The innovations introduced by these two airlines such as individual in-flight entertainment systems at each economy class seat have led to new levels of competition. With both airlines, innovation is an on-going process.

Branding has to have substance. In other words, the airline must offer a good deal for the passenger, otherwise branding alone is not enough. It must be sustainable, genuine, and change the experience of a passenger. A sustainable brand is a consistent brand. Genuine means that the brand promise should be something that the airline can realistically deliver. Branding also plays an important role in the establishment of expectations. Take the case of the understanding between airlines and customers. Customers need to realize that they cannot get "something for nothing". Certain elements of a network carrier cost money, for example, the availability of a global network, the interline system, and the delivery of lost baggage. Moreover, passengers do not make a careful distinction and tradeoff among the different products, for example, a lower fare on a new entrant that is accompanied by not just the lack of frills such as food and entertainment but also high-density aircraft, no interlining, a limited network, and possible margins of error or risk relating to delays, loss of baggage, and so forth.

Finally, we all agree that the Internet has changed the balance of power by providing customers more options and more information. In the new environment branding must now play an even more strategic role by enabling an airline to develop "trust-based" marketing that provides sufficient information for customers to select the right products or services to meet their needs. While the role of branding is still to convince the customer that the airline has and will deliver true value, branding will also help an airline to avoid margin-destroying competition.

Focusing the Board of Directors While the above comments are related to the need for disciplined airline management, there is a need for airlines boards to take their responsibilities more seriously.

Boards need to understand the issues and to recognize that re-validation of their airlines' strategic plans may be warranted. Examples include: an airline's movement away from its core business; unsustainable expansion into global markets; management decisions that produce a higher ROI in the short term but could be a disaster in the long term—an upstart airline acquiring old airplanes to lower initial investments; continuously changing marketing tactics that may produce short-term gains but totally confuse the passenger on what the airline stands for; and increasing complexity that increases costs but not necessarily revenue.

Here are some examples of the actions taken by the boards at the more successful airlines:

(a) They monitored carefully the behavior of senior management during economic booms, recessions, and times of extreme uncertainty such as now. They monitored management's vision, the relevance of numbers, and strategic thinking. They communicated their clear expectation that management learn faster, act quicker, and adapt constantly. They created the right team spirit with openness to cross-functional analysis for dealing with "grey and fuzzy" situations. They ensured that management paid attention to the execution of strategy—the discipline of getting things done. According to Larry Bossidy, the Chairman of Honeywell International and past Chairman of AlliedSignal, the key is to select the right people who in turn can work together to develop an effective strategy that must be linked to an efficient operating process.[4] One critical success factor of Southwest is the integration of people, strategy, and operations. It is the execution part of the strategy where many businesses fail. In the case of Southwest, it is not just execution but execution culture that leadership has developed.

(b) They required management to strengthen the airline's balance sheets. Strong balance sheets provide not only comfort and stability but lower financial costs and greatly increase bargaining power.

(c) They created and empowered internal groups (a highly disciplined mix of the "old", "young", and "irreverent") to re-invent the business without re-inventing the wheel. The "why not" group can play an important role in discussing such issues as the ability of low-cost airlines to provide trans-Atlantic service. The experienced sages can point out, for example that, (1) although technically feasible, it could be difficult because the scope of matching fares is much greater, and (2) the higher price (compared with intra-North American fares) may be out of the range of impulse discretionary travel.

(d) They directed management to manage for value by removing obstacles such as: legacy frameworks (network, fleet, labor contracts); business complexity (network linkages discussed in Chapter 1); government influences (routes, fares, leadership, subsidies); heavy reliance on the positive outcome of short-term actions that may be detrimental to long-term strategies.

(e) They provide oversight for the organizational structure. For example, while strategy may be developed from analyses of the business environment, the proper implementation and execution of the strategy depend very much on the organization structure. For example: how is the decision-making process assigned (for instance, the degree of centralization); how is performance evaluated; and how are employees in charge of implementing the strategy compensated? A successful organizational structure enables the efficient deployment of knowledge throughout an organization, the implementation of proper checks and balances, and the environment and facility to develop and implement innovative strategy.[5] Another aspect of the organization structure that needs the attention of the board is the degree of decentralization. Too much decentralization may in fact be detrimental. A certain amount of control must always be held by senior management and, in turn, by the board.

(f) The boards of successful companies have selected management that have a consistent and open management style to achieve harmony with labor by taking the initiative to understand employee values; that are articulate in explaining and discussing the intricacies related to such issues as cabotage traffic rights and foreign ownership of national airlines and their impact on labor; and that create an environment where, for example, aircraft and crew schedulers as well as pilots "can walk a mile in each other's shoes".

(g) Finally, they selected management that has intuition, vision, and passion.[6]

Labor

Wage Levels There are many reasons behind the structural and fundamental problems facing the US airline industry. Figure 3.1 listed, for example, six factors that are driving the shape of the airline industry. One of the major areas of focus of the management of conventional airlines deals with major labor issues (wage levels, structure, and contract provisions). To illustrate the first concern, management point to the difference in labor costs of conventional airlines and low-cost airlines. Figure 5.5 shows unit labor costs for major US airlines. America West's unit labor costs are about 50 percent less than US Airways. Management's second concern relating to labor wages is that unionized labor are more connected with their counterpart at competitive airlines than with their own management and that labor uses these connections to benchmark wages and work rules.

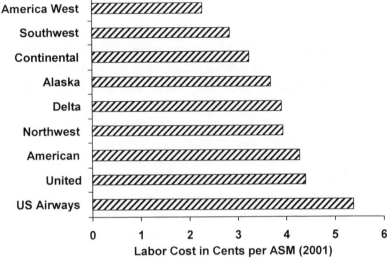

Figure 5.5 Labor Costs of Major US Airlines
Source: Constructed from data extracted from *The Airline Monitor*, September 2002.

Contract Provisions The second concern relates to restrictive contract provisions such as scope clauses that have limited the percent of regional jets in the fleet. Some analysts claim that these conditions have not only made it more difficult to match capacity to demand in selected markets but that they also have led to overcapacity. Management is also concerned about the "snap back" provisions in contracts. In the past labor has made concessions in exchange for equity, salary freezes, and contract provisions. However, except for the tradeoff of equity for concessions, these tradeoffs have usually been temporary with "snap back" provisions. Consequently, labor costs increase as soon as the financial condition of the airline improves. Finally, although pilots are a unique segment of the organized labor group (due to a unique set of skills that cannot be easily replaced), high-cost contracts negotiated with pilots have led to similar contracts at other airlines. They have also encouraged other labor groups within the same airline to negotiate for higher wages.

Awareness of the Business Environment On a broader level, managements' concern is that labor does not recognize the realties of changing market conditions and their impact on airlines. The increasing competitive pressure from the rapid growth of low-cost, low-fare airlines is only the beginning. Further down the road are the potential changes in the regulatory policies with respect to consolidation and ownership restrictions. Such changes could lead to deeply integrated alliances, mergers and acquisitions, foreign ownership of national airlines, and cost reduction through such strategies as outsourcing and sub-contracting. Furthermore, the issue is not just the need for reforms but rather the timing, the extent, and the permanency of reforms.

Inflation vs. Wages Labor has its own set of concerns. The first is that salaries have not increased at rates proportionate to inflation. Take one example, the increase in the salary of Boeing 727 captain (in the 12th year—the highest pay level at American) adjusted for inflation. Figure 5.6 shows that over a 25 year period, the wages of this one captain increased at a rate lower than the rate of inflation. The hourly rate was computed using a set of assumptions such as the amount of flying taking place during day vs. night, longevity, mileage, and weight of aircraft. Also this chart does not take into consideration profit sharing or stock based compensation.

The information contained in Figure 5.6 does not take into consideration the fact that a captain on a Boeing 727 flying for a conventional airline will move up to larger aircraft, with each step up the ladder accompanied by higher wages. Such a progression will not take place at the low-cost, low-fare airlines since they operate a single family of aircraft. This wage-rate/aircraft-size issue is one key area where labor and management need to find a win-win solution by balancing higher wages with higher productivity.

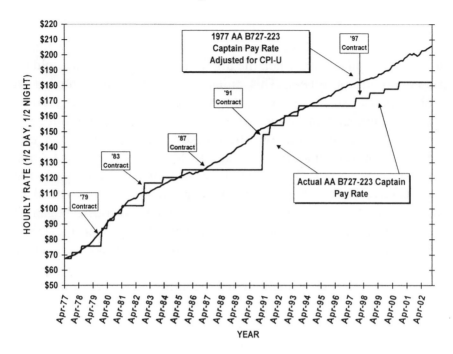

Figure 5.6 Comparison of AA B727-223 Captain Pay Rates to CPI-Urban (1977-2002)

Source: American Airlines' Allied Pilots Association.

Pilots' Salaries Labor's second concern is the press coverage given to what is considered to be "high" pilot salaries, particularly the United 21.5 percent increase negotiated in 2000. Not many know that the pilots at United did not receive any pay increase from 1994 through 2000. If one computes the cumulative average increase in the pay of a 12- year Boeing 727 Captain at United from August 1994 through April 2000, it works out to about 3.5 percent per year. This belief on the part of management and the public about inflated pilots' salaries will eventually compel both labor and management to sit down together and completely re-evaluate the worth of a pilot and the methodology that should be used to compensate them, while still grandfathering the rights of the existing pilot population.

Productivity   Labor's third concern relates to productivity. According to labor, management produced inefficient schedules during the regulated era when the higher costs of operations could be passed on to consumers. This situation led the pilot contracts to contain clauses such as duty rig, trip rig, minimum day, and average day to ensure that the pilots were not penalized for unproductive schedules. After deregulation, the clauses relating to unproductive time did not get removed, a situation that increased costs. Some labor groups claim that instead of working to eliminate the unproductive time, management at some airlines chose instead to chase the high-yield traffic. When the higher yield traffic began to decline, it became difficult for management to cover the costs associated with the strategy or the unaddressed unproductive labor cost issues.

One example of inefficient scheduling of aircraft and crews relates to the distribution strategy to maximize one's position of displays on a reservation system or to provide the shortest time through a hub to a destination. This strategy is beginning to change since passengers using the Internet are beginning to look for schedules with the lower fare rather than the shortest trip time. I applaud the "de-peaking" strategy announced by American to increase the utilization of aircraft, crews, ground personnel, and gates. For example, prior to the implementation of this strategy, American used four gates to handle 18 flights a day into and out of Hartford, Connecticut. Southwest uses two gates to accommodate 17 departures.

Negotiation Practices Labors' next concern is that management negotiates in "bad faith". For example, according to labor, management does not want to share the profits. Northwest labor gave significant concessions in the mid-1990s. The company made over two billion dollars but provided less than a 5 percent increase in wages. In the case of United, the pilots felt that they made excessive concessions in 1994 in return for equity in the company. And in fact, by 2000, their financial position was below industry average.

Then, instead of resolving the issue with pilots, management announced its desire to acquire US Airways, a decision that had not been approved by pilots whose seniority could have been affected by the acquisition.

Obviously, there are legitimate concerns on both sides. Labor knows that its success depends on the success of the airline and vice versa. At airlines where there is this mutual trust, respect, understanding and accountability, there is a win-win situation. There is clearly a need for a strategy that produces a win-win situation and that requires having a shared mindset, a topic that will be addressed later.

Airports

Airports are an integral and important component in the air transportation value chain. However, airports work within a different framework due to the differences in their ownership, control, focus, and operating environment. Here are some examples.

Since many airports in the world remain government owned or at least government controlled, the incentive to control costs is limited. Therefore airports raise charges to cover costs and make a profit. Airlines generally do not have any choice other than to pay the higher charge even though they may not be in a position to pass the increase on to their customers. When traffic is down, airports can raise rates while airlines are not able to raise fares. Second, the focus of airport management appears to be on revenue generation more than cost reduction. Since many airports derive a significant percentage of their income from non-aviation related sources; their strategies for handling passengers are not always consistent with the strategies of airlines. For example, some international airports want passengers to "linger" and experience the breadth and depth of the products and services available at their facilities. Airlines, on the other hand, want facilities and services that enable them to process passengers rapidly and cost efficiently. Some international airports— Dubai, for example, have developed extensive shopping facilities to encourage passengers to route their itinerary through their airports. Some others raise their profits with various charges and passenger handling fees, rather than using their commercial spaces to gain revenue. In Latin America, for example, only about 10 percent of the revenue comes from commercial retail spaces, causing the ratio of taxes and fees added to the ticket to be a very high percentage of the fare on short-haul segments. At most airports in developed countries the commercial activities provide up to 40 percent of their revenue. Another major concern relates to airport environmental issues that, at least in the United States, are passed down from national to local levels. This leads to inconsistent policies across the country and difficulties for airports in influencing and controlling opposition to airport development and expansion.

Governments

Governments are not only part of the airline industry problem but they are also part of the solution. The problems relate mostly to such issues as too much intervention in some areas and not enough in other areas, inconsistent policies, and too high taxes. Here are some examples.

General Policy Considerations Government policies, as they relate to the airline industry, not only differ from country to country but they have also been inconsistent over the years. While all are in favor of safety regulations, from the view point of economic regulations, some countries are for deregulation, some are for liberalization, and some are for tight regulation. The inconsistent aspect can be illustrated by the developments in the US and in China. In the US, the airlines were regulated until 1978, then deregulated, and now, according to some, re-regulated through the indirect methods of approving mergers and acquisitions, passengers' rights, and, more recently, who gets financial assistance and who does not. In the case of China, first, the tight control is exemplified by the CAAC (Civil Aviation Administration of China) not only operating all services but also performing the regulatory functions until the mid 1980s. Beginning in the 1990s not only were the regulatory functions separated but separate airlines were formed and new entry was allowed. The liberalized policies enormously increased demand that, in turn, led to a rapid increase in the number of airlines that, in turn, constrained the infrastructure and raised questions of safety. Even with tighter controls on entry and prices since the mid 1990s, competition led to losses, a situation that prompted the government to consolidate the competitors into three leading airlines.

It is some airlines' view that governments have not developed policies in critical areas to remove archaic rules relating to control and ownership, or to work with the public to speed up the development of critical infrastructure. With respect to the first point, the airline industry is no longer an infant industry that needs governments to protect and promote it by controlling it with regulatory tools. The emerging global environment requires regulatory authorities to relax their control and ownership restrictions. Ownership and control restrictions have led to alliances that are an artificial solution to an artificial problem. If governments allowed strong foreign airlines to own and control financially weak national airlines, it is possible that a strong foreign airline might significantly rationalize routes and reduce costs, but also capitalize on the synergies between the strong and weak airlines while maintaining the strong features of either or both brands. However, this concept can only work if the strong airlines do not dominate and bleed the weak airlines dry, leading governments to, once again, become cautious and protective.

Infrastructure Development With respect to the insufficiency of the infrastructure, most governments have not formed coalitions to overcome local and national opposition to increasing the capacity of their national aviation infrastructure—new runways, terminals, and so forth. The length of delay—relating to the go-ahead for Terminal 5 at London-Heathrow and the second runway at Tokyo-Narita—is intolerable. The exceptions are few. The German and the French governments were able to facilitate the expansion of Munich Airport and Charles de Gaulle Airport, respectively. It is time, for example, for the United States, the United Kingdom, the Japanese, and the Brazilian governments to overcome short-term politics and facilitate the expansion of their congested airports and the development of new airports. Governments also need to balance the required environmental compliance standards with a reasonable level and structure of costs. For example, what form should Chapter 4 noise standards take given that future noise reductions may come at the expense of retiring early expensive aircraft?

While safety and security should continue to be very high priority areas of governments, the three groups (government, management, and crew) need to come to a common understanding pertaining to flight duty regulations and policies. For example, regulatory agencies need to evaluate the regulatory rules applicable to flight crew duties so that airlines can improve productivity without sacrificing safety or the quality of life for crew members. Second, while security of passengers, employees, and property is of utmost importance, it is necessary to ensure that the process does not become bureaucratic. For instance: how much data on each passenger needs to be transmitted to the airport of destination—such as the name of the agent who performed the security check at the originating airport. While decisions involving mismatch between actual passengers on board and the manifest or between passengers on board and checked bags are vital, other less critical areas that may not affect safety but may have a significant effect on performance should be left up to the "discretion of the captain".

Financial Support Should governments bail out losing airlines without *realistic* business cases? If airlines need to be bailed out for strategic reasons, then these reasons need to be articulated. If, for example, Olympic is required to fly Athens-New York nonstop or Iberia is required to fly Madrid-Tokyo to attract investments in Greece and Spain, then the economic benefit to the country should be factored into the "declared" subsidy for the airline. However, if the government does pay a subsidy for operating a particular route, then it should demand full accountability. Moreover, even if governments provide subsidies, should not they require such decisions as divestment of non-core assets and significant reductions in headcounts? In the US, the Air Transportation Stabilization Board (ATSB) was set up to provide liquidity to carriers who could not get credit. The ATSB has supported and rejected applications of both large and small carriers.

Taxation Government taxes are another area of major concern of the airline industry. There are many examples. Government taxes in the US domestic short-haul markets have been going out of sight (for example, $40 on a $200 ticket). Probably the only taxes higher are on fuel which is another significant cost for the airlines. The Canadian government currently charges a C$24 security processing fee per return ticket—one of the highest in the world. This is a very high percentage of the fare charged by low-cost carriers like Westjet on their short-haul routes and has demonstrably hurt the traffic levels for that carrier. Governments should be sharing the security cost burden after the events of September 11th, 2001. Protection from terrorism should be part of the national defense budget.

As seen in Figure 4.5 in the previous chapter, short-haul air transportation is a substantial part of the airline industry. Such high taxes could explain some short-haul carriers' interest in long-haul markets. Taxes could be contributing to a redistribution of travel. Going to a completely different example, the imposition of considerable taxes on foreign-built aircraft is preventing Aeroflot from acquiring more Western aircraft. Take yet another case of taxes. It is known that carbon dioxide created by aviation activity is harmful to the environment. About 3.5 percent is caused by the aviation industry. In the UK, the government collects a tax called the "Airport Passenger Duty". It was passed, presumably to pay for environmental damage even though it does not carry such a label. Should governments, instead, provide incentives for airlines to become more green?

Shared Mindset

The key to dealing with the multi-faceted challenge of the airline industry is the meeting of the minds among a broad spectrum of stakeholders: management, government, labor, infrastructure providers, distributors of airline services, and consumers. Let us first consider the case of management and labor. For its part, organized labor must accept sustainable market reforms but only if (a) they help management develop and implement sustainable competitive edge, and (b) ensure that, management does not waste the concessions to introduce artificially low fares to maintain or possibly even gain market share. On the part of management, it must consider joint ownership and decision-making (although the governance relating to ownership at United did not solve the problem), performance-related pay, variable compensation to reflect variable conditions, layoff procedures, and, most importantly, respect for labor.[7] Finally, they must not overlook the customers. It can work, exemplified by the experience of Southwest where management has been able to enjoy high labor productivity by means of open and honest communications and labor has enjoyed the benefits of "no layoffs", even during bad economic times.

With respect to infrastructure, airports must find new and innovative ways to work with communities to find ways to provide needed facilities on a timely and reasonably priced basis. Governments need to maximize the benefits and minimize the risks of commercialization and the privatization of airport authorities. While airports need to be commercialized, they need to be mindful that it is the aviation activity that brings commercial activity (shops and other amenities) to the airport in the first place. Airport owners need to share excess profit with airlines through reductions in landing fees and handling charges. Unfortunately, (a) financial reports are not readily available, (b) there is no consistency from airport to airport, and (c) even if owners of airports admit that they have earned extra profit, they may decide that profit generated from non-aviation activities belongs to the owners. With respect to conflicting strategies discussed above, it may require some airports to implement dual track systems, fast track for one segment of passengers and a slower track for another segment that wishes to take advantage of the "full" services provided by an airport.

At the national level, airlines, airports, and governments need to simplify the process of passenger movement (the traditional customs/immigration and now additionally imposed security checks). This would reduce costs and make air travel more convenient. At the local level, a much closer working relationship is needed among the airlines, airports, and local planning boards—for example, if low-fare airlines are to expand their services at airports, between 40 and 80 miles from the commercial and business areas, then planning boards need to improve the access and egress from these. Consequently, governments need to encourage the development of secondary airports and provide incentives for airlines to move their operations to such locations. Islip MacArthur Airport in New York has a 7000-foot runway, insufficient for Southwest to make a nonstop transcontinental trip (even with its newest Boeing 737).

Local governments need to work with airport authorities to extend the length of runways to reduce congestion at neighboring airports as well as provide opportunities for new entrants. Stansted Airport in London is choking. It is already handling 15 million passengers per year and the high demand for air service is not integrated with either the airport infrastructure or ground access. It was recently reported in the British press that on a Monday evening in September 2002 (not considered to be the peak season), 20 flights were scheduled to arrive around or after midnight causing passengers to miss the last Stansted Express to London. A taxi ride to Central London priced at 80 pounds sterling is hardly consistent with the needs of passengers traveling on low-fare tickets on Ryanair, GO, and Buzz.[8]

Turning to some global examples, as was mentioned in the previous chapter, in order for Thai Airways to compete effectively with Malaysia and Singapore, the airline must form a close working relationship with the government and the airport. Consequently, if the new Suvarnabhumi Airport in Bangkok is to be successful, it must have sufficient airport facilities (runway capacity and gates) and ground access. Moreover, the government needs to provide an open skies regime, enabling Thailand to achieve in South East Asia what Dubai has achieved in the Middle East. On a positive side of the shared mindset, China is growing very rapidly, not just because of its economy and declining fares, but because of the support of the government (consolidation, airport construction, air traffic management, and travel facilitation—visa and currency). The consolidation policy of the government will help the three surviving airlines become strong enough to attract strategic alliance partners and investments.

For the airline industry to deal with the multi-faceted challenge, the essential key ingredients are (a) governments, management, organized labor, and infrastructure providers must have a shared mindset, and (b) each of these four entities must be individually disciplined, while still navigating collaboratively toward achieving common goals.

Learning from Successful New Entrants

Figure 3.1 in the third chapter showed six primary factors reshaping the global airline industry. Although management's failure to adapt was listed as number one and the availability of low-fare options as number two, these two factors are related. A large part of management's failure to adapt to these factors is due to the low-cost, low-fare airlines. Given the significant impact that the "low-costs" have already made and the potential further inroads this segment is likely expected to make, a few leading airlines have reached the conclusion that the survival strategy is not to copy the low-cost, low-fare airlines, but rather to learn that their products provide mutual value—both for the consumer and for an airline.

Figure 5.7 shows the four key success factors of low-fare airlines. Although different carriers have achieved success based on different factors, few carriers (probably just one—Southwest) appear to have met all the criteria listed in Figure 5.7. It is this achievement, meeting all four criteria, that explains the enduring success of Southwest during the past 30 years.

Long-Term Cost Minimization

The first lesson to be learned from the experience of low-fare airlines is the understanding that costs are more critical than revenue. Management at an airline, conventional or new entrant, has less control over revenue than over costs. While revenue can take a nose dive overnight management cannot reduce costs to the same degree and at the same rate. Consequently, the original and unchanging focus of this segment of the industry is to have the lowest costs all the time in order to better hedge the revenue risk management. For these airlines, cost reduction is not an event or an activity, it is a culture.

A number of strategies are used to achieve low costs. The first key strategy in this area is the use of a single type of aircraft. This strategy keeps crew costs and maintenance costs down as well as achieving higher utilization of aircraft as the same aircraft can be used on different routes. The second strategy is the use of innovative industrial techniques to achieve high productivity of resources: employees, aircraft, and facilities (for example, fast turnaround of aircraft and rapid processing of passengers at airports during the check-in, enplaning and deplaning processes). The third strategy is to use straight forward planning, and low-cost production and marketing methods. Examples are the use of simple cross-functionally-integrated organizational structures (planning), lower cost and less congested airports (production), and lower cost channels of distribution (marketing). Another example is JetBlue's strategy to configure all of its reservation agents to be able to work from their homes—saving money for the airline and keeping the turnover rate low because many workers may prefer this.

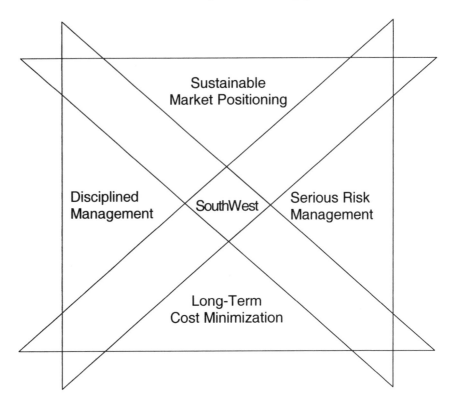

Figure 5.7 Key Success Factors of Low-Fare Airlines

Disciplined Management

A key strength of Southwest has been the discipline of management, exemplified by: "sticking to its knitting"; focusing on just two principles (low fares to attract price sensitive passengers and high frequency to attract business travelers); and measured growth. Southwest has not deviated from its basic strategy for the past 30 years. It is unlikely they will offer assigned seating, two-class service, meals that require galleys, in-flight entertainment, and so forth. It is also unlikely that they will sell cargo that might increase turnaround time, and there is even a question whether they would fly internationally, or at least outside North America. The carrier has increased the number of cities served from 3 in 1971 to 57 in 2001, an increase of less than two cities per year on the average. They always choose the new markets they enter with great care and advance planning.

Figure 5.8 shows the increase in frequency per city since 1971. The main insight from this analysis is that rather than increasing at a rapid rate the number of cities served, Southwest concentrates instead on increasing the frequency at each city. This strategy is also evident from the information presented in Figures 5.9 and 5.10. Figure 5.9 shows the number of daily departures in the top ten cities planned by Southwest in its April 2003 schedule. The number ranges from about 182 in Phoenix, Arizona to 79 in San Diego, California. Figure 5.10 shows the number of departures in the bottom 10 cities. Even in this scenario, the lowest frequency is 6 departures per day out of Houston Intercontinental Airport in Texas. Conventional airlines do not have such a distribution of departures across their network. They have very high number of departures out of their hubs and then very few out of the remaining cities served.

Finally unlike some conventional airlines or some new entrants, Southwest has always planned to grow at a relatively moderate rate. Management spends an inordinate amount of time analyzing new routes and then fully cross-functional teams make every effort to plan a successful entry into a new market. Just as an enormous amount of effort is devoted to planning service in a new market, just as much energy is spent in the execution part of the strategy. This insight is important in that it is one area where the conventional airlines fail—for example, in the execution part of the strategy to introduce a low-fare subsidiary or a new in-flight entertainment system.

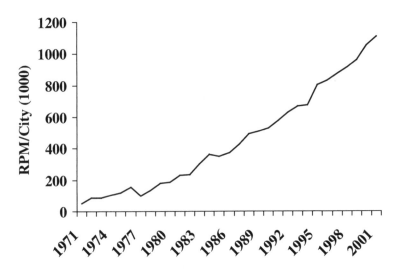

Figure 5.8 Southwest Airlines' Focus on Frequency
Source: Constructed from data provided by Southwest Airlines.

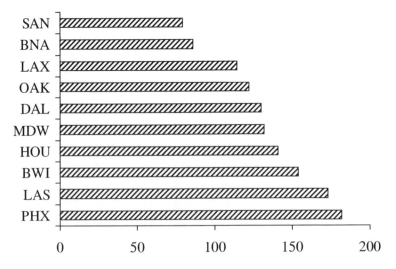

**Figure 5.9 Southwest Airlines' Top Ten Cities: Number of Daily
Flights, April 2003 Schedule**
Source: Constructed from data provided by Southwest Airlines.

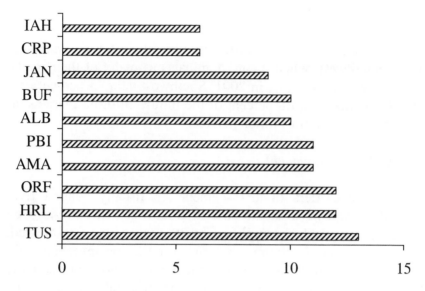

**Figure 5.10 Southwest Airlines' Bottom Ten Cities: Number of
Daily Flights, April 2003 Schedule**
Source: Constructed from data provided by Southwest Airlines.

Sustainable Market Positioning

As discussed above, most low-cost, low-fare airlines have done a
better job differentiating themselves from others than have the
conventional airlines. In Europe, for example, easyJet provides
service from conventional airports such as London-Gatwick whereas
Ryanair serves more remotely-located airports such as Hahn that is
located about 100 kilometers from Frankfurt. Similarly, Southwest
does not provide assigned seats, meals, or a two-class service
whereas AirTran does. Southwest does not provide an in-flight
entertainment system whereas JetBlue does. The one common
denominator among all carriers in this segment is that they all cater
to the basic needs of the time-sensitive and/or price sensitive
travelers. Finally, the successful carriers have kept their product and

strategy consistent, exemplified by the 30-year experience of Southwest.

In addition to the differentiation characteristic mentioned above, Southwest, when it can, dominates as many of their markets as possible. For example, in 2001, Southwest held 73 percent of the intra-Texas markets, 68 percent of the intra-California markets, and 42 percent of the intra-Florida markets.[9]

Serious Risk Management

Most of the carriers in this segment are fiscally conservative—especially Southwest. In the case of Southwest, the airline avoids unwarranted growth, exemplified by its management's knowledge that capital markets require a reasonable return on investment. One aspect of managing the risk is to outsource all non-core activities. Although this is not an insight for conventional carriers, it is a constraint for them due to the restrictions contained in some labor contracts.

As with conventional carriers, most low-cost, low-fare airlines have diversified their networks. However, the degree of diversification developed by Southwest is quite conservative. For example, Southwest does have a significant percentage of passengers that "unofficially" connect. The unchanging primary goal is to provide nonstop service. Schedulers look at what connections are made automatically and then supplement those with additional connections by adjusting the schedules very slightly.

Southwest has a good strategy for managing its business through cycles. It has a very high ratio of nonstop vs. connecting traffic, a good spread between business markets and leisure markets, a good product for both segments (higher frequency for business travelers and lower fares for leisure travelers). It flies to a greater number of secondary airports where there is less competition. It achieves a dominant position wherever possible to reduce competitive incursions. And, as already mentioned, the airline strives for low costs all the time not just during downturns, steady growth (not very high even during upturns) and only modest reductions during downturns. Finally, Southwest maintains a very low debt and a very strong liquidity. For example, its debt-to-equity ratio is 34 percent compared to 71 percent for the industry.[10] The airline also maintains a favorable debt to fixed asset ratio.

Profiting from Multi-Industry Insights

Flexibility

In addition to the insights of successful low-cost, low-fare airlines, conventional airlines can also benefit from the insights of some other businesses just like the US auto manufacturers (such as Ford) did by introducing flexible manufacturing—a more efficient manufacturing system used by their Japanese rivals.[11] This allows them to react very quickly to the changing needs of the customer by switching models and building the cars, trucks, or SUVs that are in demand. If the use of robots and sophisticated production control electronics allows this in auto manufacturing—a hard goods business, then the airlines should adopt some of these processes to allow for the scheduling of their very valuable aircraft assets closer to the departure date and time, thus also meeting changing demand volumes and avoiding fare cuts to attract marginally profitable passengers.

The Power of the Brand

The hotel industry in general, and Marriott in particular, is an example of the power that still resides in a well-managed brand. They have five in total and though they are all clearly Marriott properties, they have notable characteristics which represent different value propositions and they are clearly distinguishable to the customer. Some are meant for overnight or short stays; some for stays of many weeks. Some are full service, with staff, pools and restaurants; others are very lean in design and lack extra features.

This enables the parent company to both clearly understand their costs and then to carefully control them. Customers understand what they are getting and pay only for what they want. This could be a model for the airlines who could more clearly identify exactly what the customer will get at a given price point and therefore what the customer's expectations should be. A customer whose expectations are consistently met becomes a happy customer.

Cost Control and Mass Merchandising

Wal-Mart has become the world's number one retailer—a global powerhouse with 800,000 employees and revenues of more than US\$ 200 billion. Suppliers dread the trek to their headquarters in Bentonville, Arkansas because they know the company will drive a very tough bargain to ensure it keeps its costs down. Combined with an incredibly sophisticated management information system that gives overnight reporting on every item sold everywhere in the world, the company manages on a just-in-time basis, and maintains its thousands of stores in low-cost locations outside of cities and towns. Target Stores, by comparison, offer a slightly more sophisticated buying experience for a slightly higher price. Their stores are a little more spacious and attractive, and have a little more customer service a nd reduced check-out times. From both of these companies, the airline industry could usefully learn about the power of strategic purchasing, the merits of powerful management information tools and choosing your markets carefully, and how to merchandise profitably to the masses.

Figure 5.11 shows the changing mode of audience targeting. Successful businesses have moved from mass marketing of the brand, to direct marketing of the product, to managing expectations relating to customer benefits. It is this last area where businesses are beginning to focus on strategies to "manage" customers—catering to the best customers and "firing" the unprofitable ones. Actually, it is not so much "firing" as finding new cost-effective ways to sell their products. Clearly everyone is not looking for the lowest price by booking through the Internet. Some people still want to go through a book store and not only browse but also, in some cases, enjoy a cup of cappuccino.

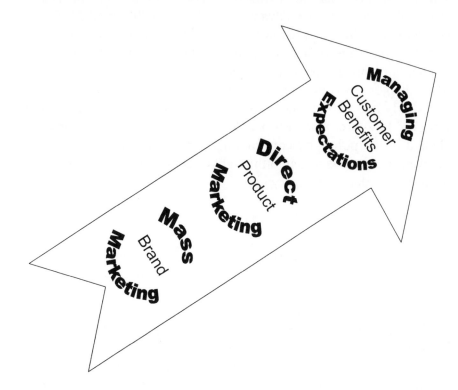

Figure 5.11 Changing Mode for Targeting Audience

However, when a certain segment of passengers is clearly unprofitable, and moreover, is likely to remain unprofitable, then the airline needs to either drop the segment or find a way to transform it to profitability. An example would be high frequency travelers who always buy tickets in the lowest bucket through advance purchase and off-peak days. While airlines are reluctant to drop passengers during difficult times, these passengers attain high privileges with low fares. Some airlines are just beginning to allocate air miles based on the amount of money spent and provide reduced or no credit for miles accumulated at the highly-discounted fares.

If airlines are going to go for wallet share instead of market share, it is necessary to focus on a core product that provides a sustainable competitive advantage and reasonable returns for the shareholder. This requires not only that the product be costed and priced to meet the needs of the marketplace, but that the strategy to implement it meet the peculiarities of our industry mentioned in the first chapter. Airlines need to stop trying to be everything to everybody and define themselves. The problem is the choice—high fare vs. low-fare, business vs. leisure, O&D vs. connecting. A change in the policy by conventional airlines to reduce the level of the walk-up fares is a step in the right direction. The above principle needs to be adopted to everything.

The Customer as a Profit Center

While companies in the past have normally measured profitability by products, divisions, territories, and so forth, a new concept being put forward is to look at a company as a portfolio of individual customers. For a large airline that number is in the millions. Based on the experience of the Royal Bank of Canada[12] and Barclays Bank, it is possible to calculate the profitability of each and every customer. These banks have over 10 million customers. It is possible today because of the power of data warehousing products such as those from Teradata. From this profitability view and other important characteristics (life time value, frequency of travel, and status in the frequent flyer plan), companies can segment and micro-segment their customers to develop strategies to maximize the profitability of every customer.

Before implementation of the above strategy a wealth of information on each passenger is required. The more detailed the information about events, costs, channels, and so forth, the more precise in measuring one can be. That said, airlines generally do not have great, detailed information in relevant areas. It becomes a process of starting with what they do have and evolving precision (and adding detail) over time. The beauty of data warehousing is that detailed data can be loaded to the data warehouse once and used many times (that is, for profitability measurement, target marketing, pricing analysis, revenue management, and operations). Information about passengers, products, services, events, and operations can be used to understand a customer and what his or her contribution is to the business.

This approach requires us to address the issue of cost allocations, always a problem in the airline industry. Different divisions of airlines are constantly arguing about the allocation of costs used to determine profitability by product, region, fleet, route, or schedule. This is an evolutionary process and agreeing is the key to determining how resources are or should be used to maximize the profitability of each and every passenger, the new "profit centers".

There is then the question of what organizational changes are required. Likewise, what culture changes are required to ensure such an approach is successful? This is an enormous challenge if an airline is to manage its entire business based on the profitability of each and every customer. The organization does not need to go through all this before any benefits are realized. Trying to gain an understanding of detailed customer profitability may only first involve making marketing-based customer management decisions. This is a tactical approach to move in the right direction without having to first re-engineer the whole company. An airline can start the process by identifying who their most valuable customers or markets are and executing three tactical programs: high value customer retention, product re-pricing, and product-feature charge management (see Figure 4.10 in the previous chapter). These three tactical (and Customer Relationship Management or CRM-related initiatives) can provide measurable returns in a very short period of time. In the case of banks, examples include asking certain customers to keep higher balances or charging certain customers monthly fees or fees based on transactions.

Table 5.1 profiles three fictitious passengers. Two of these airline customers (Rob and Ella) both generate total profits that are the same in a given period, but the profitability of each of these two passengers is generated in different ways.[13] The third customer, John, is marginal at best even though he uses the airline often. John is included in the example with the idea of showing a customer that might be "fired". Conversely, the airline marketing group might use the data to push John to the airline's website, and find ways to process John and his extra baggage more efficiently.

Tables 5.2 and 5.3 break out the P&L for one round-trip flight (Houston to Seattle) that all three of the passengers flew on the same day. The indirect costs are identical in this table. The direct or behavior-based costs vary significantly. Most of the information is self evident even though some of it was computed using formulas that are not shown explicitly.

Here are the business profiles of the three passengers. Rob Smith periodically travels for business which includes an annual trip to Europe—not sufficient for elite level for the frequent traveler program. He works for a Houston engineering firm. He makes his reservations through the local office of a major top ten travel agency. He makes his reservation early but changes it often, resulting in considerable re-booking activity. He has formed a habit of calling the airline reservation center to confirm flights and check for weather. Although he is a seasoned user of computers for business activity, he has not shown interest in the use of the Internet. He has received special treatment from the airline Frequent Traveler Center in the past when, for example, he was re-routed on a competitive airline. He always calls to insure full credit to his account. Now, he even contacts the call center to check on his mileage status and to ask about the availability of any bonus deals. He enjoys his conversations with airline personnel and makes a game of getting the most he can from his preferred airline.

Ella Jones is a registered nurse who is contracted by a Houston pharmaceutical firm to help field sales personnel "open doors" in the offices of cardiologists to promote a new heart medicine. She is very comfortable with her favorite airline's website. She is able to book in advance to obtain moderately low fares and she knows the schedule and usually sticks with the original reservation. She packs light, does not check baggage, and uses the kiosks at the airport to avoid lines at the check-in counter. By getting through the airport quickly she has more time to relax in the airline's airport lounge during her journeys. She accumulates significant frequent traveler miles and tracks her total mileage on the airline's website. She belongs to the lowest elite level of the airline's program.

John Doe buys low priced tickets from a number of major online travel agencies since he is expected to cover those expenses out of the flat rate he is paid as a customer service engineer. He is part of a medium-sized technical engineering firm which has few travel policy guidelines. He regularly flies to Atlanta using whichever nonstop carrier is the cheapest. He often checks in an extra equipment case that is used for the specialized repair work he does. John tends to call reservations to check on the current weather, airport delays, or to change his flights after they have been ticketed online. John is an elite member of the frequent traveler at the lowest tier.

Here are some examples of tactics that could be used by the airline to improve the profitability of each of these three hypothetical customers using some of the information contained in Tables 5.1-5.3. Rob needs to be encouraged to use less costly channels to communicate with the airline. Reservation and Frequent Flyer Center agents should see a flashing signal on their screen when Rob calls that instructs them to suggest the airline's website and even to go so far as to ask him to pull up the site, review the features he needs for tracking bonuses and flight cancellations, and save it as a "favorite" Internet settings on both his work and his home computers. Prompted by their reservations computer, airport passenger agents should begin to recognize Rob as a good customer, and give him name recognition and personal attention as time permits. Flight attendants should list Rob as an ideal passenger to make personal contact with when they have time. Rob likes personal contact. He likes to put a face to the service he is buying. A high touch approach to Rob during the flight may have long-term benefits in increases to his annual travel representing wallet share growth.

Ella is a consistently profitable client and while it would seem there is little to do to insure that she spends more on the airline, a few tactics may help. She should be considered for personalization on the airline's website so that she is personally recognized when she logs into the site. At that point, she may be asked to complete a short survey on the services she uses such as the kiosk check-in. In return, she would be awarded with a credit for some services at the airline's airport lounge. Dialogue with Ella should continue using web personalization tactics. She should receive a personal greeting where possible at any of the airport lounges that she uses since that is one area where high touch contact is possible. If agents learn more about Ella (type of nursing she does, her travel patterns, competitors she uses, and so forth), they should enter this information into her frequent flyer profile for future use by other front-line agents.

It is not apparent to John how much cost is generated to the airline when he uses an online agency. He should be encouraged to use the airline's website. Trained reservations sales agents may make a call to John to review his travel needs. Using a professional agent dialogue, it may be possible to include John's company as a small business account. Employees of the firm could make reservations over a special website with a minimal discount (3-5 percent) thus keeping a higher share of competitive market travel on the airline. John may also receive a special brochure that points out the best way to check his tool kit so that he is in compliance with security codes. Perhaps a certificate for a third bag free after X number of trips would help cement more of his business and push John to fares that are incrementally higher knowing that the level of service beats competitors.

The main point of these examples is first to convince the reader that it is possible for an airline to think about profitability by each and every Passenger Name Record (PNR). Banks that have over 10 million customers have already proven that it is possible to develop these methods. The second point is to relate the message that even if customers are unprofitable at first, they could be made profitable by encouraging the passenger to change his or her behavior or by having the airline change its practices with respect to a particular customer.

At first glance, individual attention may appear to be a costly process. However, within the hundreds of thousands of high-frequency travelers, there may be a thousand John Does—passengers who have accumulated enough miles to be in the elite-level status but whose profitability is questionable. The many John (and Jane Does) can be identified by mining the data warehouse by query-based proven sets of performance indicators. Cost effective marketing tactics can then be enacted to direct marginal passengers to more cost-effective channels. Thousands of buying units—small business accounts—may be identified efficiently and effectively to enhance market share. This technique represents a major breakthrough related to such time-honored sales techniques as prospecting, and qualifying. The results of customer-centric marketing using new technologies will be a compelling reason for airlines to adopt the radical changes in corporate culture.

Table 5.1 An Example of Individual Customer Profitability by Quarter (dollar amounts represent profit per ticket)

Rob Smith	$259.60	Ella Jones	$260.53	John Doe	($31.69)
Ticket Houston to London - Business Class	$335.13	Ticket Houston to Chicago - Coach Class	$44.77	Ticket Houston to Atlanta - Coach	$5.24
Ticket Houston to San Francisco - Coach Class	($29.79)	Ticket Houston to New Orleans - Coach Class	$52.08	Ticket Houston to Atlanta - Coach	($13.71)
		Ticket Houston to New York - Coach Class	$45.26	Ticket Houston to Atlanta - Coach	($11.55)
Ticket Houston to Tulsa	($13.84)	Ticket Dallas to Mexico City - Coach Class	$41.62	Ticket Houston to Atlanta - Coach	($14.00)
		Ticket Houston to New York - Coach Class	$37.64	Ticket Houston to Atlanta - Coach	$2.11
Ticket Houston to Seattle - Coach Class	($31.90)	Ticket Houston to Seattle - Coach Class	$39.16	Ticket Houston to Seattle - Coach Class	$0.22

Source: Example developed by Andy Tellers, Rick Volz, and Monica Smith of Teradata, November 2002.

Table 5.2 An Example of Itinerary Detail: Houston to Seattle Round Trip—Revenue*

	Rob	Ella	John	Comments
Revenue	$397.36	$380.57	$355.57	
Base Fare	$377.18	$344.95	$215.34	Ella booked this trip four months in advance. Rob and John purchased 14 day advance fares.
Add Collect Fees				
Excess Baggage			$40.00	
Change Fee			$100.00	
Cabin Pet Fee				
Unaccompanied Minor				
Pre Paid Ticket Advice				
Paper Ticket Fee	$20.00			Rob feels more confident with a paper ticket.
Two Seat Fee				
Upgrade Fee				
Fuel Surcharge				
Airline Partner FF Revenue				
Non Airline FF Revenue				
Programs				
Airport Lounge Services				
Club Memberships		$27.02		Lounge revenue is allocated based on 7.4 trips per year per member based on $200 annual membership.
Gift Certificates				
On Board (duty free, alcohol, headsets)				
Air Miles				
Vacation Packages				
Adv Purchase Fares - NIR	$0.18	$8.60	$0.23	Ella's Net Interest Revenue reflect the Airline's use of revenue for 1/3 year.
Award Tkt Revenue (From OA)				
Cargo				
Base Charge				
Special Services				

*: Scenario is based on all three customers flying Round Trip on the same flights same days from Houston International (IAH) to Seattle (SEA).

Source: Example developed by Andy Tellers, Rick Volz, and Monica Smith of Teradata, November 2002.

Table 5.3 An Example of Itinerary Detail Continued: Houston to Seattle Round Trip—Direct Expenses

Direct Expenses/Behavior Based Distribution	Rob	Ella	John	Comments
Commission/Override	($11.32)		($6.46)	Rob books with large national agency. John books with a leading online agency.
GDS Fees	($27.00)		($9.00)	Rob rebooked his reservation twice before purchase.
Credit Card Fees	($9.43)	($7.76)	($5.38)	Rob and John use premium level credit cards.
Reservation Processing	($47.16)	($0.64)	($11.79)	Rob made a total of four calls to Reservations regarding this PNR.
Airline Web Site		($3.40)		Ella conducted her transactions and FT mileage follow up using the airline website.
Ticket Processing	($5.07)	($0.63)	($0.63)	
Foreign Exchange				
Check in				
Kiosk		($2.54)		Rob and John use traditional check in modes.
Counter	($11.48)		($13.66)	
Baggage Handling	($3.02)		($6.04)	
Gate	($4.14)	($1.13)	($1.13)	Rob stopped at gate to request aisle seat.
Frequent Flyer				
FF Processing	($13.47)	($0.33)	($0.51)	Rob called the FF Program Center to check mileage.
Partner Airline Fees				
Award Tickets Partner Airlines				
Mileage Liability	($7.55)	($11.33)	($11.33)	Based on $0.002 /mile for mileage liability. Ella and John receive 50% bonus miles. Does not factor displacement.
Airport Lounge		($24.23)		Fully allocated cost.

Source: Example developed by Andy Tellers, Rick Volz, and Monica Smith of Teradata, November 2002.

Table 5.4 An Example of Itinerary Detail Continued: Houston to Seattle Round Trip—Indirect Expenses

By equipment per flight			Rob	Ella	John
Flight	Fuel Costs		($54.40)	($54.40)	($54.40)
	Landing Fees		($6.05)	($6.05)	($6.05)
	Catering Costs		($3.22)	($3.22)	($3.22)
	Labor				
		Flight Crew	($51.50)	($51.50)	($51.50)
		Cabin Crew	($44.33)	($44.33)	($44.33)
Airport	Labor				
		Ground Handling	($12.03)	($12.03)	($12.03)
		General Airport Staff	($8.56)	($8.56)	($8.56)
		Maintenance	($9.08)	($9.08)	($9.08)
		Aircraft Service	($2.99)	($2.99)	($2.99)
		Security	($5.20)	($5.20)	($5.20)
		Cargo Crew	($0.53)	($0.53)	($0.53)
	Maintenance				
		System	($6.02)	($6.02)	($6.02)
		Line	($4.33)	($4.33)	($4.33)
Aircraft Ownership					
	Hull Insurance		($5.06)	($5.06)	($5.06)
	Lease Costs		($14.27)	($14.27)	($14.27)
	Debt Interest		($11.02)	($11.02)	($11.02)
	Depreciation		($4.66)	($4.46)	($4.46)
Corporate Overhead			($15.84)	($15.84)	($15.84)
	Sales and Marketing		($8.20)	($8.20)	($8.20)
Risk Provision					
	Agency Default		($2.88)	($2.88)	($2.88)
	Fraud		($3.21)	($3.21)	($3.21)
	Operational				
		Weather	($4.59)	($4.59)	($4.59)
		Strike	($5.32)	($5.32)	($5.32)
		Mechanical	($6.33)	($6.33)	($6.33)
	Injuries, Loss, and Damage				

Source: Example developed by Andy Tellers, Rick Volz, and Monica Smith of Teradata, November 2002.

Conclusions

All members of the air transportation value chain must accept change and take deliberate actions to transform bureaucratic and outdated practices. All members must develop a shared mindset and partner across the value chain to produce a reasonable return for all. Airline management must develop new business models by first analyzing the firm's cost structure, and then by leveraging a broad spectrum of technology as well as its own competencies and resources—people, property, and knowledge. Just imagine a major US airline that has Southwest's cost per available seat mile and American's revenue per passenger mile! In addition to strategic management of costs, an airline, in the end, perhaps, simply provides distinctive value and does both in an increasingly uncertain environment. Managing uncertainty requires strategic thinking (some aspects of which have been discussed in this and the previous chapter), nonlinear thinking and scenario planning—topics to be discussed in the next chapter. It is nonlinear thinking that we can look forward to for the opportunities from the very markets we serve, and the markets still to come.

Notes

[1] Information provided by Justin Pettit of Stern Stewart Research, New York, January 2002.

[2] Horan, Hubert, "Is the traditional Big Hub model still viable?", *Aviation Strategy*, July-August 2002, p. 9.

[3] *Airlines International*, October-November 2002, pp. 18-20.

[4] Bossidy, Larry and Ram Charan, *Execution: The Discipline of Getting Things Done* (New York: Crown Business, 2002).

[5] Brickley, James A., Smith, Clifford W. and Jerald L. Zimmerman, *Designing Organizations to Create Value: From Strategy to Structure* (New York: McGraw-Hill, 2003).

[6] Forward, David, "Dumb and Dumber", *Airways* , November 2001, pp. 17-21.

[7] Greenwald, Gerald, *Lessons from the Heart of American Business: A Roadmap for Managers in the 21st Century* (New York: Warner Books, 2001).

[8] Chesshyre, Tom, "Terminal mayhem at Stansted", *The Times*, 14 September 2002, Travel Section, pp.1-2.

[9] Data provided by Southwest Airlines, December 2002.

[10] "Low Cost Airlines", A Report by the Center for Asia Pacific Aviation, Sydney Australia, September 2002, p. 13.

[11] Flexible manufacturing is a system that can save billions for large auto manufacturers by re-equipping assembly lines with robots that can be controlled by computers to switch the production of a vehicle from one to another to match customer demand.

[12] Shelden Larry and Geoffrey Colvin, "Will This Customer Sink Your Stock", *Fortune*, 30 September 2002, pp. 126-132.

[13] Developed by Andy Tellers, Rick Volz, and Monica Smith of Teradata, a division of NCR.

Chapter 6

A Whirlwind Tour of Some Radical Scenarios

The multi-faceted challenge for the global airline industry in the 21st Century involves identifying and delivering value to its stakeholders: customers, employees, and shareholders. Despite all the recent interest in low-fare, or low-frill air travel, let us not forget that customers do exist that depend on a global network and that are less sensitive to price. Meeting this challenge has not been easy given the complexity of the framework in which the airline industry operates. Nevertheless, in the future shareholders will insist that they receive a commensurate return for the risk they take. Most conventional airlines are not likely to meet this challenge if they continue with the mindset of incremental changes and conventional wisdom. Although most of the new low-cost, low-fare airlines have performed well, a number of them could also go the same way as People Express in the US and Debonair in the UK. Even the veteran low-cost airline—Southwest—the most consistently successful and the granddaddy of all new entrants, cannot stand on its laurels. All must continue to re-invent themselves, at least to maintain, if not improve, their lean capability at all levels of the enterprise.

The multi-faceted challenge can be met. It requires, however, thinking that is nonlinear and radical—the subject of this last chapter. Although, it is not claimed that any of these scenarios will come to fruition, thinking about them and other similar scenarios will be helpful for three reasons.

(a) It would help us to question if we have reached a plateau—for example aircraft speed? It may help us to determine if we have gone too far—for example, providing in-flight entertainment and services which are high-cost to the airlines but free to the customer? It may help us to determine if we have not gone far enough—for example, the speed and efficiency of processing passengers at airports. We need to think about the next breakthroughs, at least in the first and third areas and align costs and revenues in the second.

(b) Thinking about such scenarios may reduce the risks of being unprepared for dramatic changes.

(c) Thinking about such scenarios may trigger the reader's thoughts into conventional and unconventional strategies to seize opportunities—opportunities that might create value for the three core shareholders.

Are we at a plateau?

It is remarkable to think about what the global commercial airline industry has achieved since the early "modern" aircraft of the 1930s (twin-engine, all metal, low-wing monoplanes) that began to offer reliability of operations and cabin comfort, even if the price was prohibitive for most people.

(a) The introduction of jet aircraft reduced flying times, improved creature comfort and reduced the price of air transportation (especially for passengers traveling on charter airlines). Today, passengers flying in intercontinental markets have a choice of cabin configuration (economy, premium economy, business, and first), a broad spectrum of services and facilities at airports (electronic check-in, executive lounges, "fast-track" security lines, high-street type of shopping), and fares that range, in a market such as New York-London, from about US$250 to US$7,000 for a nonstop round trip operated from conventional airports such as New York-Kennedy and London-Heathrow.

(b) In terms of accessibility, we are flying nonstop in long-haul markets such as Taipei-New York with an Airbus 340 (7,788 miles) and New York (Newark)-Hong Kong with a Boeing 777 (8,055 miles), as well as with high frequency in such shuttle-type markets as Rio de Janeiro-Sao Paulo with 132 flights a day each way.

(c) At the other end of the spectrum for short-haul, thin markets, we have mega hub-and-spoke systems at airports such as Chicago and Atlanta where we connect passengers from small cities to a wide variety of domestic and international destinations.

(d) We have low-cost, low-fare airlines worldwide that offer different price-service options in the short- and medium-haul markets.

(e) In the area of passenger services, leading conventional airlines in the US, have reduced their distribution costs from about 20 percent of revenue to about 10 percent.

(f) We have improved the check-in process by installing self-service machines at airports and reduced the costs of call centers by installing automated re-booking and confirmation systems.

The question is: Have we reached a plateau in marketing and operations? The answer would appear to be yes, if measured by conventional wisdom, and risk aversion. However, the answer might be a resounding no if one is willing to think about some suggestions made in the previous chapters and some radical scenarios suggested in this chapter. The marketplace is being totally reshaped by tidal wave forces of technology. These include aircraft, the Internet, operations, new planning tools (the unthinkable size of data warehouses and the capability to mine this data), as well as a consumer revolt to past price-service options, and changing demographic patterns. It is the mindsets of key stakeholders in the air transportation value chain—airline management, labor, civil aviation authorities and airport operators—that are not keeping up with the speed and intensity of the approaching hurricanes.

In some markets within Europe, Ryanair has fares as low as 10 pounds sterling, less than the price of a good lunch in Central London. Southwest has fares as low as US$19 between Los Angeles and Las Vegas, again, less than the price of a reasonable evening meal in Los Angeles. Does this mean that the price of airline tickets could be headed toward zero in some markets? Before one hastily says "of course not", or "don't be silly", just think about the direction of rates for telephone calls, the rates charged to send messages via the Internet, and the fees charged for doing research on the Web.

The point here is not that we should forget about business passengers' requirements for full-service capability in long-haul intercontinental markets. It is also not to question how would the Ryanair types cover their costs and provide a reasonable return to their shareholders. The point is the need to think in a nonlinear way to re-invent the business for all carriers, those offering low-frills in the mostly short- and medium-haul, point-to-point markets to those offering high frequency service in global markets. To begin mindset change, one must first look for breakthroughs with nonlinear thinking.

Where is the next breakthrough?

Aircraft

Let us start with aircraft. There already exist aircraft that could change the structure of the industry. One area where substantial progress in the near future seems unlikely is higher speeds from completely new designs. In late 2002, Boeing announced it will stop development on the Sonic Cruiser (at least for the time being), designed to fly near the speed of sound. Airlines did not appear to think the 20 percent reduction in (long-haul intercontinental) travel time would justify the potential increase in premium fares.

Let us take the case of ultra long-range aircraft. Just as the structure of the trans-Atlantic market was changed by the availability of smaller-capacity twin-engine aircraft and the liberalization of the markets on both sides of the Atlantic, a similar situation could take place in the trans-Pacific markets with the availability of ultra long-range and high-capacity aircraft (such as the Airbus 340-500 and the Boeing 777-200LR in the first case and the Airbus 380 in the second case) coupled with management's decision to take some risks. Some markets are ready for nonstop service, by-passing the traditional stop or connection at the Tokyo-Narita Airport. The ultra long-haul aircraft can be cost-effective if it can complement the smaller capacity narrow-body at one end and the very high capacity at the other end, particularly if there is crew commonality all the way through the entire fleet. Such a restructuring would have a significant impact on the aviation activity at the Tokyo-Narita Airport as well as the development of existing and planned hubs in South East Asia. Similarly, the use of smaller-capacity long-range aircraft could change the structure of the industry if airlines such as Air India start operating nonstop flights between their large second-tier cities such as Ahmadabad, Bangalore, Chennai, Hyderabad, and Kolkata and the secondary airports surrounding the greater London area such as Stansted Airport.

Though the manufacturers are struggling to produce passenger-friendly long-haul regional jets at an attractive cost, work goes on. The long-haul domestic and international routes are crying out for such an aircraft which will lower plane-mile costs relative to the narrow-body families, and allow smaller markets such as San Diego-New York (Stewart) to be connected. A lack of spaciousness and comfort can be an issue with such aircraft in long-haul applications.

Consider the scenario of a totally new four-seat jet that can be purchased for under a million dollars and a changed infrastructure allowing it to fly in and out of almost 3000 airports in the United States. Think of the challenges to conventional airlines if thousands of such aircraft came into the marketplace through fractional ownership or through new concepts such as air-taxis. What if the fare level of such air taxis was comparable to the recently reduced walk-up fares? What if in short- and medium-haul markets the low-cost, low-fare airlines took almost all of the low-end customers of the full-service airlines and the air taxi and fractional ownership businesses took almost all of the top end customers?

Community-Sponsored Airline Operations

Who knows better the transportation needs of a city, the city itself or an airline considering service there? What if regional carriers were given economic incentives by large second-tier cities such as San Antonio, Texas to provide nonstop service to key markets? Imagine if cities started putting out proposals for bids from airlines to provide a certain level of service. What if a few dozen cities in the US followed such a strategy? Imagine also if the regional carriers partnered with cities instead of with larger airlines. Finally, imagine if airline scheduling departments sought out input from the cities served.

Alliances

As one industry executive said, alliances among international carriers are an artificial solution to an artificial problem. If governments were to remove restrictions around ownership and control, there would be less desire to have alliances. That is one possibility. If this scenario takes place it would most certainly change the market structure and conduct of the global airline industry. This appears possible within a few years, as the US and Canada reopen talks on removing the rules surrounding cabotage and allowing airlines from either country to fly anywhere they wish within both countries.

There is a second scenario, however, at the other end of the spectrum. Suppose, with the approval of civil aviation authorities and labor groups, alliances operated a jointly-owned fleet in major intercontinental markets, with each partner flying its own brand at each end of the major route. This scenario assumes the existence of a joint fleet operated under a common certification of aircraft, pilots, maintenance, security, and so forth. Under such a scenario, computer programs could suggest the optimal routing of high-capacity (such as the Airbus 380), alliance-branded aircraft in global networks. Utilization of aircraft could optimize such aspects as stage lengths, departure and arrival time preferences, time zones, and airport curfews and slots.

International Services by Low-Cost, Low-Fare Airlines

The conventional wisdom is that low-cost, low-fare airlines will not be able to duplicate their business model across the North Atlantic. Given that Southwest and JetBlue are already providing transcontinental services within the US and easyJet is already flying in medium-haul markets such as London-Athens, how much of a stretch is it to imagine service between London and New York. The US-UK is the largest trans-Atlantic market and New York-London is by far the largest trans-Atlantic segment, large enough to have an O&D market sufficient for low-fare airlines to fly, say between New York's Islip MacArthur Airport and London Stansted. Such a service does not need to be no-frills. Simple meals could be served that do not require a galley. Simple entertainment could be provided that is cost-effective for 6-7 hour flights, or passengers could be encouraged to bring their own hand-held audio-video systems.

Let us take this already radical scenario one step further. What if easyJet offered service between London and New York and between London and Toronto that connected with service offered by JetBlue and Westjet? Conversely, what if Westjet or JetBlue flew between Toronto and London and between New York and London and connected with easyJet out of London. There would not have to be an official strategic alliance. Passengers would simply connect after clearing government formalities, or use a bonded in-transit lounge.

Airport Passenger Processing Staff and Systems

Currently, passengers are processed at airports either by the staff of individual airlines or by their strategic partners. At a number of airports where airlines have limited activity such business practices do not appear to be cost-effective. What if at small and medium-size airports, passengers were handled by dedicated and airline-trained airport staff. Overlay software is already in use connecting the systems of different airlines. The same concept could apply to check-in machines, in other words, a common set used by passengers traveling on different airlines. It is already in practice.

This business practice is likely to be resisted according to the rules of conventional wisdom based on such objections as access to confidential information, different operating systems and platforms, and the loss of brand equity. As radical as the idea may be, there are two supporting arguments. First, there are ATMs where a customer of one bank can conduct business using cards issued by a variety of other banks. Second, as to the issue of branding, while it is true a few global airlines have reached a level of brand equity that sets them apart from other airlines, airline brands in general rank low when compared to other businesses. While most airlines put a lot of faith in the value of their brands, it might be valuable to think about a study conducted by the firm of Interbrand and BusinessWeek in the year 2001. This study identified and ranked the world's most valuable 100 brands.[1] Despite the limitations of this study pointed out in the note at the end of this chapter, there was not a single passenger airline in the top 100 global brands. The only airline to even make the list was the express courier Federal Express and they achieved a rank of 86.

In The Year 2020 ...

How might things look 17 years from now? The industry might muddle along as it has for many years, or it just might undergo a dramatic transformation. Imagine Southwest Airlines continuing its growth pattern of the last 30 years for the next 17. At just 9 percent compounded annually, that would bring its fleet of aircraft to about 1500, and might give it almost complete control of the North American airspace, especially if a NAFTA-type agreement is reached permitting any carrier from the US, Canada, and Mexico to operate anywhere it wants in the three countries. With an annual compound growth of 10 percent, its fleet would grow to almost 1800 airplanes! It could result in the demise of all the other low-cost low-fare carriers which were unable to match all the key aspects of Southwest's business model. It might also result in the disappearance of numerous "conventional" airlines which lost their North American internal traffic, and had to survive on whatever international traffic they could muster from other countries or which connected to them at North American gateways of Southwest's choosing.

Overly dramatic? Impossible to imagine? Just think of some of the American automobile manufacturers of the last few decades—the Plymouth division of DaimlerChrysler, the Oldsmobile division of General Motors, American Motors, Bricklin, DeLorean, Studebaker, Packard, Stutz, Cunningham, Cord, Duesenberg, and Auburn. All gone. Does it seem more likely now?

Conclusions

The purpose of these scenarios was to give a general introduction to several potential breakthroughs, not to provide a deep discussion or a comprehensive list of possibilities. Rather, the purpose was simply to provide a few examples that could at least make the stakeholders sit up and take notice, if not inspire them.

The global airline industry has adapted to many developments in the past—the introduction of the jet aircraft, deregulation/liberalization, the Internet, and the phenomenal growth of the low-cost, low-fare airlines. However, at the beginning of the second century of flight, stakeholders need to manage not just the confluence of the major forces that have already been discussed in previous chapters but some new ones that could be unleashed by technologists (aviation, telecommunications and information) as well as by civil aviation authorities who are under pressure to change focus from protecting the flag carrier to protecting the consumer interest.

The one major difference between the first and the second century of flight would most likely be this. Whereas, during the first century demand was driven by supply, in the second century supply will most likely be driven by demand. For example, are we willing to accept low organic growth (1 or 2 percent) in some mature markets but take some calculated risks in some developing markets (5 to 10 percent). In the new environment where supply is driven by demand, airlines at both ends of the cost and service spectrums can prosper. The key to success is whether an airline can identify an un-served or under-served market and deliver the right product, be it low point-to-point fares in the US, good service in India, or global network service in Germany or Singapore. I hope that the information provided and the challenges and opportunities addressed in this book have stimulated your inquisitiveness and motivated your thinking about the future.

Note

[1] "The 100 Top Brands", *BusinessWeek*, 6 August 2001, p. 64. For a brand to be considered, it had to have a value greater than US$1 billion. The other selection criteria were: (a) the brand had to be global in nature, deriving 20

percent or more of its sales from outside its home country; and (b) there had to be publicly available marketing and financial data on which to base the valuation. This condition did exclude some big brands such as VISA, BBC, Mars, and CNN.

Index

About the Author

Nawal Taneja has more than 30 years of experience in the airline industry. As a practitioner, he has worked for and advised major airlines and airline-related businesses worldwide in the areas of strategic and tactical planning. His experience also includes the presidency of a small airline that provided scheduled and charter service with jet aircraft, and the presidency of a research organization that provided consulting services to the air transportation community worldwide. In academia, he has served as Professor and Chairman of the Aerospace Engineering and Aviation Department at the Ohio State University, and an Associate Professor in the Flight Transportation Laboratory of the Department of Aeronautics and Astronautics of the Massachusetts Institute of Technology. On the government side, he has advised civil aviation authorities in public policy areas such as airline deregulation, air transportation bilateral agreements, and the management and operations of government-owned airlines. He has also served on the board of public and private organizations.